MW00535332

ADVANCED POWER ELECTRONICS CONVERTERS

IEEE Press
445 Hoes Lane
Piscataway, NJ 08854

ADVANCED POWER ELECTRONICS CONVERTERS

PWM Converters Processing AC Voltages

EUZELI CIPRIANO DOS SANTOS JR.
EDISON ROBERTO CABRAL DA SILVA

IEEE PRESS

Published by John Wiley & Sons, Inc., Hoboken, New Jersey.
Published simultaneously in Canada

For general information on our other products and services or for technical support, please contact our Customer Care Department within the United States at (800) 762-2974, outside the United States at (317) 572-3993 or fax (317) 572-4002.

Wiley also publishes its books in a variety of electronic formats. Some content that appears in print may not be available in electronic formats. For more information about Wiley products, visit our web site at www.wiley.com.

Library of Congress Cataloging-in-Publication Data is available.

ISBN: 9781118880944

Printed in the United States of America

10 9 8 7 6 5 4 3 2 1

CONTENTS

PREFACE

This book deals with a new methodology to present an important class of electrical devices, that is, power electronics converters. The common approach to teaching converters is to consider each type individually, in a separate and isolated fashion. The direct consequence is that the learning process becomes passive since the power electronics configurations are presented without consideration of their origin and development. Since the teaching process is based on the topology itself, students do not develop the ability to construct new topologies from the conventional ones.

A systematic approach is taken to the presentation of multilevel and back-to-back converters, instead of showing them separately, which is normally done in a conventional presentation. Another special aspect of this book is that it covers only subjects related to the converters themselves. This will give more room for exploring the details of each topology and its concept. In this way, the method of conceptual construction of power electronics converters can be highlighted appropriately.

While presenting the basics of power devices, as well as an overview of the main power converter topologies in Chapter 2, this book focuses primarily on configurations processing ac voltage through a dc-link stage. This text is ideally suited for students who have previously taken an introductory course on power electronics. It serves as a reference book to senior undergraduate and graduate students in electrical engineering courses. However, due to the content in Chapter 2, it is expected that even students who the lack knowledge of power devices and basic concepts of converters can understand the subject.

Although the primary market for this text is heavily academic, electrical engineers working in the field of power electronics, motor drive systems, power systems, and renewable energy systems will also find this book useful.

The organization of the book is as follows: Chapter 1 is the introductory chapter. Chapter 2 presents the basics of power devices as well as an overview of the main power converter topologies. Chapter 3 provides a brief review of the main power electronics converters that process ac voltage; additionally, it furnishes the introduction to the power blocks geometry (PBG), which will be used to describe the power converters described in this book. In fact, this chapter brings up a compilation of the topologies explained throughout this book. The fundamentals of PBG and its correlation to the development of power electronics converters are presented in a general way. Multilevel configurations are presented from Chapters 4–7. Neutral-point-clamped, cascade, flying capacitor, and other multilevel configurations are presented in Chapters 4–7, respectively. Chapter 8 deals with techniques for optimization of the pulse width modulation (PWM), considering the fact that the number of pole voltages is higher than the number of voltages demanded by the load. After describing many topologies throughout Chapters 2–7, highlighting the circuits themselves, as well as PWM strategies in Chapter 8, Chapter 9 handles control actions needed to keep a specific variable of the converter under control. Chapter 9 is strategically placed before the presentation of the back-to-back converters (Chapters 10 and 11) due to their need for regulation of electrical variables. Single-phase to single-phase back-to-back converters are presented in Chapter 10, and the final chapter deals with three-phase to three-phase back-to-back converters.

Euzeli Cipriano Dos Santos Jr.
Edison Roberto Cabral Da Silva

CHAPTER 1

INTRODUCTION

1.1 INTRODUCTION

Power electronics may be considered a revolutionary field in electrical engineering because of the new insights obtained during its development. This has actually been the case from the beginning, when mercury arc rectifiers and thyratrons were employed in grid-controlled circuits. After this first generation of power devices and converters, power electronics with silicon power diodes and thyristors was developed to overcome many of the problems of the first generation, such as the operation in low efficiency. As mentioned in Reference 1, the so-called power electronics, with gas tube and glass-bulb electronics, was known as industrial electronics, and the power electronics with silicon-controlled rectifiers began emerging in the market in the early 1960s.

The different definitions of power electronics lead to the same concept or idea: that the control of power flow between an apparatus that furnishes electrical energy and another one that demands electrical energy. For instance, the definition given in References 2 and 3 say, respectively: " … power electronics involves the study of electronic circuits intended to control the flow of electrical energy. These circuits can handle power flow at levels much higher than the individual devices ratings … " and " … power electronics deal with conversion and control of electrical power with the help of electronic switching devices."

Power electronics involves several academic disciplines creating a complex system, including semiconductor physics, control theory, electronics, power systems, and circuit principles. The comprehensive aspect of power electronics makes the presentation of its contents difficult. The interdisciplinary nature of power electronics requires the integration of the practices and assumptions of all the academic disciplines involved, as well as calling for significant prerequisites on the part of the students enrolled for the course. Figure 1.1 illustrates this by analogy, with the prerequisite skills needed for a power electronics course being shown as the roots of a tree, the various power electronics devices as the trunk, and the resulting technologies and applications (power quality, renewable energy systems, etc.) as the branches.

Since the dawn of solid-state power electronics, the use of semiconductor devices has been the major technology to drive power processors. A comparison

---

Advanced Power Electronics Converters: PWM Converters Processing AC Voltages,
Forty Fifth Edition. Euzeli Cipriano dos Santos Jr. and Edison Roberto Cabral da Silva.

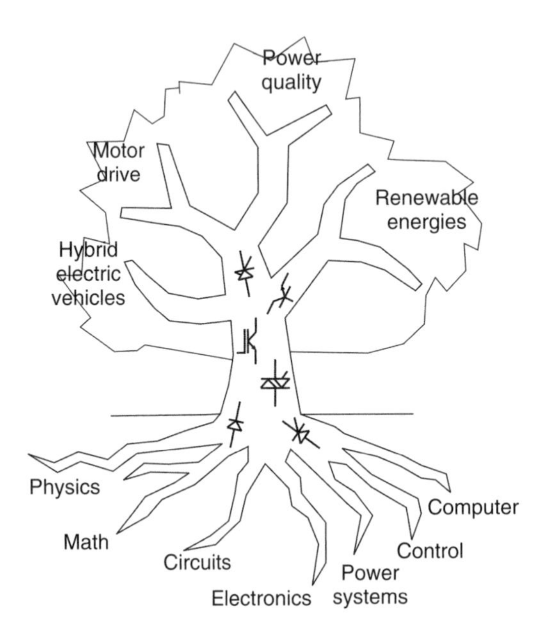

Figure 1.1 Interdisciplinary nature and new insights obtained from power electronics.

of the semiconductor devices formerly used in controlled rectifiers with new technologies underlines this dramatic development. In addition to the improvement of power switches, there has also been great activity in terms of circuit topology innovations.

A power electronic converter is the centerpiece of many electrical systems. Common applications include, but are not limited to, motor drive systems, renewable energies, robotics, electrical and hybrid vehicles, and circuits promoting power quality. These applications have required considerable research worldwide to develop semiconductor devices, configurations that process ac and dc variables, control and diagnosis, fault-tolerant systems, and the like.

In addition to the technical side mentioned already, the educational aspects have considerable importance, as students usually consider power electronics courses to be particularly difficult, perhaps because of their interdisciplinary nature. Achieving student motivation is thus a fundamental task of educators involved in the field of power electronics.

In this context, this book discusses a novel methodology for presenting an important set of power electronics converters, that is, topologies that process ac voltage. The common approach to teaching converters is to consider each type individually, in a separated and isolated manner. The direct consequence is that the learning process becomes passive as the power electronics configurations are presented without any consideration of their origin and development. Since the teaching process is based on the topology itself, students develop no ability to construct new topologies, different from the conventional ones. Section 1.2 outlines this new methodology.

1.2 BACKGROUND

Although presenting the basics of power devices as well as an overview of the main power converter topologies in Chapter 2, this book focuses primarily on configurations processing ac voltage through a dc-link stage. This book is ideally suited for students who have already taken an introductory course in power electronics. It also serves as a reference book to senior undergraduate and graduate students in electrical engineering courses. However, students can easily manage despite the lack of knowledge of power devices and basic concepts of converters, because they are explained in Chapter 2.

Systems with power electronics conversion have been used to guarantee grid and load requirements in terms of controllability and efficiency of the electrical energy demanded, especially in industrial applications. Power electronics topologies convert energy from a primary source to a load (or to another source) requiring any level of processed energy.

Classifications of the power electronics topologies can be done in terms of the type of variable under control (i.e., ac or dc), as well as the number of stages of power conversions used, as observed in Fig. 1.2. Figure 1.2(a) shows, in a general way, many of the possibilities related to energy conversion. Figure 1.2(b) highlights a direct ac–ac conversion, which converts an ac voltage (v_1) with a specific frequency (f_1) to another ac voltage with a different (or same) voltage (v_2) and frequency (f_2); this converter is normally called a cycle converter. Figure 1.2(c) depicts the ac–dc or dc–ac conversion, while Fig. 1.2(d) shows a dc–dc converter. Even admitting

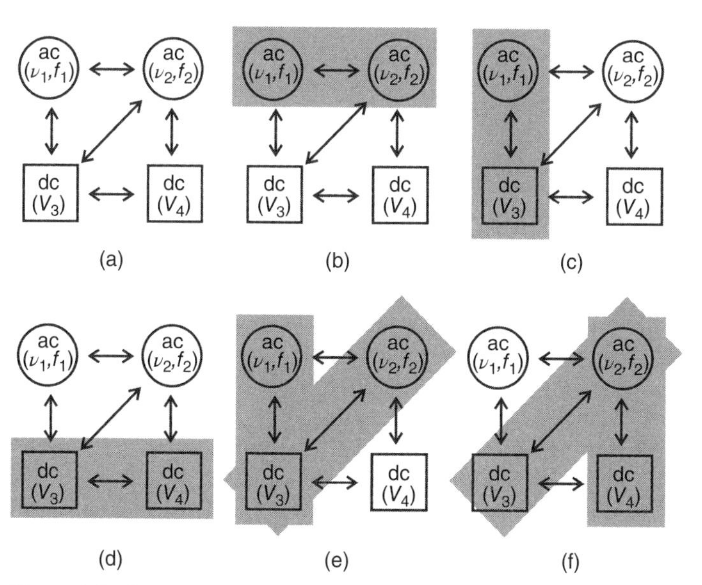

Figure 1.2 Power conversion: (a) all possibilities of conversion, (b) cycle converter, (c) rectifier or inverter, (d) chopper, (e) ac–dc–ac, and (f) dc–ac–dc.

that Fig. 1.2(e) and 1.2(f) could be considered as extended versions of the previous cases, those conversion systems (ac–dc–ac and dc–ac–dc) are presented in Fig. 1.2 because of the large use in different applications.

Special attention is given to the conversion systems presented in Fig. 1.2(c) and 1.2(e), dealing with configurations that process ac voltage (at input and/or output converter sides) with one dc stage. A systematic approach is taken for the presentation of those configurations, instead of just showing them separately, as is normally done in a conventional presentation. Another aspect of this book is that only the subjects related to the converters themselves will be considered, which means that the contents dealing with either ac filters or transformers will be omitted. This will give more room for exploring the details of each topology and its concept. In this way, the method of conceptual construction of power electronics converters can be highlighted appropriately.

1.3 HISTORY OF POWER SWITCHES AND POWER CONVERTERS

Configurations of power electronics converters have provided an attractive alternative for the applications needing energy processing, considering the acceptable level of losses associated with the conversion process itself, as well as improvement in reliability. As previously mentioned, power electronics converters must control the power flow, which means that the development of the devices used in those converters is crucial to guarantee the expected features. In this section, a historic view of the power electronics devices will be furnished, highlighting the main events that contributed to the current development.

The history of power electronics predates the development of the semiconductor devices employed nowadays. The first converters were conceived in the early 1900s, when the mercury arc rectifiers were introduced. Until the 1950s the devices used to build power electronics converters were grid-controlled vacuum tube rectifier, ignitron, phanotron, and thyratron. There were two important events in the power electronics development: (i) in 1948, when Bell Telephone Laboratories invented the silicon transistor, with applications in very low power devices such as in portable radios and (ii) in 1958, when the General Electric Company developed the thyristors or SCR, first using germaniums and later silicon. It was the first semiconductor power device.

Besides these two events, many developments have been achieved in terms of switching development. Between 1967 and 1977, the gate turnoff (GTO) (gate-controlled switch) and gate-assisted turnoff thyristor (GATT) (gate-assisted turnoff switch) were invented. Power transistors, MOSFETs (metal oxide semiconductor field-effect-transistors), MCTs (MOS-controlled thyristor) and IGBTs (insulated-gate-bipolar transistors) have been invented since the end of 1970s. In addition, it is worth mentioning that the area of power electronics was deeply influenced by microelectronics development, and the history of power electronics is closely related to advances in integrated circuits to control switching power supplies.

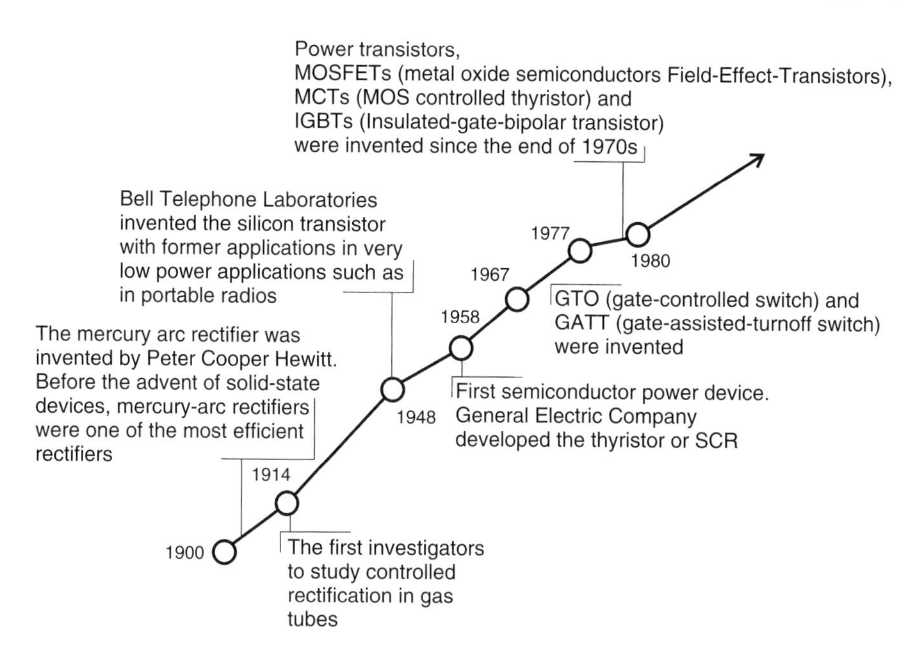

Figure 1.3 Timeline of historical events in the power electronics devices evolution.

Figure 1.3 depicts the timeline showing the development of power electronics devices.

An important chapter in the history of power electronics converters was the development of switching power suppliers. In 1958, the IBM 704 computer, which was developed for large-scale calculations, used as a switching power supplier the primitive vacuum tube-based switching regulator. But the revolution in power supplier concepts came in the late 1960s, when the switching power supplies replaced the linear ones. In a linear power supply, regulated dc voltages are obtained from the ac utility grid throughout the following sequence of steps: (i) 60 Hz power transformer, to converter 120 ac voltage at the primary transformer side to low voltage at secondary transformer side; (ii) such voltage is converted to dc with a simple diode rectifier; and (iii) a linear regulator drops the voltage to a desired value. Indeed, it is possible to identify many problems related to this technology, such as low efficiency (50–65% of the power is wasted as heat), and it was heavy and large (mainly due to the low frequency transformer, heatsink and fans to deal with the heat). The advantages are that it has a very stable output voltage and the conversion system is noise-free.

To overcome the disadvantages of the linear regulators, General Electric published a design of an early stage switching power supply in 1959.

The concept of switching power suppliers is very different from linear regulators. Instead of conducting power 100% of the time (i.e., turning excess power into heat), the switches and passive elements are connected to rapidly turn the power on and off. Unlike linear regulators, the ac utility voltage is converted directly to dc voltage, and the gating signal controls the time of the switching, regulating the average voltage desired at the output converter end.

Another important development in power electronics configurations was the controlled rectifiers, especially with the production of the silicon-controlled rectifier (SCR or thyristor). Such a device allowed the control of high power by just changing the signal applied to its gating circuit with higher efficiency rather than the older technology of employing a mercury arc rectifier.

1.4 APPLICATIONS OF POWER ELECTRONICS CONVERTERS

The range of applications for power electronics converters is so large that it goes from low power residential applications to high power transmission lines. Many of those applications can be considered as traditional ones (e.g., rectification circuits and motor drive systems). On the other hand, a few emerging applications have generated wide interest (e.g., renewable energy systems). A brief discussion matching the power electronics converters with those applications will be introduced here, with the details of those applications being presented throughout the chapters.

Figures 1.4 and 1.5 summarize some examples that demonstrate the presence of power electronics in a wide range of applications. Figure 1.4(a) shows schematically the application of power electronics in hybrid/electric vehicles. From the power

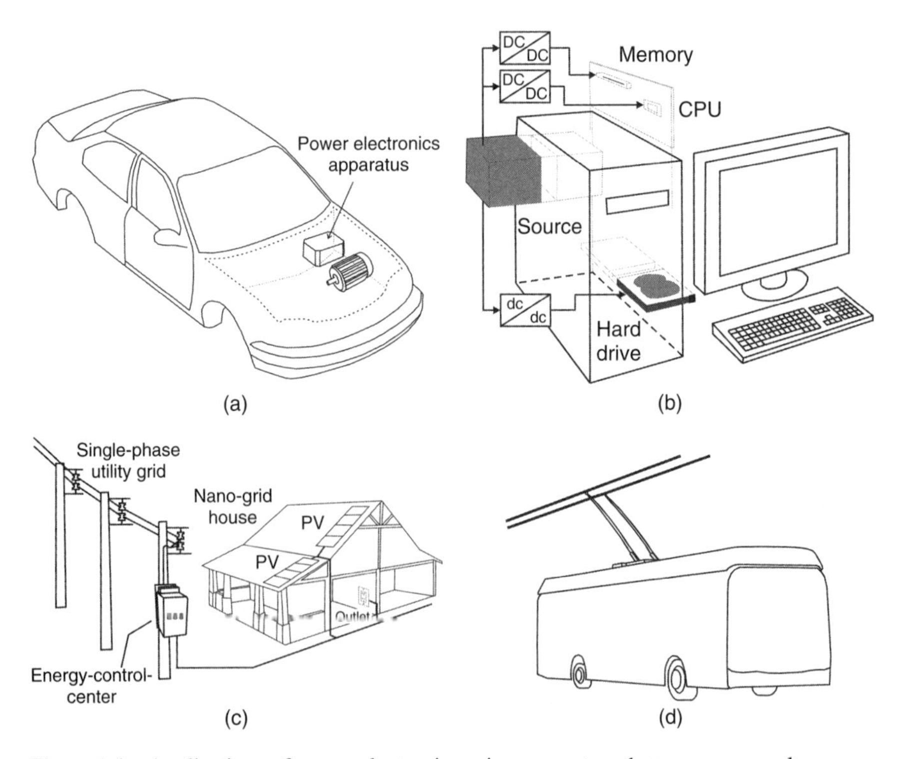

Figure 1.4 Applications of power electronics using converters that process ac voltage.

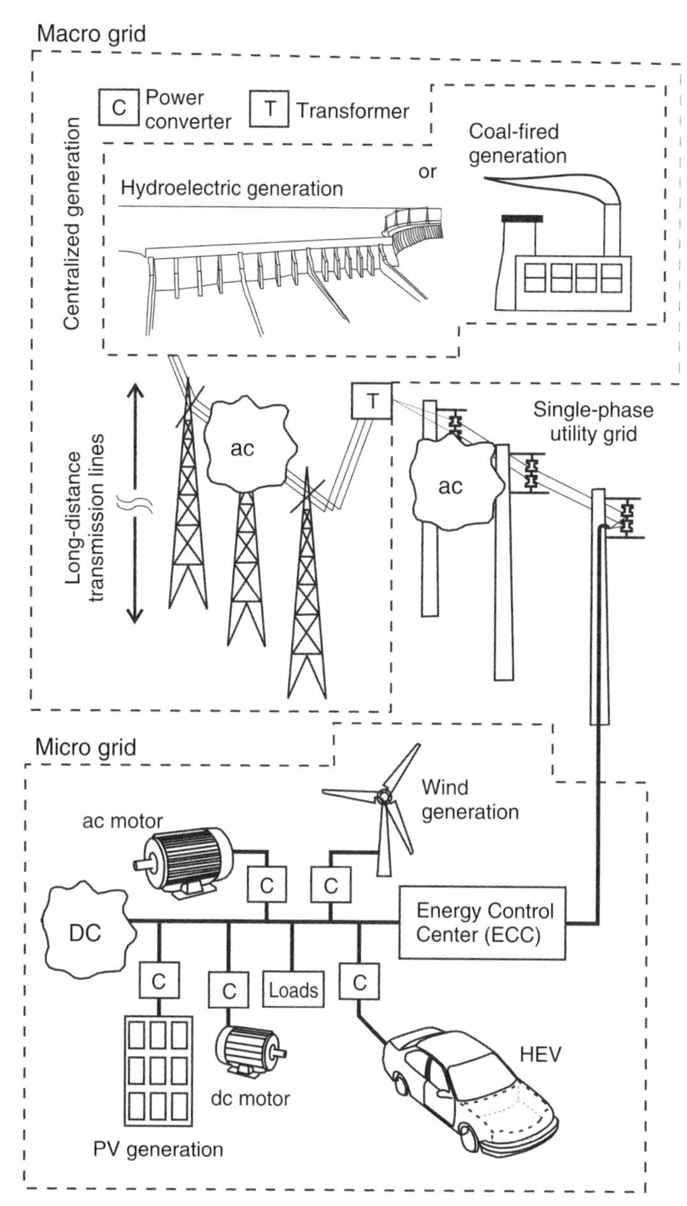

Figure 1.5 Application of power electronics in a distributed generation system. C stands for converter.

electronics point of view, the hybrid and fully electric automobiles differ one from another, mainly due to the power ratings of the inverters used. While a typical inverter rating is about 50 kW for the hybrid vehicle, the inverter rating for a fully electric vehicle is about 200 kW. The inverter motor drive system that furnishes energy to the power-train is by far the most important power electronics system used in this kind of application, but the battery charge and other peripheral systems are also crucial. The main features expected in this application are high efficiency performance, compact on-board energy storage, and low manufacturing cost for market competition with conventional thermal-engine vehicles.

Desktop and laptop computers can be considered as systems with on-board distribution schemes where different dc bus voltages are required. Inside these equipment can be found many power electronics converters, as seen in Fig. 1.4(b). An ac–dc converter produces a dc voltage bus from an ac utility grid, which will be employed by different dc–dc converters to supply the microprocessor, disk drive, memory, and so on. In the case of laptops, a battery charger is added with a power management system to control sleep modes, which guarantees extension in battery life via power consumption reduction.

Figure 1.4(c) shows the application of the power electronics converters in renewable energy systems, which nowadays is a hot topic in the political agenda of many industrialized countries, mainly due to environmental issues and as an alternative way to establish a decentralized generation system. It is worth mentioning that, besides the advantages of renewable energy, this kind of system presents a high price energy generation, especially when it is compared to conventional sources such as hydroelectric power and coal. In this sense, power electronics converters must deal with efficiency, reliability, and cost reduction, in order to make those alternative sources of energy more competitive.

Figure 1.4(d) shows a trolley bus, which is an electric bus that receives electrical energy directly from overhead wires (generally suspended from roadside posts) by using spring-loaded trolley poles.

A well-defined traditional power distribution system has a radial topology and unidirectional power flow to feed end-users. However, in the last few years, there has been research and development in replacing this paradigm by a new and complex multisource system with active functions and bidirectional power flow capability. In this new scenario, the utility grid is supposed to guarantee load management and demand side management, as well as using market price of electricity, and forecasting of energy (e.g., based on wind and solar renewable sources) in order to optimize the distribution system as a whole.

A microgrid can be defined as a localized grouping of electricity generation, energy storage, and loads that are normally connected to a traditional centralized grid (macrogrid), as seen in Fig. 1.5. Figure 1.5 shows a microgrid with a dc bus, where the power converters (represented generically by the letter C) interface distributed sources and loads with the dc bus. The point of common coupling (PCC) between micro- and macrogrid can be disconnected, which means that the microgrid can then operate autonomously. In this case, an island detection system is necessary, which safely disconnects the microgrid. The interface between micro- and macrogrid is possible due to advances made in the power electronics

The important equipment in this scenario is the Energy-Control-Center (ECC), consisting of a bidirectional ac–dc (or dc–ac) power conversion converter used to interface the utility ac grid and dc bus. The multiple dispersed generation sources and the ability to isolate the microgrid from a larger network would provide highly reliable electric power.

Another important area in which power electronics is becoming more and more common is in aerospace industry. Many loads classically powered by hydraulic networks were replaced by electrical power loads (e.g., pumps and braking). Besides facing the common challenges, the power electronics converters must deal with harsh environment constraints in terms of temperature, low pressure, humidity, and vibrations.

1.5 SUMMARY

Following the introduction, this chapter presents in Section 1.2 the background of the book, highlighting the type of configurations that this book will deal with (i.e., dc–ac and ac–dc–ac converters). Section 1.3 gives a brief history of the power electronics devices and power electronics converters, focusing on the development of switching power suppliers and SCR rectifiers. Finally, some applications are considered in Section 1.4 to show the wide range of applications of power electronics converters. Readers can find further discussion from References 4 to 13.

REFERENCES

[1] Liserre M. Dr. Bimal K. Bose: a reference for generations [editor's column]. IEEE Ind Electron Mag 2009;3(2):2–5.

[2] Rashid MH. *Power Electronics Handbook: Devices, Circuits, and Applications*. San Diego, CA: Elsevier; 2007. p 1–2.

[3] Bose BK. *Power Electronics and Motor Drives: Advances and Trends*. San Diego, CA: Elsevier; 2006. p 1–3.

[4] Blaabjerg F, Consoli A, Ferreira JA, van Wyk JD. The future of electronic power processing and conversion. IEEE Trans Ind Appl 2005;41(1):3–8.

[5] Gutzwiller FW. Thyristors and rectifier diodes - the semiconductor workhorses. IEEE Spectr 1967;4:102–111.

[6] Elasser A, Kheraluwala MH, Ghezzo M, Steigerwald RL, Evers NA, Kretchmer J, Chow TP. A comparative evaluation of new silicon carbide diodes and state-of-the-art silicon diodes for power electronic applications. IEEE Trans Ind Appl 2003;39(4):915–921.

[7] dos Santos EC Jr, Jacobina CB, da Silva ERC, Rocha N. Single-phase to three-phase power converters: state of the art. IEEE Trans Power Electron 2012;27(5):2437–2452.

[8] Goldman A. *Magnetic Components for Power Electronics*. Kluwer Academic Publishers; 2002.

[9] IBM Customer Engineering Reference Manual: 736 Power Supply, 741 Power Supply, 746 Power Distribution Unit; 1958. p 60–17.

[10] Karcher EA. Silicon controlled rectifiers in monolithic integrated circuits. Electron Devices Meeting, 1965 International; vol. 11; 1965. p 6–17.

[11] Mohan N, Undeland TM, Robbins WP. *Power Electronics: Converters, Applications, and Design*. 3rd ed. John Wiley & Sons; 2002.

[12] Rashid MH. *Power Electronics: Circuits, Devices, and Applications*. Prentice Hall; 1993.

[13] El-Hawary ME. *Principles of Electric Machines with Power Electronic*. Wiley: IEEE Press; 2002.

CHAPTER 2

POWER SWITCHES AND OVERVIEW OF BASIC POWER CONVERTERS

2.1 INTRODUCTION

The basic principles and characteristics of the main power switches are presented in this chapter. Furthermore, an overview of the principal power electronics converters is furnished, highlighting the main characteristics for each type of topology. Semiconductor power devices are the center piece of the power electronics converters. While the knowledge about such devices is crucial to design a power converter with specific characteristics for a given application, the study of different topologies brings up new possibilities to improve the energy process system.

Before dealing specifically with pulse-width modulation (PWM) converters processing ac voltage, which is the core of this book, this chapter considers a large variety of power conversion possibilities. The converters studied in this chapter include dc–dc, dc–ac, ac–dc, and ac–ac converters, as well as voltage-, and current-source converters.

As the objective is to furnish an overview of the different types of switches and converters, a deep analysis of the converters described in this chapter is omitted. However, the following chapters present a systematic description of both power converters, called Power Block Geometry, and the advanced PWM converters processing ac voltage (e.g., multilevel converters and back-to-back converters) to the smallest detail.

This chapter is organized as follows: Section 2.2 presents the ideal characteristics of the major power switches available in the market, highlighting their static and dynamic features; Section 2.3 shows the real characteristics of such semiconductor devices, sorted in terms of dynamic characteristics; Section 2.4 describes basic power electronics converters, and finally, Section 2.5 summarizes the chapter.

Advanced Power Electronics Converters: PWM Converters Processing AC Voltages,
Forty Fifth Edition. Euzeli Cipriano dos Santos Jr. and Edison Roberto Cabral da Silva.

2.2 POWER ELECTRONICS DEVICES AS IDEAL SWITCHES

The design and construction of power semiconductor devices focus on how to improve their performance toward the hypothetical concept of an ideal switch. Figure 2.1 shows the power switches and the year of development of each switch. The performance of a given power switch is normally measured by its static and dynamic characteristics. An ideal switch must have the following characteristics: (i) infinite blocking voltage capability, (ii) no current while the switch is off, (iii) infinite current capability when on, (iv) drop voltage equal to zero while on, (v) no switching or conduction losses, and (vi) capability to operate at any switching frequency.

There are different ways to classify a power switch. In this book, two different ways are considered: static characteristics and dynamic controllability. The main

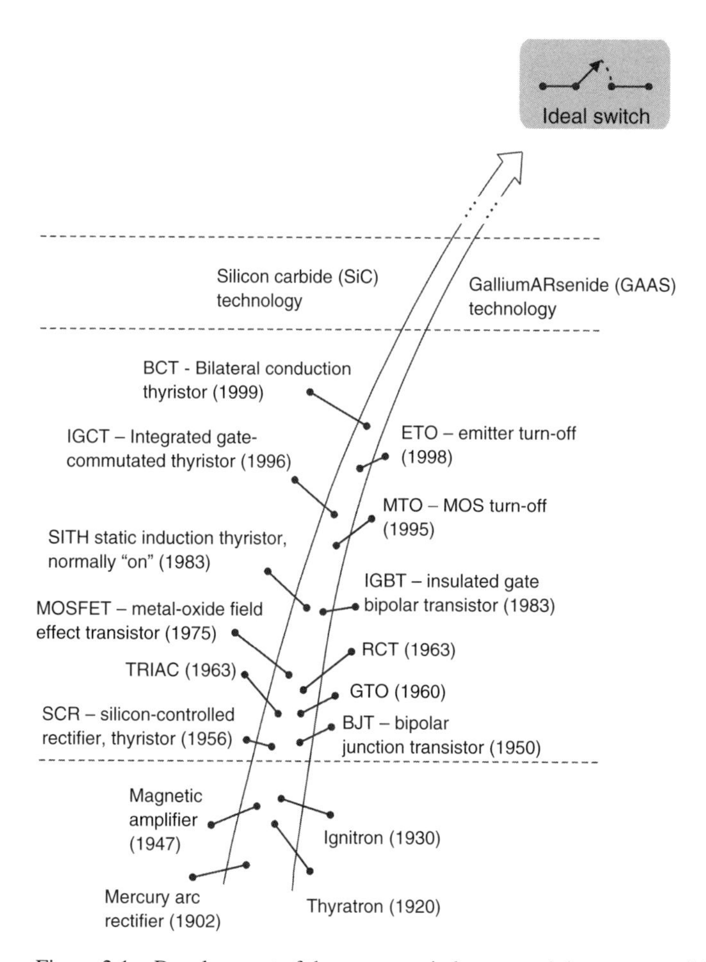

Figure 2.1 Development of the power switches toward the concept of the ideal switch.

figures of merit for the static characteristics are the graphs describing the current versus voltage behaviors ($I-V$). On the other hand, for the dynamic features, the capability to change the states on and off through either an external signal (gating command) or by the variables of the circuit in which the switch is connected is considered. For example, there is no gating signal to turn on and off a diode, its conduction or blocking depends upon the voltage and current imposed by the circuit.

2.2.1 Static Characteristics

The static characteristics of the power semiconductor devices are related to their ability to either conduct or block one or two polarities, as shown in Fig. 2.2.

Figure 2.2(a) and 2.2(b) shows the ideal $I-V$ characteristic for the blocking and conducting states, respectively. Figure 2.2(c)–2.2(g) depicts the voltage versus current ($I-V$) ideal characteristics for the main devices found in the market.

The semiconductor devices can, therefore, operate with either unidirectional or bidirectional (UniC or BidC) current, and with either unidirectional or bidirectional voltage (UniV or BidV). For example: (i) diode, bipolar junction transistor (BJT), and insulated bipolar junction transistor (IGBT) are unidirectional voltage and current type of devices, (ii) SCR is an unidirectional current and bidirectional voltage device, (iii) TRIAC and bidirectional controlled thyristor (BCT) are bidirectional in current and voltage, and (iv) MOSFET is unidirectional in voltage and bidirectional in current.

The characteristics presented in Fig. 2.2(c)–2.2(g) play an important role for the specification and design of the power electronics converters. These graphs are in fact approximations of the real $I-V$ characteristics of the device. For example, Fig. 2.2(c) is an approximation of the real characteristics of a power diode, which is presented later in this chapter.

2.2.2 Dynamic Characteristics

The dynamic characteristics of a specific device are related to the behavior of voltage and current when there is a change either from conduction to blocking state or from blocking to conduction. Such a change is known as commutation or switching. The commutation (or switching) from the blocking state to the conduction state is referred to as either turn-on or conduction. The commutation from the conduction state to the blocking state is referred to as either turn-off or blocking. For the $I-V$ characteristic curves (shown in Fig. 2.2), the commutation process corresponds to going from the operating point on an axis to another one. For example, for a switch UniC/UniV with direct voltage as in Fig. 2.2(c), the blocking procedure makes the variables of the switch (voltage and current) go from the I axis to the V-axis.

In fact, the commutation process can be spontaneous or controlled, as shown in Fig. 2.3. Four cases are presented as follows:

- Spontaneous conduction (SC)—see Fig. 2.3(a);
- Spontaneous blocking (SB)—see Fig. 2.3(b);

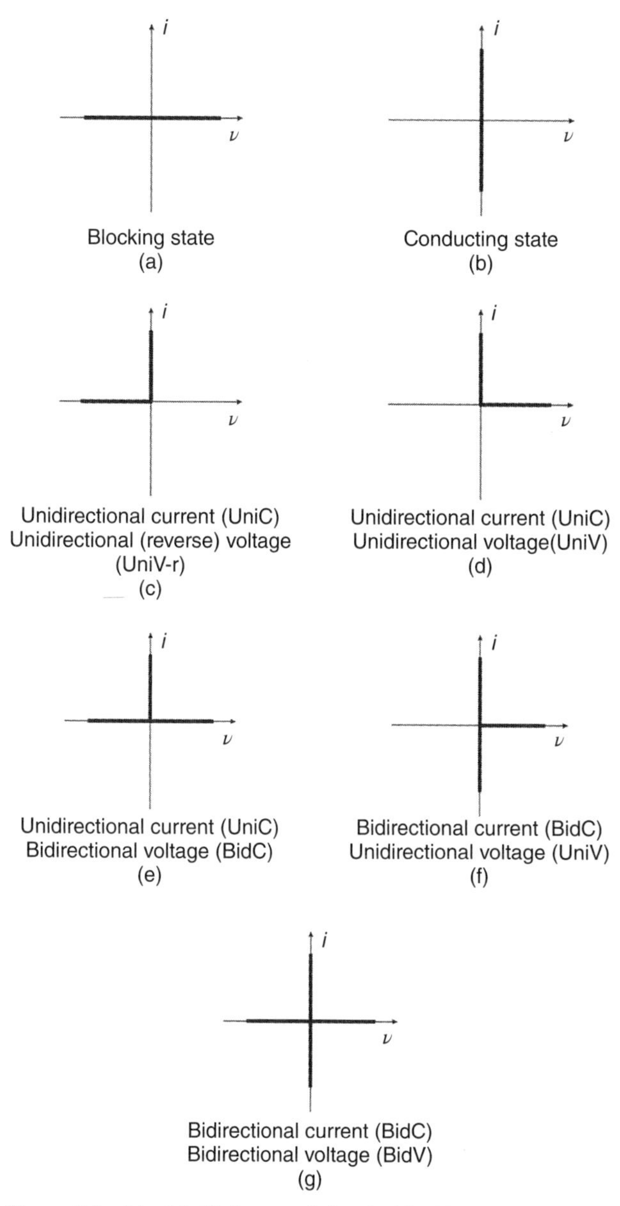

Figure 2.2 Ideal I–V characteristics for blocking and conduction states of the power switches available in the market.

- Controlled conduction (CC)—see Fig. 2.3(c);
- CB—see Fig. 2.3(d).

In the case of spontaneous commutation (i.e., conduction and blocking), the change of state is defined by the variables of the power circuit, while for the controlled

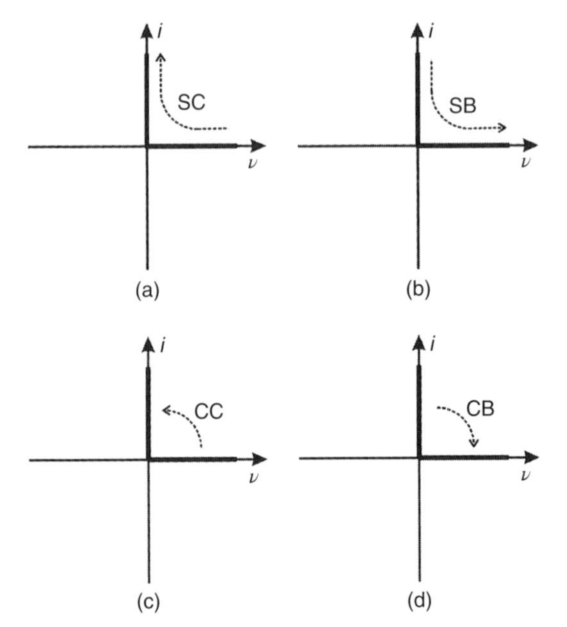

Figure 2.3 Dynamic characteristics of an ideal device considering its commutation process. (a) Spontaneous turn on, (b) spontaneous turn off, (c) controlled turn on, and (d) controlled turn off.

commutation, the change of state is guaranteed via a gating signal (command signal).

Figure 2.4 shows the behavior of the voltage and current in time-domain (left side) and dynamic characteristics (right side) for SC, SB, CC, and CB. The points (P1, P2, and P3) along with the axes in Fig. 2.4 illustrate the behavior of the variables when the switching process occurs for both spontaneous and controlled commutation. For example, in Fig. 2.4(a) P1 indicates the operation point when the switch is blocked, which means that such a device is submitted to a negative voltage while its current is zero. The operation point P2 shows the device's behavior while its voltage has been reduced. Finally, P3 shows the values of voltage and current when the switch is turned on. This commutation process occurs, for example, in a rectifier circuit with diodes where the circuit itself allows the state change from blocking to conduction. A similar analysis can be done for SB in Fig. 2.4(b).

On the other hand, a controlled commutation device must have a control electrode, usually called a gate or base, in addition to the two main terminals. A control signal applied to the gate or base, while the voltage applied between the main terminals is positive, results in change of state in a desirable manner, that is, from one axis to another, as shown in Fig. 2.4(c) and 2.4(d).

In terms of dynamic characteristics, the ideal power devices can be classified as follows:

- SC/SB;
- CC/SB;
- CC/CB;
- SC/CB.

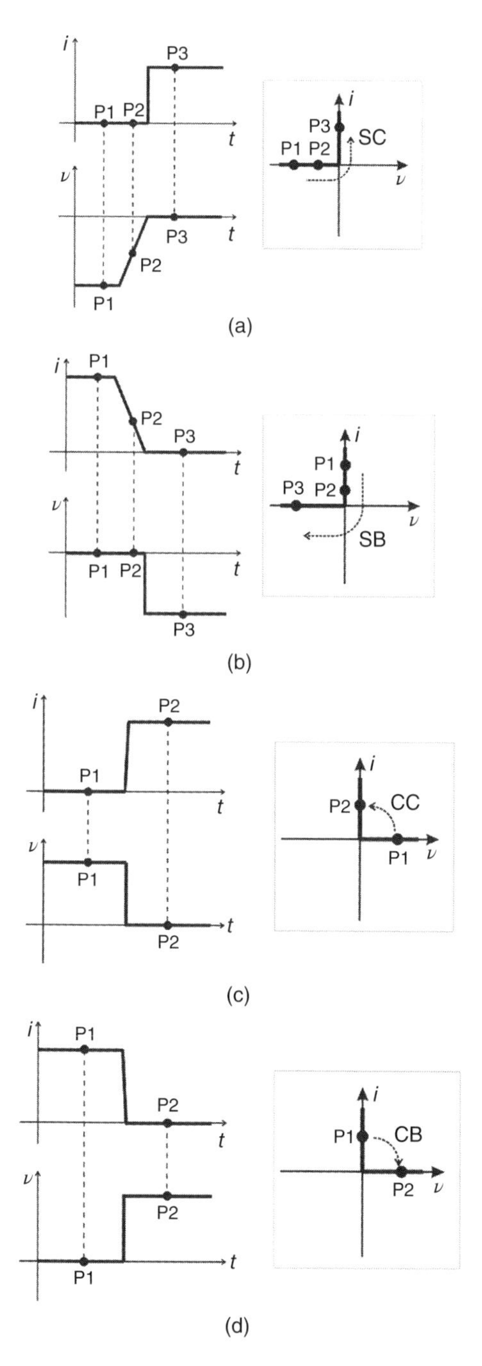

Figure 2.4 Comparison on time-domain and dynamic characteristics during the commutation process. (a) SC, (b) SB, (c) CC, and (d) CB.

2.3 MAIN REAL POWER SEMICONDUCTOR DEVICES

The most common power electronics devices are diodes, thyristors, and power transistors. Usually they have two power terminals (anode/cathode, or collector/emitter or drain/source) and one or more command terminals (gate/ base). Unlike the ideal characteristics presented earlier, real devices have practical limits for rated voltage and current, as well as for their operation frequency. Such limits are normally specified by the datasheet furnished by manufacturers. Therefore, the device characteristics and their specification are crucial to choose a particular power device instead of others.

The most common nonideal device characteristics are

(a) *The Forward and Reverse Voltage Capability.* The main limiting ratings are as follows:

  i. *Forward Blocking Voltage.* The maximum repetitive forward voltage that can be applied to the power terminals of the device (normally from anode to cathode, from collector to emitter, from drain to source) so that the device blocks the current flow (blocking state) in the direct sense, unless commanded to turn on.

  ii. *Reverse Blocking Voltage.* The maximum repetitive reverse voltage that can be applied to the power terminals of the device (from cathode to anode, for instance) so that the device blocks the current flow in the reverse sense.

  iii. *Maximum Peak Nonrepetitive Forward and Reverse Voltage.* The maximum nonrepetitive forward and reverse voltages, respectively, under transient conditions.

  iv. V_{dc} and V_R. Maximum continuous direct (forward) and reverse blocking voltages, respectively. This is the maximum dc voltage that the diode can withstand in reverse-bias mode on a continual basis.

  v. *Forward Voltage Drop.* This is the instantaneous value of the drop voltage, which is normally dependent on the temperature.

(b) The current capability while the device is on (conducting) is junction temperature-dependent. The main limiting ratings are the following:

  i. *On-State Current.* It is the average value of the conduction current.

  ii. *On-State Root Mean Square (RMS) Current.* It is the RMS value of the conduction current.

  iii. *Peak Repetitive Forward Current.* It is the maximum repetitive current that can flow through the device.

  iv. *Peak Surge Forward Current.* It is the maximum nonrepetitive forward current that can flow through the device under transient conditions.

It should be noted that when blocked, the device still conducts a leakage current that can be forward or reverse, depending on the device state.

(c) The switching is not instantaneous, so there are limits in the switching frequency, as follows:

  i. *Turn-On Time.* It is the time required to complete the turn-on process.

ii. *Turn-Off Time or Recovery Time* (t_{rr}). it is the minimum interval of time required from the instant the conduction current is decreased to zero so that the device is capable of withstanding the forward voltage without turning on.

iii. dv/dt. It is the maximum variation rate of the forward voltage that can be applied to the device in the blocking state without starting a nonprogrammed turn-on.

iv. di/dt. It is the maximum variation rate of the forward current during turn-on that can be applied to the device; higher di/dt than the one specified by the manufacturer may destroy the component.

(d) The maximum switching frequency depends on the recovery time of the device.

(e) *Power Losses*. There are three main components of losses in a semiconductor device: (i) switching losses (turn-on and turn-off), (ii) conduction losses, and (iii) reverse conduction losses.

The most used devices in industrial applications are considered in the following section. Such switches are sorted in terms of their dynamic characteristics.

2.3.1 Spontaneous Conduction/Spontaneous Blocking

The conventional diode and the Schottky diode have the characteristic of SC and SB, as presented in the sequence.

The Conventional Diode The conventional diode is a silicon p–n junction device with two terminals, anode (A) and cathode (K), that conducts current from A to K and blocks the reverse current. The rating of the voltage goes up to 9 kV with 4.8 kA and 5 kV with 13 kA. Its symbol is presented in Fig. 2.5(a) and its real I–V characteristic is given in Fig. 2.5(b). The diode conducts when $v_{AK} > 0$, the current being limited by the external circuit, its typical direct voltage drop is 0.7 V. Generally speaking, it can be said that they recover their reverse blocking capability when the forward current i_A goes to zero and a reverse voltage is applied across its terminals, for an interval of time longer than the reverse recovery time t_{rr} obtained in its technical data sheet. In reality, after reaching zero, the current reverses its direction, reaching a reverse peak called "peak reverse recovery current" that is comparable to the forward current. Snubber

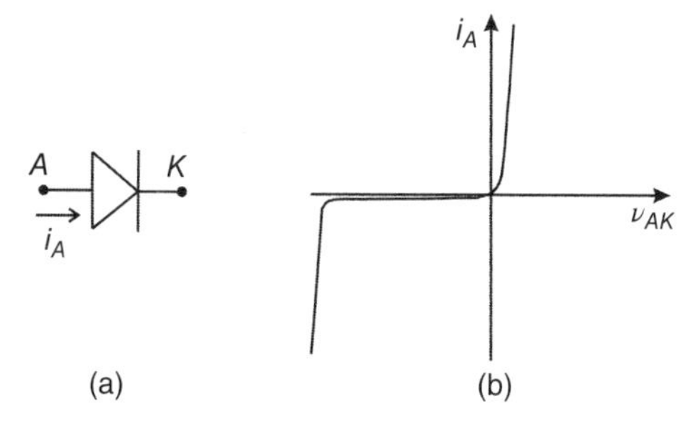

Figure 2.5 The diode: (a) symbol and (b) I–V characteristic.

circuits are essential for the adequate protection of the diode. The snubber circuit will protect it from overvoltage spikes, mainly due to the junction capacitance and leakage inductance from terminals and circuit connections. The basic snubber circuit is composed of a capacitor in series with a resistor connected in parallel with the diode. The size of the snubber circuit is reduced for the fast ($t_{rr} < 1$ μs) and ultrafast recovery diode ($t_{rr} < 100$ ns), with ratings reaching: (i) 600 V, 30 A and $t_{rr} = 50$ ns, and (ii) 1200 V, 120 A, $t_{rr} = 85$ ns. Recent development in the silicon carbide material allows reducing the diode reverse recovery time up to 16 ns.

The Schottky Diode Unlike the standard diode presented previously, the Schottky diode is formed by a metal–semiconductor junction (the p material is replaced by metal). It has a lower conduction drop voltage (typically 0.5 V) and faster switching time than the standard diode (less than 100 ps), which allows its operation in higher frequency. However, it has lower blocking voltage (typically up to 200 V) and higher leakage current. Its silicon carbide (SiC) version is promising and has been already tested to withstand 1.2 kV with 60 A.

2.3.2 Controlled Conduction/Spontaneous Blocking Devices

The main devices in this group are the silicon-controlled rectifier (SCR) and the TRIode AC (TRIAC).

The SCR The SCR is a silicon p–n–p–n device with three terminals: two power terminals, anode (A) and cathode (K) through which the current flows, and a control terminal, named the gate terminal (G). It is unidirectional in current and blocks in both forward and reverse directions (UniC/BidV). The device is triggered by a positive gate current pulse. When the gate is not triggered the SCR blocks the current in the forward direction even with $v_{AK} > 0$. If $i_{GK} > 0$ while $v_{AK} > 0$, the SCR conducts the current from anode to cathode imposed by the circuit in which the SCR is inserted. Its voltage drop in conduction is from 1 to 4 V. Once in conduction, the SCR behaves like a diode and it can only be turned off when the anode current becomes zero. After the current reaches zero, a reverse voltage should be applied across its terminals in order to accelerate the capability of forward blocking. Note that the gate loses control over the SCR during its conduction. The symbol and real I–V characteristics are shown in Fig. 2.6(a) and 2.6(b), respectively. The direct and reverse voltages across its terminals are symmetrical in the conventional SCR (there are SCRs with asymmetrical voltage characteristics, the ASCR) and can reach 5 kV with 5 kA and 12 kV with 2.3 kA, or even 5 kV with 8 kA. Unexpected turn-on can occur due to a high dv/dt. For these reasons a snubber circuit with a capacitor can be used to avoid both triggering by dv/dt and overvoltage spikes, a resistor that limits the current peak, and an inductor that limits the di/dt rate.

The TRIAC The TRIAC is a thyristor that operates as two SCRs monolithically integrated connected in antiparallel. Such a device can conduct and block in both forward and reverse directions (BidC/BidV). It has two power terminals, T_1 and T_2, and one gate G. Its symbol and its I–V characteristic are presented in Fig. 2.7(a) and

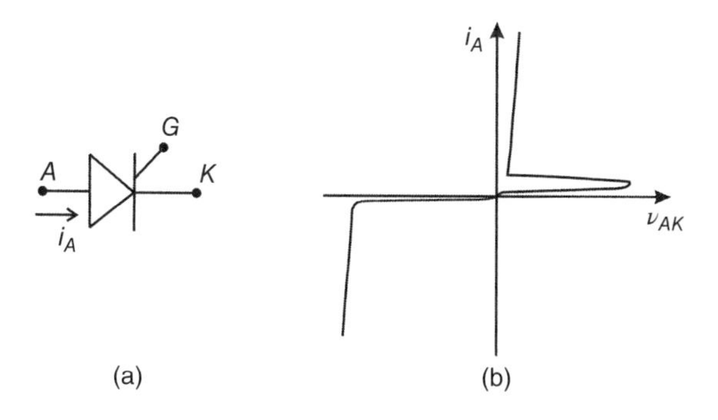

Figure 2.6 The SCR: (a) symbol and (b) I–V characteristic.

2.7(b), respectively. When $v_T = v_{T1} - v_{T2}$ is positive (quadrant I) and the TRIAC is turned on by a positive gate current pulse. But when the v_T is negative (quadrant III) it is turned on by a negative gate current pulse. The operation in quadrants II and IV is also possible but the gate triggering is less sensitive in these cases. Also, the dv/dt is poorer than that of SCRs.

Other CC/SB devices are the light-activated SCR (LASCR), the reverse conducting thyristor (RCT), which functions as an SCR with an inverse-parallel diode (BidC/UniV), and the BCT, in which two SCRs are also connected in antiparallel, but differently from the TRIAC, each one acting independently with its independent gate control (BidC/BidV).

2.3.3 Controlled Conduction/Controlled Blocking Devices

The main CC and CB devices are (1) basic transistors, like the BJT and the metal oxide semiconductor field-effect transistor (MOSFET); (2) basic thyristors, like the gate turn-off (GTO); (3) mixed transistors, like the IGBT; and (4) mixed thyristors, like the MOS-controlled thyristor (MCT) and the Integrated gate commutation thyristor (IGCT).

Bipolar Junction Transistor (BJT) Unidirectional in current, the BJT can be either n–p–n or p–n–p and it has asymmetric blocking, only withstanding some tens of reverse blocking voltage (UniC/UniV). Its symbol and static characteristics

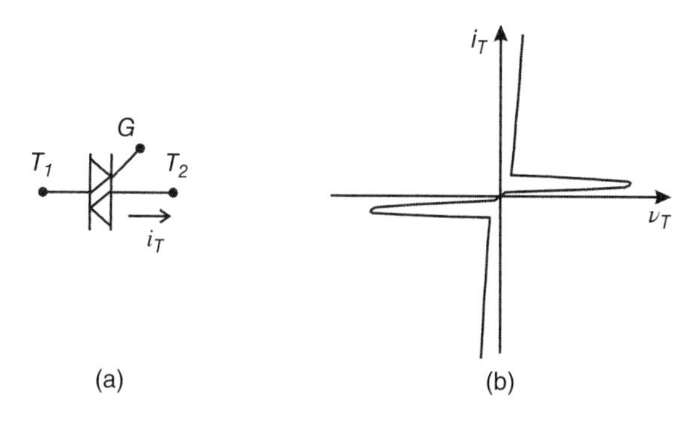

Figure 2.7 The TRIAC: (a) symbol and (b) I–V characteristic.

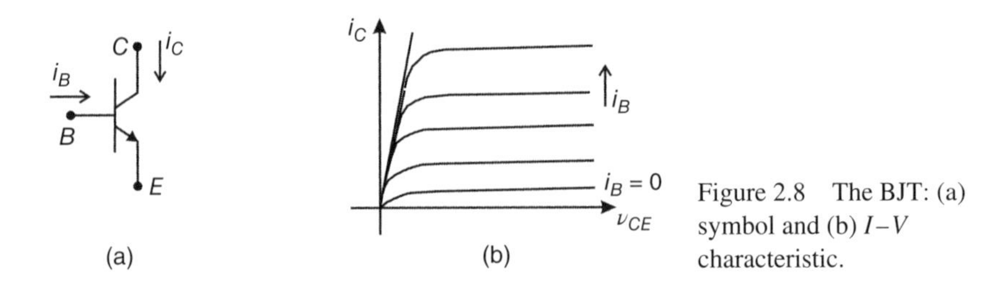

Figure 2.8 The BJT: (a) symbol and (b) $I-V$ characteristic.

are shown in Fig. 2.8. It has three terminals, the collector (C) and emitter (E) through which the current (i_c) flows and the base (B) that controls the amplitude of i_c. Unlike lower power BJT devices, the power devices operate in the quasi-saturated region, which reduces the stored charge, avoiding the long recovery time obtained when operated in the saturated region. As its dc current gain (h_{fe}) is much lower than its signal level counterpart (in general as low as 5), the Darlington connection is more common because its dc current gain is higher. A disadvantage of this arrangement is its drop voltage and leakage current. BJTs with forward blocking voltage up to 1 kV are available.

The MOSFET Application of the metal oxide semiconductor (MOS) technology to the field-effect transistor resulted in the power MOSFET. It has three terminals, the drain (D), the source (S), and the gate (G).

Its symbol and $I-V$ characteristics are shown in Fig. 2.9. It is an $n-p-n$, or $p-n-p$, device in which its two $p-n$, or $n-p$, layers are connected through a metal, so that a capacitor is formed between G and D. The MOSFET is turned on by a voltage pulse. The MOSFET acts as a resistance while conducting and behaves like a transistor before being turned on. Similar devices have been developed under different names, depending on the manufacturer, such as the HEXFET ("hexagonal-field-effect-transistor"), SIPMOS ("Siemens-Power Metal Oxide Silicon") and TMOS ("T flowing current metal oxide silicon"). Its forward blocking voltage is in the range of 1 kV and it can operate in high frequency. Since it has an intrinsic diode in antiparallel connection, it operates with forward and reverse current and can be classified as a BidC/UniV device. It has replaced the BJT in the range of low voltage and high frequency but its disadvantage comes from its high conduction resistance that increases with the voltage.

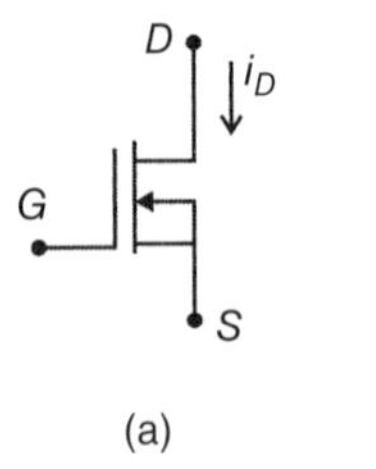

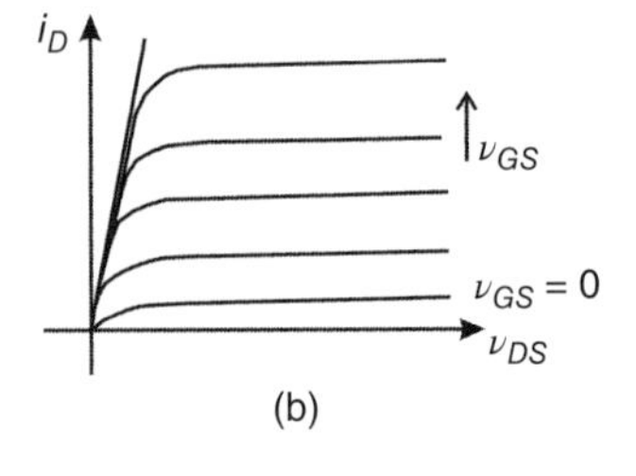

Figure 2.9 The MOSFET: (a) symbol and (b) $I-V$ characteristic.

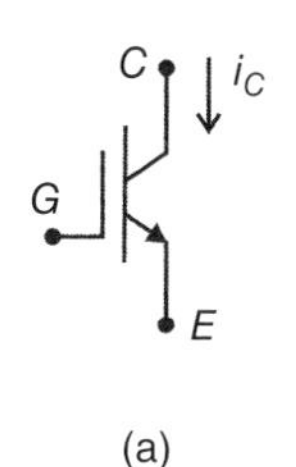

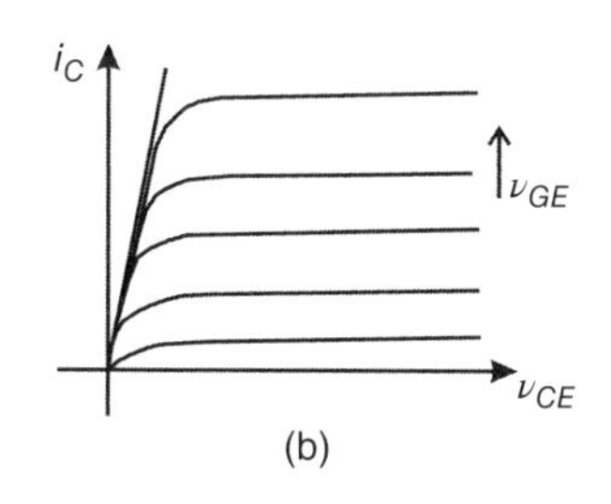

Figure 2.10 The IGBT: (a) symbol and (b) $I-V$ characteristic.

However, it has been demonstrated that SiC MOSFET can have extremely low on-resistance, thus saving energy. It has been predicted that in some applications the SiC MOSFET will replace the Si-IGBT in the voltage range of more than 1 kV.

The IGBT The IGBT promoted a great revolution in power electronics as it uses a mixed bipolar-MOSFET technology. It combines the advantages of the MOSFET and the BJT Darlington, with controlled turn-on and turn-off. Like the MOSFET it has a high input impedance and needs low energy to be switched on. Its symbol and $I-V$ characteristics are given in Fig. 2.10. The IGBT conduction drop voltage is small (from 2 to 3 V in a device of 1 kV). The conventional device is unidirectional in current and only blocks forward voltage (UniI/UniV) with a voltage of 6.5 kV for 750 A, or 1.7 kV for 3.6 kA for dv/dt of order from 50,000 V/μs to 100,000 V/μs. The type NPT ("nonpunch-through") can reach 3.5 kV and 2 kA. Its typical turn on and turn off time is from 200 ns to 1 μs. It is possible to design it to block the voltage in both directions, forward and reverse voltages. Such a device has been recently introduced in the market with the name of reverse blocking IGBT but for lower voltage (1.7 kV) and current (25 A), 1200 V/40 A, 600 V/200 A. As for the MOSFET, the SiC technology is also designing SiC–IGBT, which has reached the high blocking voltage of 15 kV.

The GTO The GTO did appear to give thyristors the option to control its turn-off. Its symbol and $I-V$ characteristics are given in Fig. 2.11 and it is, normally, a UniC/UniV device. For turning on, it only needs a small pulse of positive current at the gate. However, its turn-off current gain of the negative current pulse is typically 3–5, which means that for turning off a 6 kV/6 kA GTO, it needs a negative gate

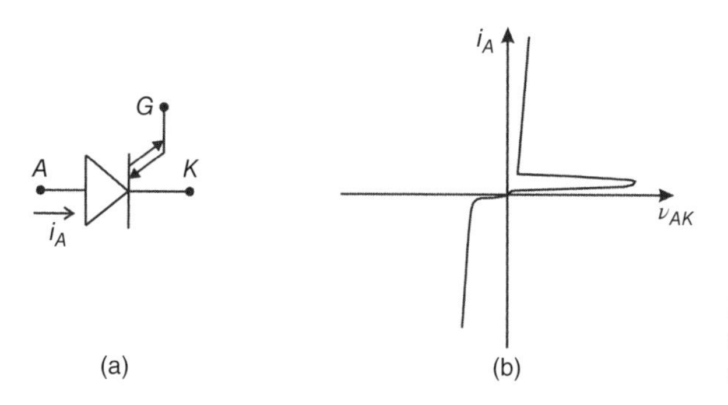

Figure 2.11 The GTO: (a) symbol and (b) $I-V$ characteristic.

current pulse of 1.5 kA. The reverse blocking GTO can withstand reverse voltage of 4.5 kV with 3 kA. Also, the GTO needs snubbers when used with inductive load.

The MCT The MCT is also called MOS-GTO and combines the characteristics of the FET integrated with the p–n–p–n structure of a thyristor. It is a UniC/UniV device. When designed, it was expected that it would be able to handle more than 200 kVA and more than 1 MVA in subsequent years. In conduction, it approximates to the SCR characteristics. Its blocking is obtained through the turn-off gate. However, even though it reached values of 2 and 3 kV and hundreds of amperes, its acceptance by the market is still undefined at the moment. Its symbol is given in Fig. 2.12(a).

The IGCT The IGCT was introduced in 1997 and it is a high voltage, high power, asymmetric blocking device, with a structure very similar to a GTO thyristor. As it is designed with a monolithically integrated antiparallel diode, it is a BidC/UniV type of device. Its symbol is given in Fig. 2.12(b). It is a device with unity turn-off current gain. This means that a 4.5 kV IGCT with a controllable anode current of 3000 A requires a turn-off negative gate current of 3 kA. So it needs a great amplification for turn off and, also, the gate driver must have an ultralow leakage inductance in order to have short duration and very large di/dt. With this purpose, the gate drive circuit is built in and such integration between the command and the device results in high turn-off speed (1 μs) and basically eliminates the problem of dv/dt found in GTOs permitting snubberless operation. In IGCTs the cathode current is diverted to the gate before any distribution of current between gate and anode is observed so that the structure p–n–p–n can be converted to a p–n–p structure. As a result, the IGCT conducts as a GTO but turns off as an IGBT, combining the characteristics of these devices. Other parameters superior in the IGCT as compared to the GTO are the conduction drop voltage and gate-driver loss. Its typical frequency is around 500 Hz it can reach 1 kHz. The device has been applied in power system installations of 100 MVA and medium power (up to 5 MW) industrial drives. Like the GTO and IGCT, the reverse blocking IGCT (RBIGCT), symmetric, has been recently developed, and can withstand 5 kV.

2.3.4 Spontaneous Conduction/Controlled Blocking Devices

Two devices can be classified as SC/CB, that is, the static induction transistor (SIT) normally-on and the Static Induction Thyristor (SITH) normally-on.

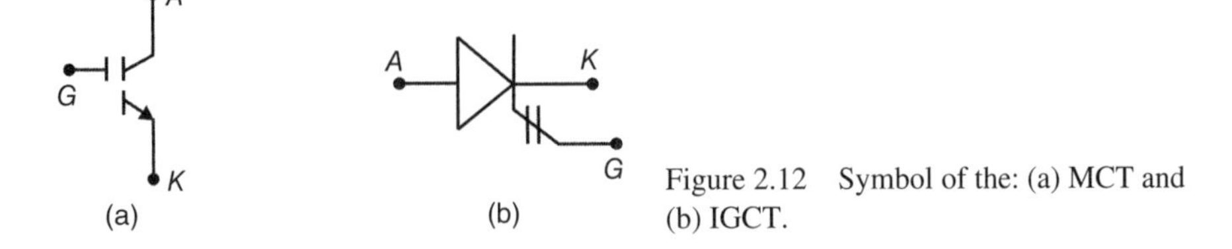

Figure 2.12 Symbol of the: (a) MCT and (b) IGCT.

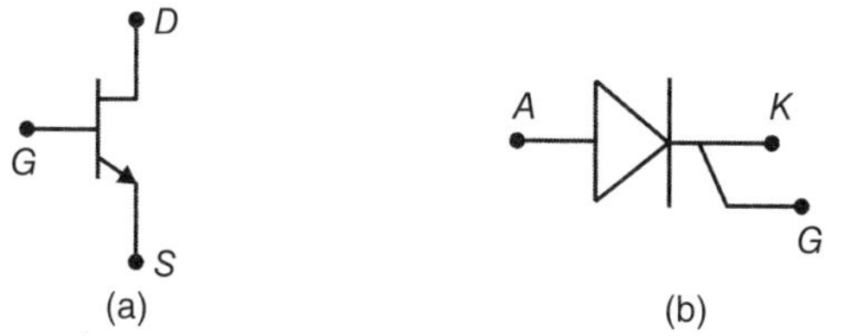

D
G
S
(a)

A K
G
(b)

Figure 2.13 Symbol of: (a) SIT and (b) SITH.

The SIT The SIT, also known as Junction Field-Effect Transistor (J-FET), is an n-type field-effect transistor, voltage driven. Its symbol is given in Fig. 2.13(a). It is a UniC/UniV device with ratings of 1.2 kV and 300 A and can operate up to 100 kHz. It conducts when $v_{GS} = 0$ and its drain current can be controlled by the gate-source voltage, v_{GS}. However, the gate-source voltage required for turning off the device is high and it is not uncommon that a voltage as high as 40 V (negative) is needed. This characteristic of normally turned on with controlled turn-off classifies the device as SC/CB (the SIT normally-off has been already developed, with limited performance). However, its high forward voltage makes it unsuitable for most power electronics applications, unless radio frequency operation is needed. Additionally, the difficulty in manufacturing it raises concerns about its mass production. This device has been shown to be more promising with the SiC technology reaching ratings of 1.2 kV/17 A and 6.5 kV/5 A. Its range of power operation is up to 50 kW. It has been shown that a SiC SIT of 125 V/2.2 kW can operate in the ultrahigh frequency range up to 450 MHz. However, the gate-source voltage required for turning this device off is about $V_{GS} = -30$ V. Although SiC SIT is a normally-on device, recently a SiC SIT normally-off has been developed with ratings of 1.2 kV and 35 A, with a turn-off V_{GS} of only −2.5 V.

The SITH The SITH, or SIThy, a device unidirectional in voltage, is a combination of an n-channel SIT and a p–n–p transistor. Its symbol is given in Fig. 2.13(b). It is a normally-on device with turn-off controlled by negative gate voltages and it does not have reverse blocking capability. It can handle currents in the range of 300 A to 2 kA with a recovery time from 2 to 4 μs and it is a device rated with 1.5 kV/300 A and has a rated frequency of 10 kHz but it is expected to be applied to power sources up to 10 MHz. Although it can operate at higher frequencies compared to the GTO, its higher drop voltage and lower current gain are its main disadvantages. As for the SIT, the complexity of the manufacture process and the high negative gate voltage for turning it off are its major drawbacks.

Exercise 2.1

The simple circuit presented below allows the power flow control between the voltage source ($V_{dc} = 1$ kV) and the resistive load (R_o), by turning-on and -off the switch S with switching frequency equal to f_s. From the ratings of the switches given in this chapter determine the appropriate power device to be employed as the switch S, considering the following conditions:

(a) $f_s = 1$ kHz and $R_o = 0.6$

(b) $f_s = 50$ kHz and $R_o = 2.5$

(c) $f_s = 200$ kHz and $R_o = 10$.

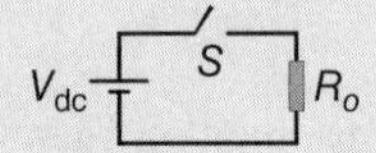

2.3.5 List of Inventors of the Major Power Switches

The list of inventors of the major power devices available in today's market is presented here:

- *Diode pn.* The first semiconductor diode, a germanium-made device, was created in 1952, 200 V/35 A, by R.N. Hall. Its counterpart in silicon was invented by Russell Ohl at Bell Laboratories, 500 V.
- *BJT.* In 1947, W. Shockley, J. Bardeen, and W. Brattain built a germanium point-contact transistor. The BJT was created by W. Shockley from Bell Laboratory in 1948 and developed in 1950 for 500 V/20 A. The first commercially available silicon devices (grown junction) were manufactured in 1954 by Gordon Teal.
- *SCR (Thyristor).* The SCR or thyristor was proposed by William Shockley in 1950. It was theoretically described in 1954 and 1955 by J.L. Moll from Bell Laboratory but it was not well accepted until GE manufactured a feasible device in 1957. Its commercial version was available in 1958 and was championed by G. E.'s Frank W. "Bill" Gutzwiller (300 V/16 A).
- *GTO.* The GTO thyristor was created in 1962 by R. Aldrich and N. Holonyak from GE. It was used until the 1970s when it was replaced by the silicon power MOSFET and IGBT. This happened because the GTO was limited to low current. For instance, in 1967 its maximum rates were 500 V/10 A. However, an improved Hitachi GTO was developed in 1981 allowing for higher voltage and current (2500 V/1000 A).
- *TRIAC.* The bidirectional triode thyristor was created by F. W. Gutzwiller, from GE, in 1963. It reached 40 A experimentally but the first commercial Triacs were the SC40 and SC45, rated 200 V, 6 A, and 10 A, in 1965 and 1966, respectively, following basic research steps developed by GE's Aldrich and Nick Holonyak (1958), and Finis E. Gentry and Tuft (1963).
- *RCT.* The RCT was created in 1970 by Kokosa and B. Tuft from GE.

- *MOSFET*. The first metal-oxide field-effect transistor was created in 1958 and reported in 1960 by D. Kahng and M.M. Atalla, from Bell Laboratory. However, the first successfully commercialized power MOSFET was from International Rectifier.
- *SITH or FCT (Field-Controlled Thyristor)*. The SITH was proposed by J. Nishizawa from Mitsubishi Electric Corporation in 1975 (700 V); in the same year a similar device that received the name of FCT was reported by D.E. Houston.
- *IGBT*. The first IGBT with substantial current ratings was developed in 1982 by J. Baliga, from GE, with a symmetric blocking voltage of 600 V for 10 A (6 kVA). It was also reported by J.P. Russel in 1983 under the name of COMFET, reaching 400 V and 30 A.
- SIT (normally "on" or JFET) was introduced by J. Nishizawa in 1975, 300 V, 2 A, with an output power of 5 W at 1 GHz, 36W at 200 kHz and 40 W at 100 MHz.
- *BCT*. The bilateral controlled thyristor was created by ABB in 1998, having voltage and current ratings of 6.5 kV and 1390 A.
- *IGCT*. The IGCT was conceived in 1993 at ABB and reported in 1996 by P.K. Steimer, H. Gruning et al. from ABB for 4.5 kV and 3 kA. It was also reported in 1996 by J. Sakano et al. from Hitachi and was able to block 4 kV.
- *MTO*. The MOS turn-off thyristor was invented in 1995 by D.E. Piccone et al. from Silicon Power Corporation, with rated values of 6000 V and 500 A.

2.4 BASIC CONVERTERS

As mentioned earlier, a power electronics converter interfaces a source and a load (or another source). The source can be either of voltage type or current type. Also, the load can be of either voltage type (e.g., a capacitive load) or current type (e.g., an inductive load). The connection between a sourcc and a load should obey the following sequences: (i) *voltage (V)–current (I)–voltage (V)–current (I)*, or (ii) *current (I)–voltage (V)–current (I)–voltage (V)*. Four connections between source and load are then possible, as shown in Fig. 2.14, which establishes the following connections, V–I, I–V, $V-V$ and I–I. Notice that, for the V–V connection as in Fig. 2.14(c), the power converter interfacing load and source are expected

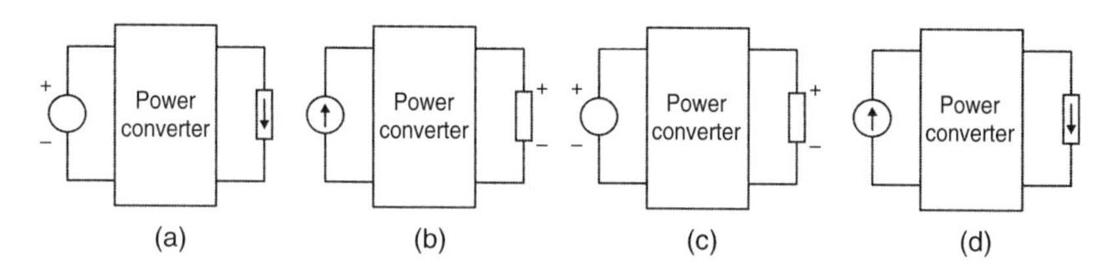

Figure 2.14 Four possibilities of connection between source and load.

to have an inductor element inside that block. The same rationale applies for Fig. 2.14(d).

Only sources and loads of different types can be directly connected, that is: (i) a voltage source can be directly connected to a current-type load and (ii) a current source can be directly connected to a voltage-type load. Such a restriction comes from the need to avoid, for example, a parallel connection between two voltage-type elements, and so preventing short-circuit between these two elements.

Except for the case of resistive load, the minimal number of switches that allows interconnecting a source and a load is two, which is called basic commutation cell. In fact such a commutation cell can be obtained with two different arrangements, as seen in Figs 2.15 (Cell I) and 2.16 (Cell II). Cell I in Fig. 2.15(a), also known as leg when it is arranged as presented in Fig. 2.15(b), can be used to connect a voltage source to an inductive load. In Fig. 2.15(c), the source is connected between points M and N while the load is connected to the point A, named as pole. Instead, in Cell II in Fig. 2.16(a), or its leg representation [Fig. 2.16(b)], the current source connected to point A allows feeding a capacitive load connected between points M and N (voltage-type load), as shown in Fig. 2.16(c). It should be noticed that while one of the switches is conducting, the other one is turned off. Note that the pole (point A) can be connected to either point M or point N, so that it can only assume the potential of those points. For this reason, both cells are said to be two-level cells.

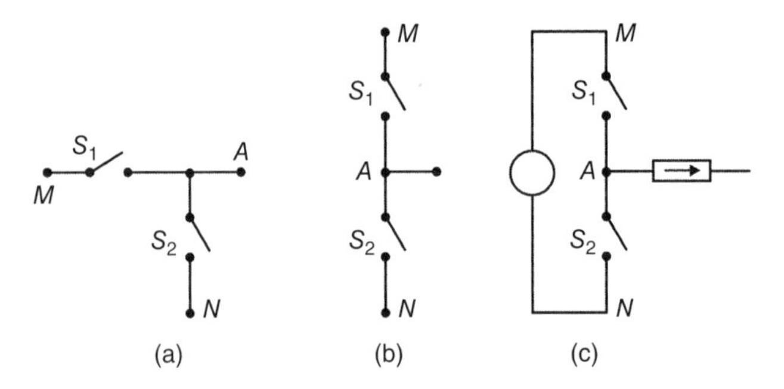

Figure 2.15 Basic cell employed to connect voltage source to current-type load.

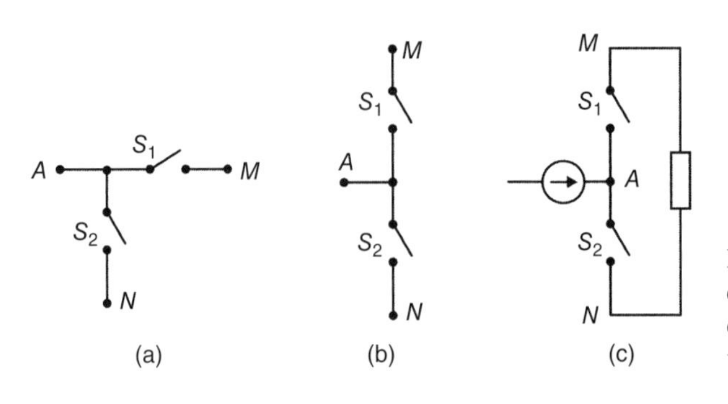

Figure 2.16 Basic cell employed to connect current source to voltage-type load.

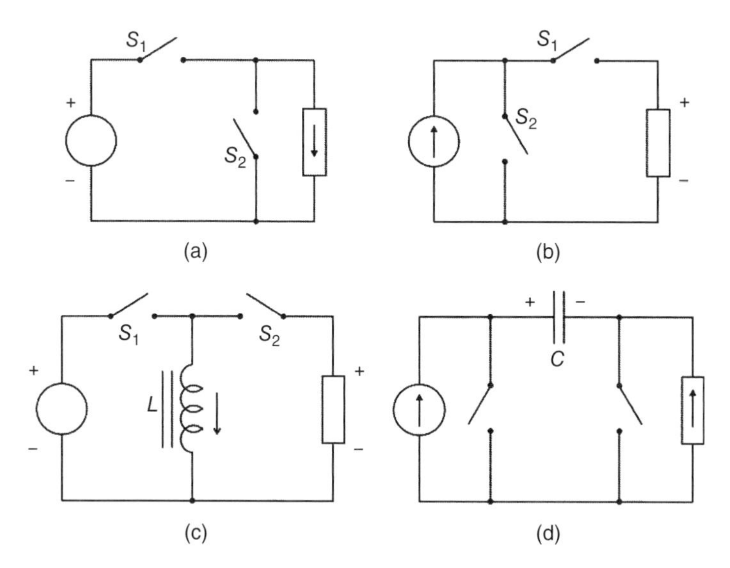

Figure 2.17 Principles for building power converters.

Figure 2.17 shows how different types of sources and loads can be connected through the basic cells presented in Figs 2.15 and 2.16. As mentioned earlier, sources and loads of the same type can only be connected through an intermediate coupling circuit. For example, to connect a voltage source to a voltage-type load, it is necessary to employ a current-type element (inductor) between the source and load [see Fig. 2.17(c)]. For duality, the voltage-type element is a capacitor that is used to connect both current source and current-type load, as shown in Fig. 2.17(d).

The basic cells permit the basic converters used to convert energy from dc-to-dc, dc-to-ac, ac-to-dc, and ac-to-ac (or simply dc–dc, dc–ac, ac–dc, and ac–ac) and also to extend them to more complex structures, in a more systematic way. This will be in fact the approach employed in this book.

Depending of the application, it is possible to define two scenarios: (i) one load supplied by one or more sources and (ii) one source supplying one or more loads. Consider first the case in which there is only one load supplied from one or more sources. Each source can be connected to either a switch (corresponding to the point M in Fig. 2.15, for example), as shown in Fig. 2.18(a)–2.18(d), or to the pole (midpoint of the leg, i.e., point A), as shown in Fig. 2.18(e) and 2.18(f). Similarly, for the second scenario, when there is only one source and the load is constituted by one or more elements, each part of the load can be connected to either a switch (point M, for example), as shown in Fig. 2.19(g) or to the midpoint of the leg (A), as shown in Fig. 2.18(h) and 2.18(i).

Notice from Figs 2.15–2.18 that the power switches, sources, and loads are represented generically. The procedure to choose first the appropriate power switch and the converter with desirable characteristics depends on the type of source and load that have been considered, as well as the requirements for a specific application.

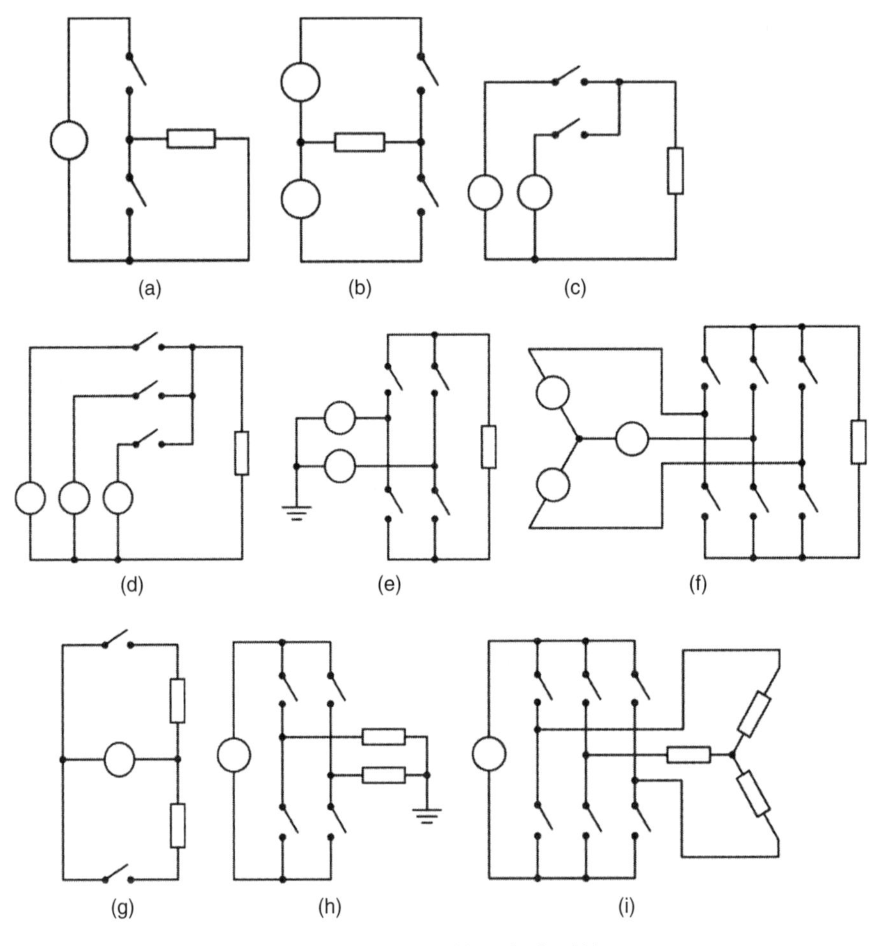

Figure 2.18 Different ways to connect source(s) to the load(s).

The following technical characteristics are used for choosing the appropriate switch and converter: (i) UniC or BidC, (ii) UniV or BidV, (iii) SC or CC, and (iv) SB or CB. Other essential features are switching frequency, voltage and current ratings, and cost.

2.4.1 dc–dc Conversion

Buck Converter The buck converter is presented in Fig. 2.19(a), which comprises (i) a dc voltage source (V_{dc}), (ii) a basic cell as in Fig. 2.15(a) with switches S_1 and S_2, (iii) energy storage component (L) due to the interface between two voltage-type elements, and (iv) a resistive load with voltage V_o. As demanded for this application, both switches S_1 and S_2 should be (a) UniC and UniV and (b) CC and CB. However, notice that S_2 must be complementary to S_1 to avoid a short circuit of the dc source (V_{dc}), which means that when S_1 is on, S_2 should be off, and when S_2 is on, S_1 should be off.

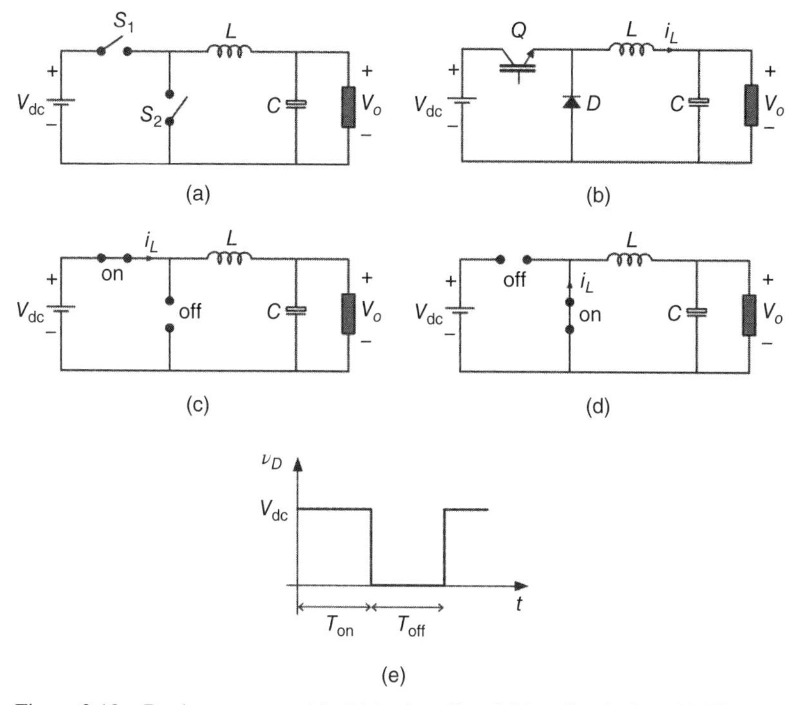

Figure 2.19 Buck converter with: (a) basic cell and (b) real switches. (c) TS 1. (d) TS 2. (e) Diode voltage.

This means that the circuit will have the same functionality if a SC/SB type of switch (i.e., diode) is employed for S_2. Then, the circuit will be as shown in Fig. 2.19(b).

There are two topological states (TSs) for the buck converter:

TS 1—The switch Q is on and conducts the inductor current, while the diode is off, as shown in Fig. 2.19(c).

TS 2—When Q is turned off, the diode conducts the inductor current, as shown in Fig. 2.19(d).

Example 2.1

Explain why a buck converter such as the one presented in Fig. 2.19 requires two power switches with UniC and UniV characteristics.

Solution

As the output voltage is dc, the load current is also dc, which leads to a dc current for I_L. During the TS 1, the switch Q will carry this unidirectional type of current (UniC) and during the TS 2 the diode will carry I_L. When both switches are turned off their current will be zero. In terms of the voltage applied to those devices, it is also UniV, as the voltage will be either zero or V_{dc} for Q and either zero or $-V_{dc}$ for D.

Figure 2.19(e) shows the diode voltage, which is the voltage applied to the LC filter. While the time domain representation—as in Fig. 2.19(e)—provides a good understanding of the topology itself, the frequency domain with the Fourier series allows identification of the frequency components. In general terms, the expansion of a periodical waveform $f(t)$ in Fourier series is given by

$$f(t) = F_o + \sum_{n=1}^{\infty} f_n(t) = \frac{1}{2}a_o + \sum_{n=1}^{\infty} \{a_n \cos(n\omega t) + b_n \sin(n\omega t)\} \tag{2.1}$$

where $F_o = 1/2(a_o)$ is the average value and

$$a_n = \frac{1}{\pi}\int_0^{2\pi} f(t)\cos(n\omega t)d(\omega t) \quad \text{for } n = 0, \ldots, \infty \tag{2.2}$$

$$b_n = \frac{1}{\pi}\int_0^{2\pi} f(t)\sin(n\omega t)d(\omega t) \quad \text{for } n = 1, \ldots, \infty \tag{2.3}$$

In the case of Fig. 2.19(e), the average value F_o in (2.1) is given by

$$V_o = \frac{T_{on}V_{dc}}{T_s} = dV_{dc} \text{ with } d = T_{on}/T_s \tag{2.4}$$

where T_{on} is the interval of time with the switch S_1 on, T_s is the switching period ($T_{on} + T_{off}$), and d is known as duty cycle.

Notice that the input voltage of the LC filter has infinite components of frequency [see equation (2.1)], which must be filtered allowing, ideally, an output voltage V_o with only the zero frequency component, as in equation (2.4).

Example 2.2

Assume the inductor voltage waveform and the information that this energy storage component is in fact a passive element to deduce equation (2.4) without the expansion in Fourier series.

Solution

The graphs below show the inductor waveforms (voltage and current) assuming a continuous conduction mode (i.e., i_L is always higher than zero). As the inductor is a passive component, the average power on this device is zero, leading to an average voltage also equal to zero. It turns out that:$(V_{dc} - V_o)T_{on} = V_oT_{off}$ and consequently: $V_o/V_{dc} = T_{on}/T = d$.

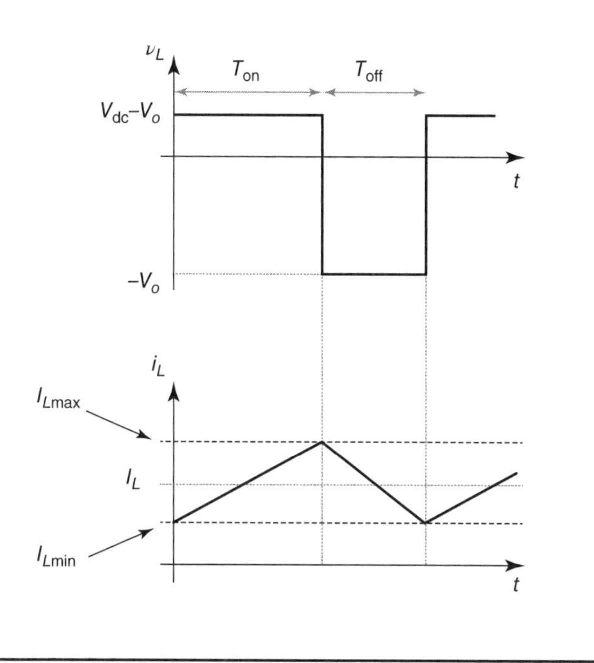

Exercise 2.2

By using equations (2.2) and (2.3) calculate the Fourier coefficients a_n and b_n from 0 to the seventh harmonic for two waveforms of the diode voltage with buck converter operating with $d = 0.5$ and 0.75. Also, consider a switching frequency equal to 50 kHz. If the LC filter in Fig. 2.19 is designed with a cut frequency equal to 20 kHz (3dB), find out if this filter can cut the switching frequency of the buck converter for both cases: $d = 0.5$ and 0.75?

Boost Converter The boost converter with the representation of the real semiconductor devices is presented in Fig. 2.20(a). Its TSs are given in Fig. 2.20(b) and 2.20(c).

TS 1—In this mode, the switch is on and the diode is off. The input voltage is then applied to the inductor with a consequent increase in its current i_L. Since D is off, the load current is provided by the capacitor C.

TS 2—This mode starts when the switch is turned off and consequently the diode is on. At this moment, the energy stored at the magnetic field of the inductor is released to the capacitor and load.

The average output voltage of the boost converter is given by

$$V_o = \frac{V_{dc}}{(1 - d)} \tag{2.5}$$

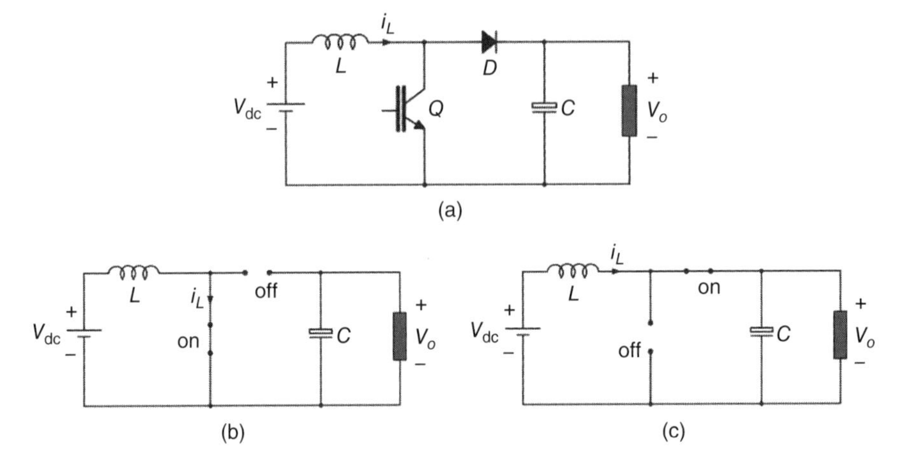

Figure 2.20 Boost converter and its topological states.

The characteristics of the boost converter are the following:

1. Its output average voltage value is greater than the input voltage source.
2. Its input current is continuous with a ripple that depends on the value of the input inductance and switching frequency.
3. The output voltage is positive.

Exercise 2.3

A boost converter as in Fig. 2.20 aims to obtain a dc output voltage of 120 V from a voltage source with 24 V. The resistive load is 120 Ω, switching frequency is equal to 50 kHz and the converter operates at continuous conduction mode ($i_L > 0$). Determine: (a) duty cycle d, (b) average load current, (c) average source current, and (d) inductor value to guarantee a current ripple less than 5%.

Other Basic dc–dc Converters In addition to the buck and boost converters presented previously, there are four other basic nonisolated dc–dc converters that can control the output voltage, as shown in Fig. 2.21. The absolute value of their average output voltage is given by

$$V_o = \frac{dV_{dc}}{(1-d)} \tag{2.6}$$

For all of the topologies in Fig. 2.21, the average of the output voltage can be either smaller or greater than that of the voltage source, depending on the value of d. The main characteristics of the configurations presented in Fig. 2.21 are presented in the sequence.

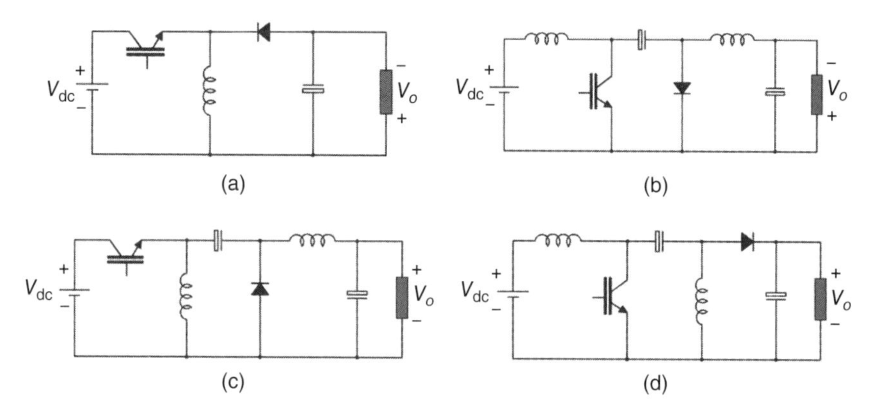

Figure 2.21 Four basic dc–dc converters: (a) buck-boost, (b) cuk, (c) zeta, and (d) SEPIC.

(i) *Buck-Boost Converter*. This converter is shown in Fig. 2.21(a) and can connect a voltage source to a voltage type load. The characteristics of the buck-boost converter are the following:

1. Both input current and the current i_D are discontinuous.
2. The output voltage is negative.

(ii) *Cuk Converter*. This converter is shown in Fig. 2.21(b) and can connect a current source to a current-type load. The characteristics of the Cuk converter are the following:

1. Both input current and output current are continuous.
2. The output voltage is negative.

(iii) *Zeta Converter*. This converter is shown in Fig. 2.21(c) and can connect a voltage source to a current-type load. The characteristics of the Zeta converter are the following.

1. The input current is discontinuous and the output current is continuous.
2. The output voltage is positive.

(iv) *Single-Ended Primary-Inductor Converter (SEPIC) Converter*. The SEPIC converter is based on the scheme of Fig. 2.21(d) and can connect a current source to a voltage type load. The characteristics of the Zeta converter are the following:

1. The input current is continuous and the current i_D is discontinuous;
2. The output voltage is positive.

2.4.2 dc–ac Conversion

The circuits that permit a conversion from dc to ac are also known as inverters. They can be voltage-source inverters (VSIs), current-source inverters (CSIs) and impedance-source inverters (ZSIs). The fundamental concepts and main

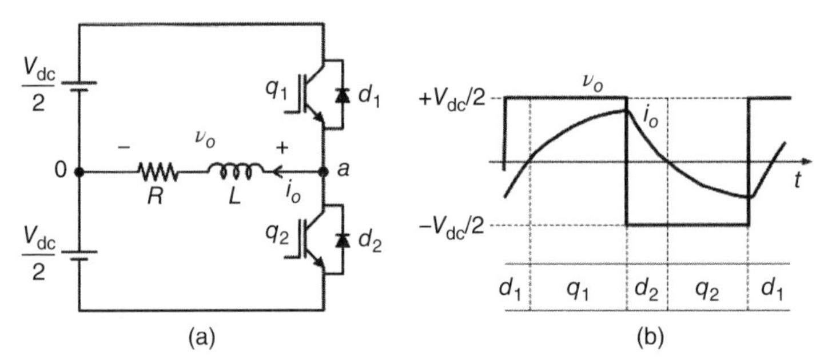

Figure 2.22 (a) Half-bridge inverter and (b) its output voltage and current waveforms for $d = 0.5$.

characteristics of the VSI and CSI are given in the sequence, while the ZSI is considered in Chapter 8 (Section 8.6.2).

Voltage Source Inverter (VSI) The simplest way to generate an ac voltage from a dc source is with the arrangement as in Fig. 2.18(b) with the load connected between the basic cell and two sources. Figure 2.22 shows the same converter with the representation of real switches.

Half-Bridge When the switch q_1 is on, the voltage source $+V_{dc}/2$ is applied to the load. The load current then increases exponentially. When q_1 is turned off (q_2 is on), the current is obliged to decrease via diode d_2, and a negative voltage is applied to the load. A similar operation but related to d_1 and q_1 starts when q_2 is turned off. When $d = T_{on}/T_s = 0.5$, a square wave is generated at the inverter output. The output voltage is depicted in Fig. 2.22(b) along with the load current. The expansion in Fourier series of the output voltage is given by

$$v_{a0} = \frac{4}{\pi}\frac{V_{dc}}{2}\left[\sin(\omega t) + \frac{1}{3}\sin(3\omega t) + \frac{1}{5}\sin(5\omega t) + \frac{1}{7}\sin(7\omega t) + \ldots\right] \tag{2.7}$$

Equation (2.7) allows calculating the amplitude of the output fundamental voltage as well as the existing odd harmonics.

The sinusoidal variation of the duty cycle d inside the switching interval T_s allows shifting the harmonics to the high frequency region. The order of the most significant ones is around the frequency modulation index, defined as

$$m_f = \frac{f_s}{f_m} \tag{2.8}$$

where f_s is the frequency of the carrier signal that defines the switching frequency (e.g., triangular waveform) and f_m is the frequency of the modulating signal (e.g., sinusoidal signal).

This technique is known as sinusoidal pulse-width modulation (SPWM). Its implementation is achieved by comparing a sinusoidal (modulating signal) with a

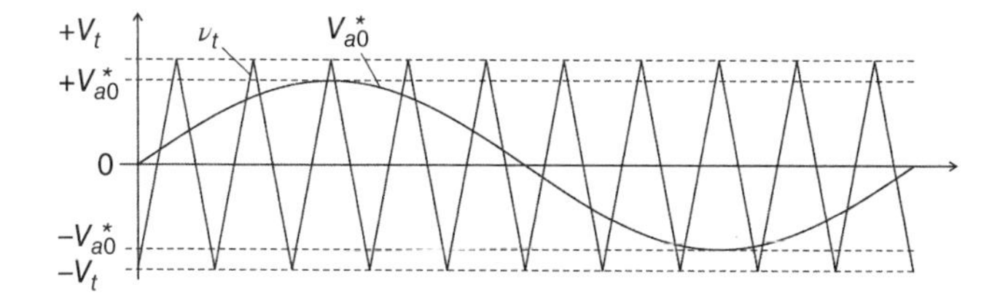

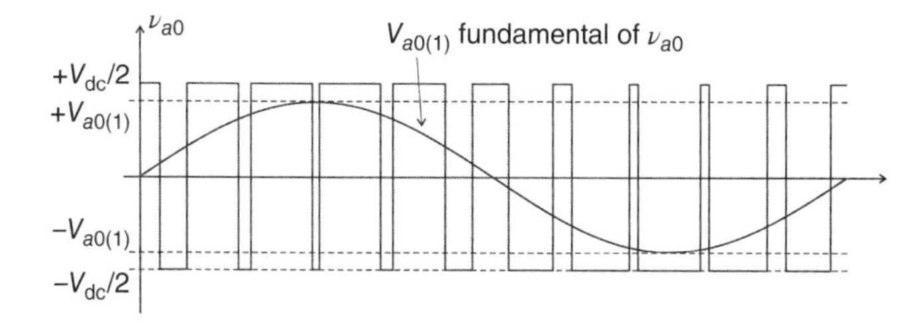

Figure 2.23 Sinusoidal modulation: reference (top) and pole voltage waveform (bottom).

triangle waveform (carrier signal), the frequency of the modulating signal being the fundamental one and the carrier frequency defining the switching frequency. The main waveforms of the SPWM are presented in Fig. 2.23 (top), while Fig. 2.23 (bottom) shows the waveforms for the output voltage.

The relationship between the amplitude of modulating signal, V^*_{a0}, and that of the carrier signal, V_t, defines the amplitude modulation index, m_a, that is

$$m_a = \frac{V^*_{a0}}{V_t} \tag{2.9}$$

For the sinusoidal modulation (SPWM), the amplitude of the fundamental voltage is given by

$$V_{a0(1)} = \frac{V^*_{a0}}{V_t}\frac{V_{dc}}{2} = m_a\frac{V_{dc}}{2} \tag{2.10}$$

This means that the amplitude of the fundamental is directly proportional to the amplitude modulation index (m_a) for a given dc bus value (V_{dc}).

Example 2.3

For a half-bridge inverter as presented in Fig. 2.22(a) supplying a resistive load of 5Ω, switching frequency of 10 kHz and $V_{dc} = 100$ V (duty cycle equal to 50%), calculate: (a) the RMS of the output voltage assuming all harmonic frequencies, (b) the RMS of the output voltage at the fundamental frequency, and (c) output power as well as voltage and current processed by each switch.

Solution

(a) By definition, the RMS value of a periodic function $f(t)$ with period T is given by

$$f_{\mathrm{RMS}} = \sqrt{\frac{1}{T}\int_0^T [f(t)]^2 dt}$$

Applying this equation for the waveform of the voltage presented in Fig. 2.22(b), gives an output RMS voltage equal to

$$V_{o\mathrm{RMS}} = 50 \text{ V}$$

(b) As observed in equation (2.7), the fundamental component of the output voltage is given by $v_{ao1} = 2V_{\mathrm{dc}}/\pi(\sin(\omega t))$, consequently the RMS value of the fundamental voltage will be given by $V_{o1\mathrm{RMS}} = 2V_{\mathrm{dc}}/\pi\sqrt{2} = 45.15$ V.

(c) The output power is given by $P_o = V_{o\mathrm{RMS}}^2/R = 500$ W. The blocking voltage and current processed by each switch is given by 100 V and 20 A, respectively.

Exercise 2.4

Repeat the last example with a converter operating with a switching frequency equal to 20 kHz and a RL load with $R = 5\Omega$ and $L = 15$ mH.

Full-Bridge The circuit of the single-phase full-bridge inverter is given in Fig. 2.24(a). The output voltage can be written as a function of the pole voltages, as

$$v_{ab} = v_{a0} - v_{b0} \tag{2.11}$$

The main waveforms of the converter are shown in Fig. 2.24(b). In this case the pole voltages v_{a0} and v_{b0} are phase-shifted by180 electrical degrees.

As in the half-bridge with duty cycle equal to 50%, the output voltage includes odd harmonics components, that is, third, fifth, seventh, and so on, as $d = 0.5$.

When the pole voltages are phase-shifted of θ degrees, as shown in Fig. 2.25(a), four of the eight modes of operation are given in Fig. 2.25(b)–2.25(e). In Mode 1, switches q_1 and q_2 conduct the increasing positive load current ($t_0 < t < t_1$), then q_2 is turned off forcing q_1 to conduct the load current along with d_3 in a free-wheeling mode (Mode 2, $t_1 < t < t_2$) until q_1 is turned off (t_2). From this moment on, Mode 3 starts with d_3 and d_4 conducting together; the load voltage becomes negative and the load current is then forced to zero ($t_2 < t < t_3$). The zero crossing determines the beginning of Mode 4, in which the current increases in the negative direction.

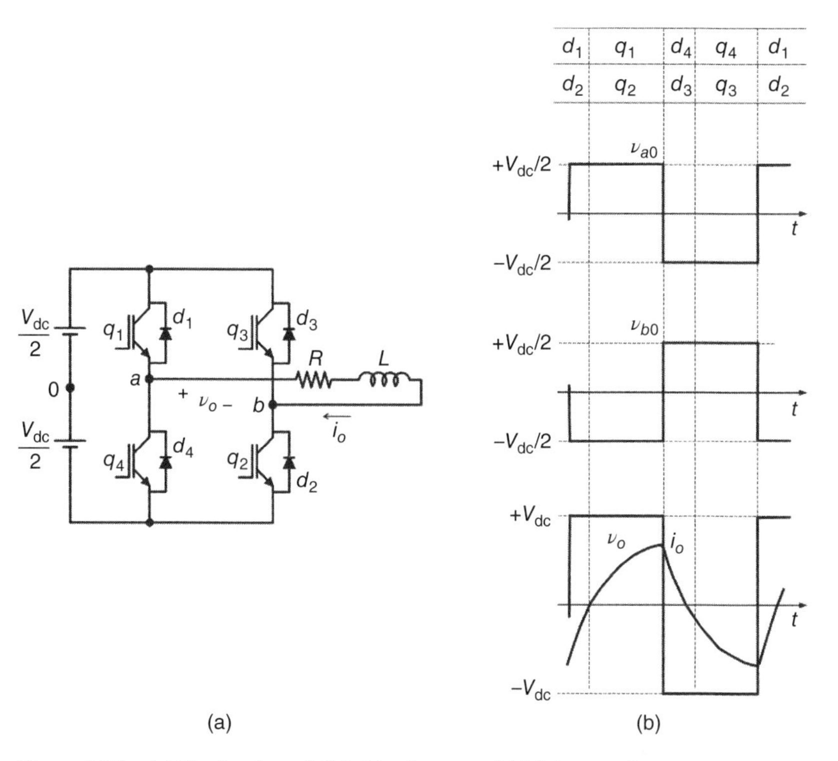

Figure 2.24 (a) Single-phase full-bridge inverter. (b) Main waveforms.

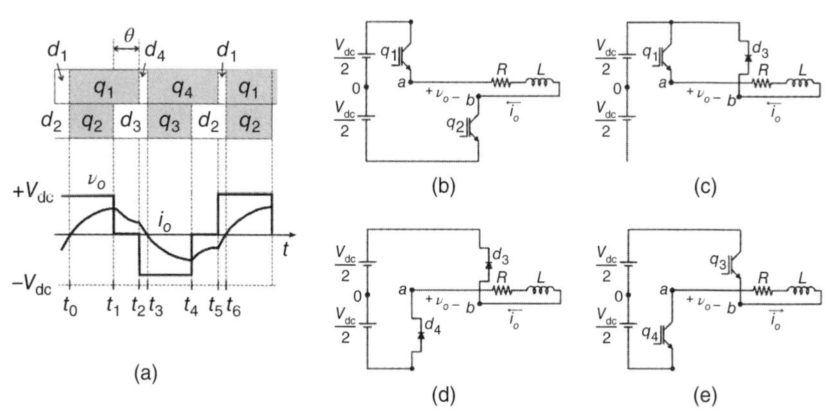

Figure 2.25 Single-phase full-bridge inverter. (a) Gating signals and main waveforms. (b) Mode 1. (c) Mode 2. (d) Mode 3. (e) Mode 4.

When q_4 is turned off, q_3 starts conducting with d_1, and so on, until the cycle is completed.

The strategy of adding the phase-shift angle θ modifies the amplitude of the fundamental and harmonics of the load voltage, which is given in equation (2.12)

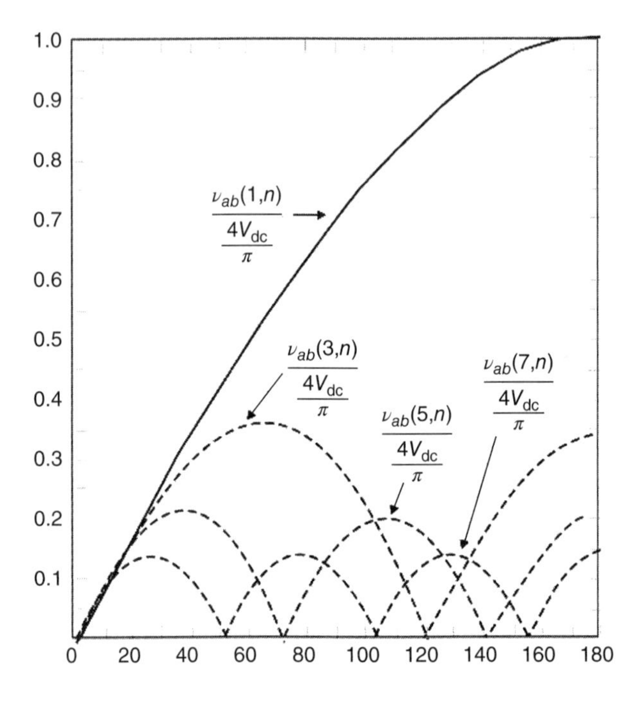

Figure 2.26 Behavior of the fundamental and harmonics of the output voltage as a function of γ.

as a function of $\gamma = \pi - \theta$. Figure 2.26 depicts the behavior of the fundamental and harmonics components of the load voltage as a function of γ. It can be observed that for $\gamma = 120°$ the amplitude of the third harmonic is zero, although the 5th, 7th, 11th, 13th harmonics are still present.

$$V_{o(n)} = \frac{4V_{\text{dc}}}{n\pi} \sin\left(\frac{n\gamma}{2}\right) \tag{2.12}$$

Another way to reduce the low frequency harmonic components for the output voltage is to apply a PWM strategy for the gating signals, as previously done for the half-bridge converter. There are different ways to generate the PWM signals for the full-bridge converter, such as: (i) *bipolar modulation*—when the sinusoidal references for pole 1 and pole 2 are in phase, (ii) *unipolar modulation*—when the sinusoidal reference voltages are phase-shifted by 180°, and (iii) a third possibility is to operate one of the legs with PWM signal while the other one is operated at the fundamental frequency; this corresponds to the unipolar modulation with clamped voltage. Details about the PWM for the full-bridge converter are furnished in Chapters 5 and 8.

Three-Phase Inverter A dc–ac three-phase converter is shown in Fig. 2.27. The gating signal generation for the power switches can be obtained as done previously by either changing γ to eliminate specific harmonics components or with a PWM approach, which will push the harmonics to higher frequencies components. First, assuming the pole voltages with phase-shift equal to 120° from each other, the

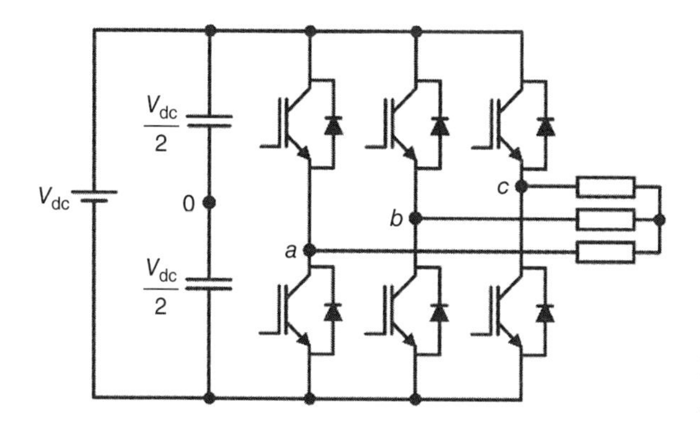

Figure 2.27 Three-phase inverter.

line-to-line voltages can be calculated as

$$\begin{aligned} v_{ab} &= v_{a0} - v_{b0} \\ v_{bc} &= v_{b0} - v_{c0} \\ v_{ca} &= v_{c0} - v_{a0} \end{aligned} \tag{2.13}$$

As the load is Y-connected, it turns out that

$$\begin{aligned} v_{an} &= \frac{1}{3}(2v_{a0} - v_{b0} - v_{c0}) \\ v_{bn} &= \frac{1}{3}(2v_{b0} - v_{c0} - v_{a0}) \\ v_{cn} &= \frac{1}{3}(2v_{c0} - v_{a0} - v_{b0}) \end{aligned} \tag{2.14}$$

Also, it can be shown that the relationship between the load neutral and the midpoint of the dc bus for a balanced three-phase system is given by

$$v_{n0} = \frac{1}{3}(v_{a0} + v_{b0} + v_{c0}) \tag{2.15}$$

The corresponding waveforms for equations (2.13)–(2.15) are shown in Fig. 2.28.

As the line-to-line voltage in Fig. 2.28(b) was obtained with $\gamma = 120°$, then neither third harmonic nor its multiples are measured in these voltages. This is an important characteristic for three-phase systems, as the natural phase angle displacement among the phases (120°) allows elimination of harmonics.

It can be seen from Fig. 2.28 that there are six modes of operation. These modes are highlighted in Fig. 2.29(a)–2.29(f). In addition to these six modes of operation, there are two other modes of operation that apply zero voltage to the loads. These are the free-wheeling configurations presented in Fig. 2.29(g) and 2.29(h), where three bottom or upper switches conduct at the same time. Note that for simplification purposes the operation modes of the three-phase circuit are presented with ideal switch.

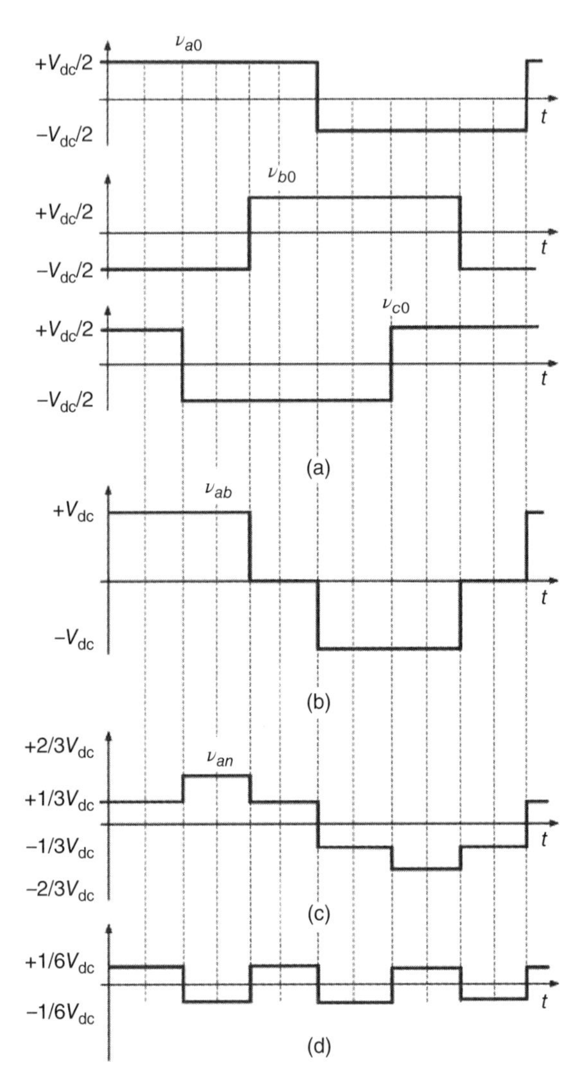

Figure 2.28 Waveforms of three-phase inverter. (a) Pole voltages; (b) line-to-line voltage; (c) load phase voltage; and (d) zero-sequence voltage.

Another way to eliminate harmonics for the output voltage is to generate a SPWM signal for the gating of the switches, as presented in Fig. 2.30, which shows three reference signals phase-shifted by 120° compared to a carrier signal. This figure shows also the pole voltages as a result of the sine-triangular comparison.

Example 2.4

Draw the waveforms of the pole voltages (v_{a0}, v_{b0}, and v_{c0}), line-to-line voltage (v_{ab}), and load phase voltage (v_{an}) considering a SPWM strategy applied to dc–ac three-phase converter with the switching frequency and modulating frequency equal to 10 kHz and 60 Hz, respectively. Also prove that $m_a \rightarrow \infty$.

Solution

As the amplitude modulation index (m_a) is infinite, from equation (2.9) it means that the amplitude of the modulating signal is infinitely higher than the amplitude of the carrier signal, which leads to a pole voltage and load voltages as presented in Fig. 2.28.

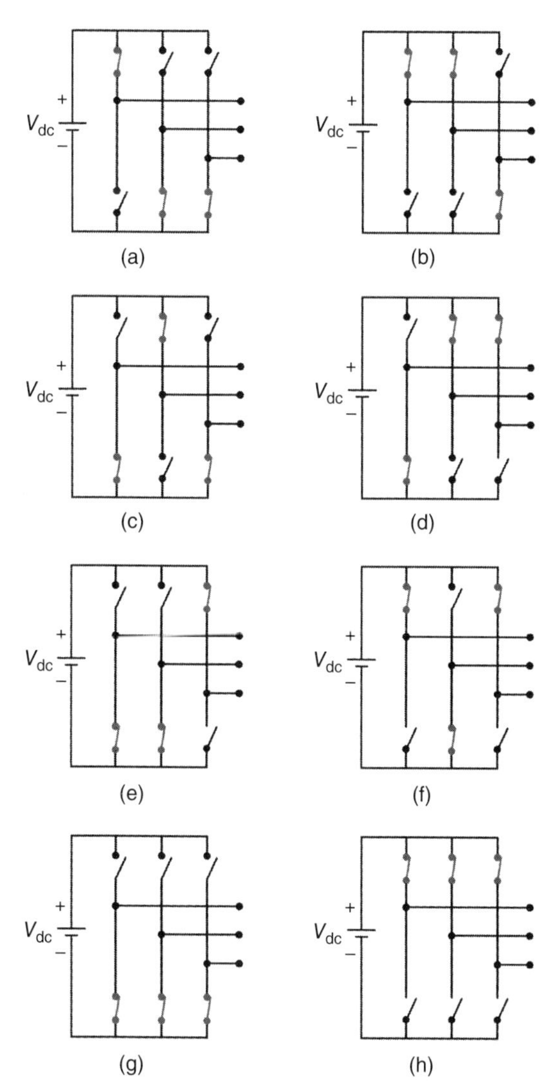

Figure 2.29 Modes of operation of the three-phase inverter.

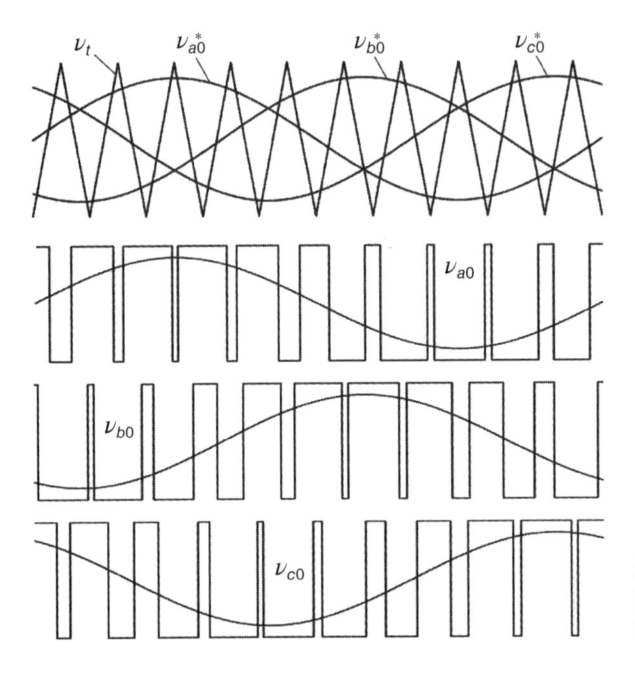

Figure 2.30 Three-phase inverter waveforms with PWM approach.

Exercise 2.5

The three-phase inverter in Fig. 2.27 ($V_{dc} = 300$ V) is operated with a sinusoidal PWM approach with a modulation index $m_a = 0.9$. The ratio between the frequency of the carrier and modulating signals (m_f) is high enough to guarantee sinusoidal load currents when the three-phase load is constituted by $R = 5\ \Omega$ and $L = 20$ mH (per phase). The frequency of the modulating signal is 50 Hz.

(a) What is the RMS value of the voltage v_{a0} at the fundamental frequency?

(b) What is the RMS value of the voltage v_{ab} at the fundamental frequency?

(c) What is the RMS value of the current in each phase?

(d) CSI

The three-phase version of the CSI is shown in Fig. 2.31. As the current source is unidirectional, the switches are of UniC type. However, due to the bidirectional load voltage type, the switches must be of BidV type. In fact, the CSI is dual to the VSI. Because of this, the line-to-line output voltage waveform of a VSI is equivalent to the line output current waveform of a CSI. Using this knowledge, the CSI can be controlled by the same PWM techniques used to control the VSI. However, practical aspects such as dead-time generation follows different rules for both types of converters.

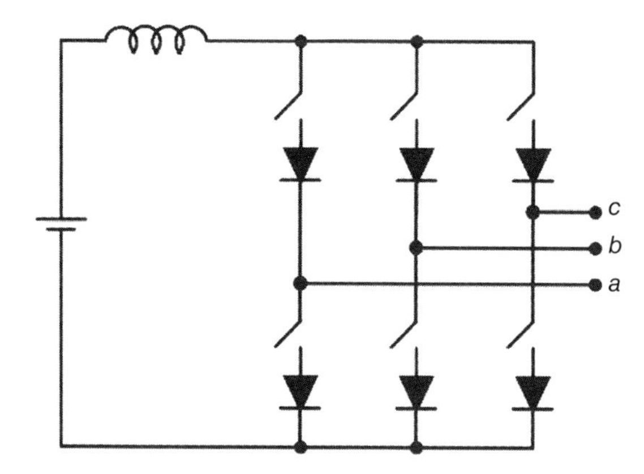

Figure 2.31 Current source inverter (CSI).

2.4.3 ac–dc Conversion

Ac–dc converters are also known as rectifiers. Such converters can be sorted as uncontrolled or controlled rectifiers. An overview about this type of converter is presented here.

Uncontrolled Rectifiers Notice that a sinusoidal waveform as those obtained on the utility grid has average voltage equal to zero. Some applications, such as dc motor drive systems, require a waveform with average voltage different from zero.

If a resistive load is employed, the simplest way to satisfy the condition of non-null average voltage is to use a power converter able to cut, for instance, the negative part of the sinusoidal waveform, as presented schematically in Fig. 2.32(a). A power switch series connected between the source and load will implement this rectifier circuit with no need to use the basic cells as in Figs 2.15 and 2.16. It is evident that the output voltage and output current are unidirectional, which leads to a device with the following characteristics SC/SB/UniC/UniV-r, that is, diode. There are two TSs defined by the variables of the circuit as there is no gating signal control. The TSs are shown in Fig. 2.32(b).

However, for an RL type of load [see Fig. 2.33(a)] the sinusoidal input voltage obliges the current to become zero somewhere between $\omega t = \pi$ and $\omega t = 3\pi/2$

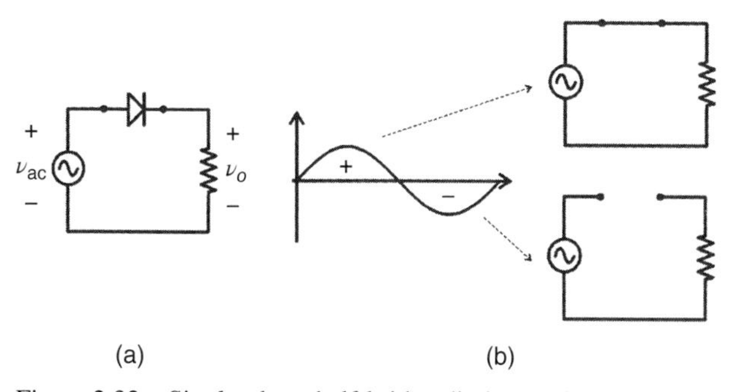

Figure 2.32 Single-phase half-bridge diode rectifier with resistive load: (a) circuit; (b) operation modes.

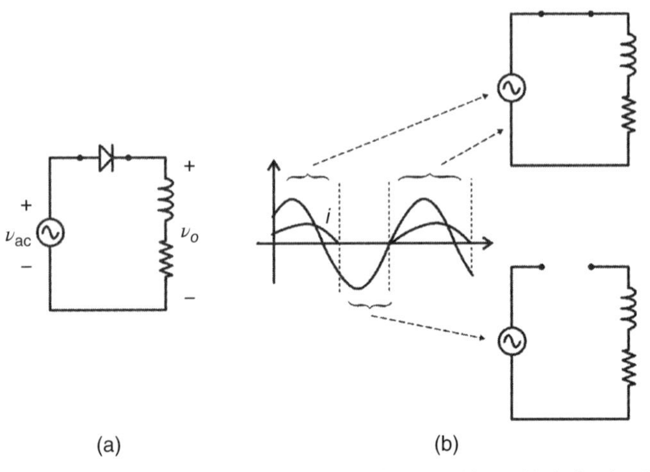

Figure 2.33 Single-phase half-bridge diode rectifier with inductive load: (a) circuit; (b) operation modes.

(depending of the load power factor), thus turning off the diode. During the diode conduction after $\omega t = \pi$, the negative input voltage is applied to the load. Even dealing with negative voltage, the diode will turn off only when both conditions are satisfied: reverse voltage and zero current [see section "The Conventional Diode"]. In this case, the average of the output voltage will vary with the load power factor, which is sometimes an undesirable condition. The output voltage varying with the power factor of the load can be avoided by using a diode in parallel with the load (free-wheeling diode), that, in fact, clamps the negative part of the load voltage at zero, as shown in Fig. 2.34. The arrangement of two diodes as presented in this figure is indeed a particular case of the basic cell shown in Fig. 2.15. In this case, the output voltage will be equal to

$$\overline{V}_o = \frac{V_m}{\pi} \tag{2.16}$$

where V_m is the amplitude of the sinusoidal waveform.

Example 2.5

Deduce equation (2.16).

Notice from Fig. 2.34 that due to the inductive load, the arrangement of diodes is in fact a basic cell (as in Fig. 2.16) constituted by diodes.

While a single-phase full-bridge diode rectifier is presented in Fig. 2.35(a), the three-phase version is presented in Fig. 2.35(b). Both circuits are also one of the cases presented in Fig. 2.18 with higher output dc voltage than the half-bridge, as in Figs 2.32–2.34.

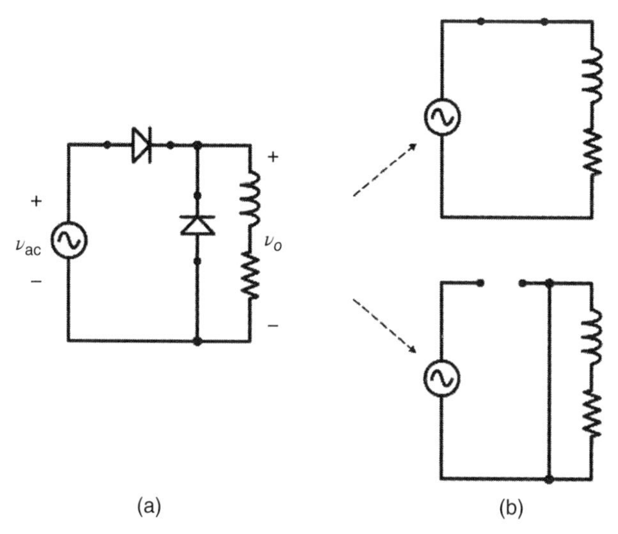

Figure 2.34 Single-phase half-bridge diode rectifier with inductive load and free-wheeling diode: (a) circuit; (b) operation modes.

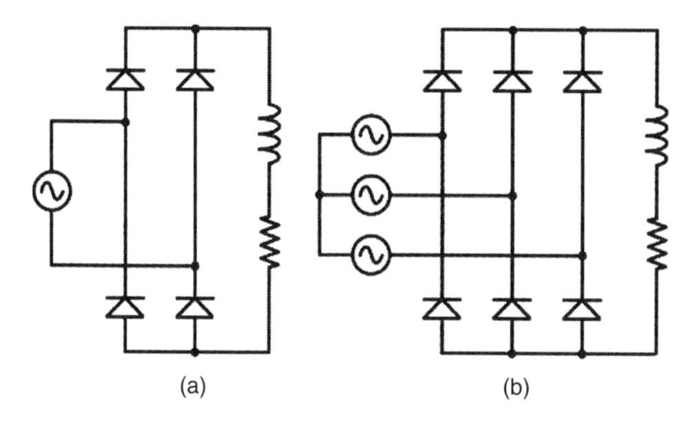

Figure 2.35 Diode rectifier: (a) single-phase and (b) three-phase circuits.

Controlled Rectifiers The controlled rectifiers can also be either half-bridge or full-bridge topologies. Their control can be obtained either by phase control or by PWM control, depending on the type of switch employed in the circuit.

Phase Control Considering a sinusoidal (bidirectional) voltage source feeding a unidirectional inductive load, as shown in Fig. 2.36(a), the type of switch for S_1 and S_2 can be given respectively by: CC/SB/UniC/BidV (SCR) and SC/SB/UniC/UniV-r (diode), as in Fig. 2.36(b). The SCR is turned on at a given angle α in relation to the moment that the SCR voltage becomes positive on its terminals. Unlike the previous rectifiers, the angle control allowed by the gating circuit of the SCR permits varying the output average voltage by changing α. This technique is known as phase control.

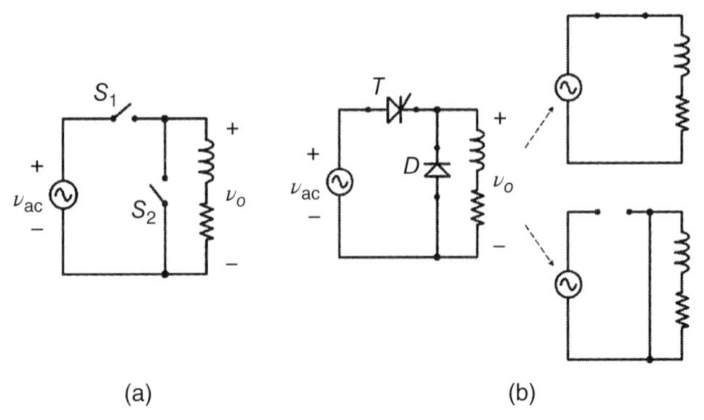

Figure 2.36 (a) Controlled ac–dc converter using a switching cell. (b) Single-phase rectifier with SCR.

The average of the output voltage ($\overline{V}_{o\alpha}$), as a function of both α and the amplitude of the sinusoidal voltage, is given by

$$\overline{V}_{o\alpha} = \frac{V_m}{2\pi}(1 + \cos\ \alpha) \tag{2.17}$$

It can be seen from equation (2.17) that when $\alpha = 0°$, the value of the output voltage is the same as in equation (2.16), as the SCR is equivalent to the diode when $\alpha = 0$.

Exercise 2.6

Deduce the expression for the average of the output voltage as in equation (2.17) of the controlled rectifier as presented in Fig. 2.36.

The single split-phase half-bridge rectifier is shown in Fig. 2.37(a), in which the sinusoidal voltage sources v_{ac1} and v_{ac2} are phase-shifted 180° to each other, as shown in Fig. 2.37(b). Suppose that the load current is continuous (i.e., always above zero) and that the SCR T_1 is turned on at a given angle α after v_{ac1} becomes positive. This voltage is then applied to the load. From $\omega t = \pi$, v_{ac1} becomes negative while v_{ac2} becomes positive allowing T_2 to be turned on when $\omega t = \pi + \alpha$. Until this time the load current is furnished by v_{ac1}, as shown in Fig. 2.37(c). From $\omega t = \pi + \alpha$ on, v_{ac2} feeds the load current, as shown in Fig. 2.37(d).

Note that when a free-wheeling diode is added as in Fig. 2.38(a), the negative part of the output waveform is clamped to zero as in Fig. 2.38(b) and the output average voltage is calculated from equation (2.17).

The three-phase and six-phase half-bridge rectifiers are shown in Figs 2.39 and 2.40, respectively. In the results presented in Figs 2.39(b) and 2.40(b) their output voltages are obtained for $\alpha = 30°$. The general expression for the average output voltage for a n-phase half-bridge rectifier considering $\alpha = 0°$ is given by (2.18).

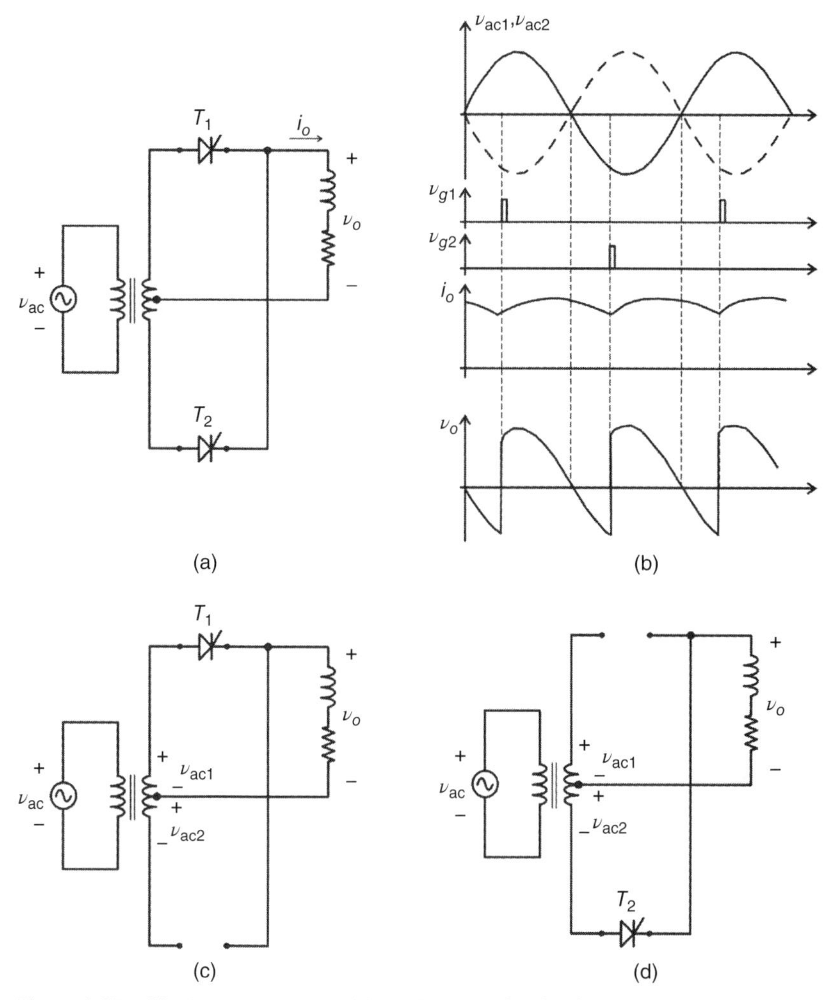

Figure 2.37 Single split-phase rectifier: (a) power circuit, (b) waveforms, and (c) and (d) modes of operation.

$$\overline{V}_o = \frac{nV_m}{\pi} \sin\left(\frac{\pi}{n}\right) \tag{2.18}$$

Examples of the full-bridge rectifiers for single-phase and three-phase circuits are presented in Fig. 2.41(a) and 2.41(b), respectively. The output voltage waveform of the full-bridge single-phase bridge is similar to that of the single split-phase half-bridge rectifier. The output voltage waveform of the three-phase bridge is similar to that of the six-phase half-bridge.

PWM Control A three-phase PWM converter with bidirectional power flow capability, sinusoidal grid current, and power factor control is presented in Fig. 2.42. Each phase can employ the boost principle so that the output voltage is higher than the amplitude of the input voltage. This topology is discussed in detail in Chapter 9.

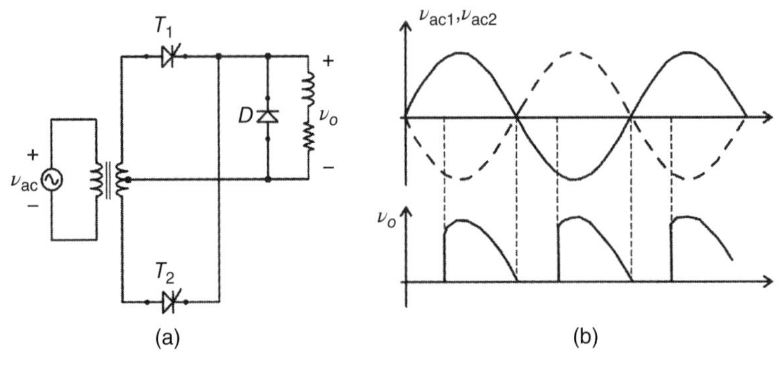

Figure 2.38 Single split-phase rectifier with a free-wheeling diode: (a) power circuit and (b) main waveforms.

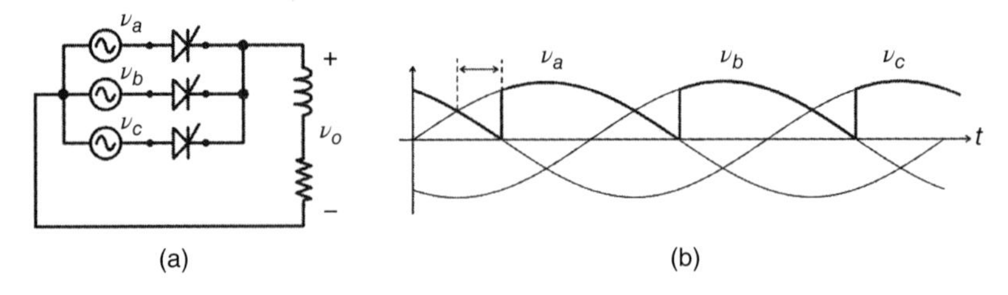

Figure 2.39 (a) Three-phase half-bridge controlled rectifier and (b) output waveform.

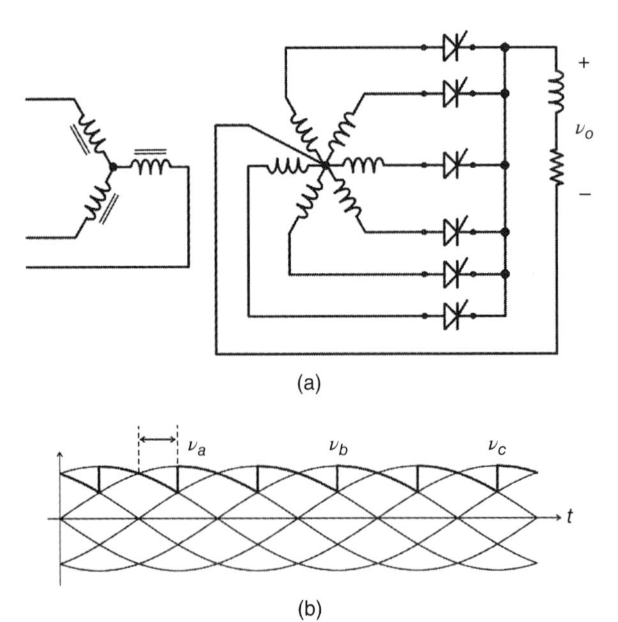

Figure 2.40 (a) Six-phase half-bridge controlled rectifier and (b) output waveform.

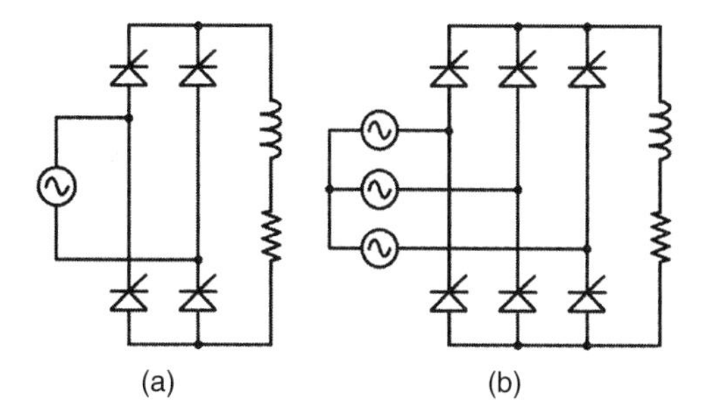

Figure 2.41 Full-bridge controlled: (a) single-phase and (b) three-phase rectifiers.

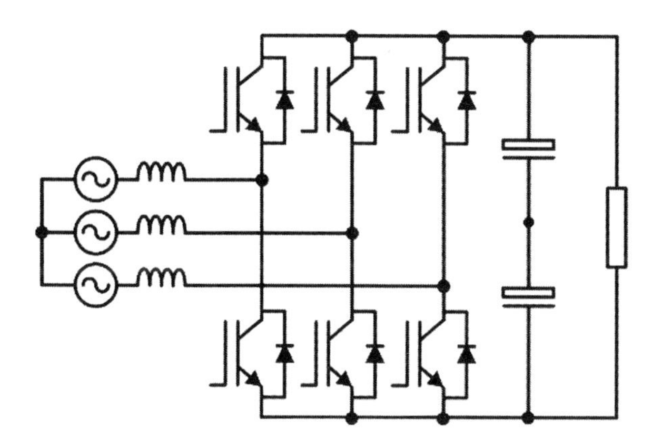

Figure 2.42 Three-phase controlled rectifier.

2.4.4 ac–dc Conversion

An example of a simple ac–ac converter is presented in Fig. 2.43(a), where the switch is a TRIAC (or two SRCs in antiparallel), for a resistive load. By varying the switch turn-on angle α, the RMS value of the fundamental component of the output voltage is varied as well. During the positive semicycle of the input voltage, when turned on, the TRIAC conducts from M_1 to M_2 and during the negative semicycle of the input voltage, when turned on, it conducts from M_2 to M_1. The voltages are shown in Fig. 2.43(b). The need for a device of BidV type is confirmed from the waveform of the voltage across the device. The RMS of the output voltage is given by

$$V_{o\alpha} = \frac{V_m}{\sqrt{2}}\sqrt{\frac{1}{\pi}\left(\pi - \alpha + \frac{\sin 2\alpha}{2}\right)} \tag{2.19}$$

The circuit shown in Fig. 2.44(a) is a single-phase ac–ac converter, referred to as a cycloconverter. The switches must be of the type CC/SB and BidV/UniC as in this figure. The positive (P) and negative (N) converters are controlled to achieve an output voltage with variable amplitude and a variable frequency, as in Fig. 2.44(b).

Another ac–ac converter is the matrix converter, which consists of an array of bidirectional switches that directly connect the load to the source. The three-phase to three-phase conversion is given in Fig. 2.45(a) with ideal switches. The practical realization of the matrix converter is presented in Fig. 2.45(b).

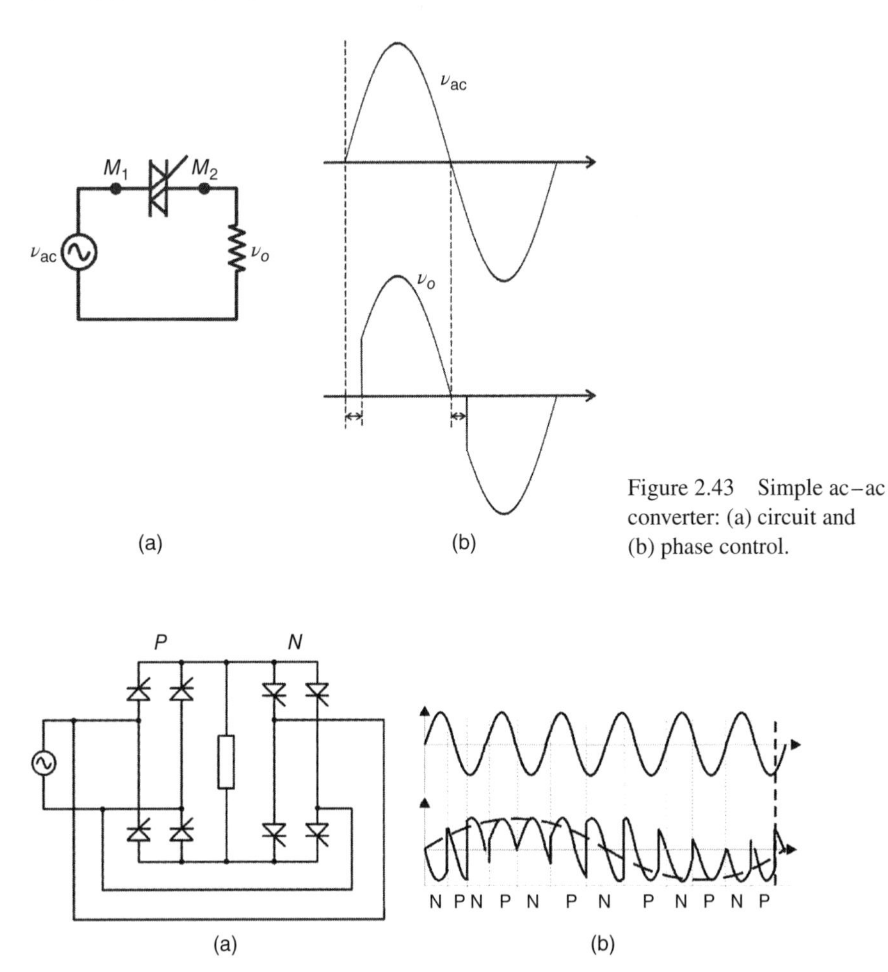

Figure 2.43 Simple ac–ac converter: (a) circuit and (b) phase control.

Figure 2.44 (a) Single-phase cycloconverter. (b) Waveforms of the circuit: (top) grid voltage and (bottom) load voltage with dashed line.

The back-to-back converter is another way to generate a controlled ac voltage from the utility grid voltage. Such a converter is also known as indirect ac–dc–ac conversion, as there is a dc link between both ac stages, as in Fig. 2.46. This converter is described in detail in Chapters 10 and 11. The ac–dc–ac converter can also be implemented with a current source configuration.

2.5 SUMMARY

This chapter first classified the static (unidirectional or bidirectional features in terms of current and voltage) and dynamic (spontaneous or controlled turn-on and/or turn-off) characteristics of the ideal power switches. Then a summary of the main

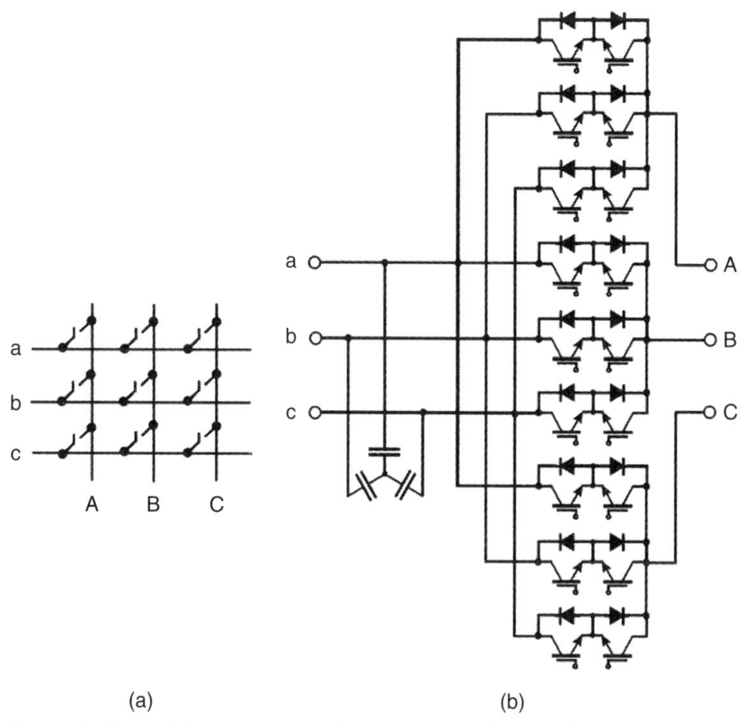

Figure 2.45 (a) Three-phase switching matrix. (b) Practical implementation of the matrix converter.

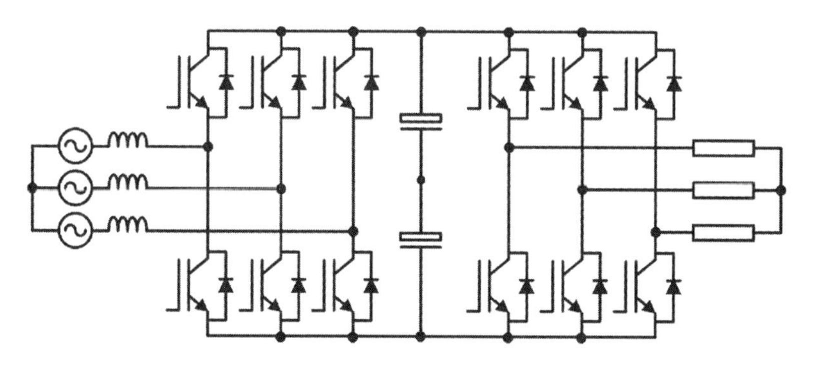

Figure 2.46 Indirect converter: ac–dc–ac conversion.

real power semiconductor devices available in the market was given, taking into account the static and dynamic characteristics. It was outlined their limit values for current, voltage, and switching frequency. The bedrock for the following chapters was given by introducing the concept of basic switching cell, from which it is possible to generate the basic converter topologies used in power electronics. By using these switching cells, a general discussion on the principles of power conversion from dc sources (dc–dc and dc–ac conversion) and ac sources (ac–dc and ac–ac) is furnished. Also the principle of operation of the main converter topologies is explained. It is expected from this chapter to provide the reader a systematic way of deciding which type of power semiconductor is the most appropriate choice for a given application based on: (i) the static and dynamic characteristic, (ii) voltage and current ratings, and (iii) on the frequency of operation. In addition to the generation of conventional power electronics topologies, the concept of cell can be applied to conceive other topologies used by researchers and by related industry. These other topologies, named in this book as advanced power electronics converters, are introduced in the following chapters. Details on the topics can be found from References 1 to 77. As will be presented from Chapter 3 onwards, this text will focus on advanced power electronics converters that process AC voltage by using PWM strategies as the gating signal control of the power switches.

REFERENCES

[1] Gutzwiller FW. *Silicon-Controlled Rectifier Manual.* 5th ed. Auburn, NY: General Electric Co.; 1972.

[2] Burns WB III, Kociecki J. Power electronics in the minicomputer industry. Proc IEEE 1988;76(4):311–324.

[3] Hower PL. Power semiconductor devices: an overview. Proc IEEE 1988;76(4):335–342.

[4] Baliga BJ. Evolution of MOS-bipolar power semiconductor technology. Proc IEEE 1988;76(4): 409–418.

[5] Chen DY. Power semiconductors: fast, tough, and compact. IEEE Spect 1987;24.

[6] Thomson-C.S.F. Le transistor de puissance dans son environement. Paris: Thomson-CSF-DSD, 1978.

[7] Ferrieux JP, Forest F, Lienart P. The insulated gate bipolar transistor: switching modes. Proceedings of European Power Electronics Conference; 1969. p 171–175.

[8] Heumann K, Papp G, Jung M. Comparative study on new power transistors with respect to high frequency inverter applications. Ibid; 1989. p 99–104.

[9] Muraoka K, Kawamura Y, Ohtsubo Y, Sugawara S, Tamamushi T, Nishizawa J. Characteristics of the high-speed SI thyristor and its application to the 70-kHz 100-kW high efficiency inverter. IEEE Trans Power Electron 1989;4(1):92–100.

[10] Akagi H, Sawae T, Nabae A. 130 kHz 7.5 kW current source inverters using static induction transistors for induction heating applications. IEEE Trans Power Electron 1988;3(3):303–309.

[11] Escaud B, Marty P. Introduction à l'étude des structures des convertisseurs statiques. Electron Ind 1983;56:65–71.

[12] Marty P, Escaud B, La commutation dans les convertisseurs statiques, Ibid., No. 60; 1983. p 69–73.

[13] McMurray W. Power electronic circuit topology. Proc IEEE 1988;76(4):428–437.

[14] Cheron Y. Analyse des contraintes subies par les interrupteurs pendant les commutations. Recherche des regles de leur utilisation optimale dans les convertisseurs. Proceedings de la Journée d'Etude S.E.E./GRECO; Toulouse; 1987.

[15] Liu K, Lee F. Topological constraints on basic PWM converters; IEEE PESC'88 Record; 1988. p 164–172.

[16] Wood P. *Switching Power Converters.* Malabar, Fl.: Robert E. Krieger Publishing Co.; 1984.

[17] Mohan N, Undeland TM, Robbins WP. *Power Electronics: Converters, Applications, and Design.* New York: John Wiley and Sons; 1989.

[18] Ledwich G. Current source inverter modulation. IEEE Trans Power Electron 2001;6(4):618–623.

[19] Hsin-Hua PL Bidirectional lateral insulated gate bipolar transistor having increased voltage blocking capability US patent 5977569 A. 1999 Nov 02.

[20] Dahono P, Kataoka T, Sato Y, Dual relationships between voltage-source and current-source three-phase inverters and its applications. Proceedings of PEDS; 1997. p 559–565.

[21] Hefner A et al. Recent advances in high-voltage, high-frequency silicon-carbide power devices. Proceedings of IEEE IAS'09; 2009. p 1–8.
[22] Bose BK. *Modern Power Electronics and AC Drives*. Prentice Hall; 2001.
[23] Ralph McArthur, Defining diode data sheet parameters; Advanced Power Technology, Application Note APTO 30 Rev.; 2003.
[24] Nishizawa T, Terasaki , Shibata J. Field effect transistor and analog transistor. Res. Inst. Electrical Comm. Tohoku Univ. Tech Rep TR-36; 1973. p 1–65.
[25] Nishizawa J, Terasaki T, Shibata J. Field effect transistor versus analog transistor (static induction transistor). IEEE Trans Electron Devices 1975;ED-22:185–197.
[26] Nishizawa J, Ohmi T, Mochida Y, Matsuyama T, Iida S. Bipolar mode static induction transistor (BSIT) - high speed switching transistor. IEDM Tech Dig 1978:676–679.
[27] Nakamura Y, Tadano H, Sugiyama S, Igarashi I, Ohmi T, Nishizawa J. Normally-off type high speed SI-thyristor. Tech Dig IEDM 1982:480–483.
[28] Nakamura Y, Tadano H, Takigawa M, Igarashi I, Nishizawa J. Experimental study on current gain of BSIT. IEEE Trans Electron Devices 1986;ED-33:810–815.
[29] Aoki S, Tsukiyama N, Hie T, Tadano H and Nishizawa J. Low loss high gain 300 V - *200* A class normally-off SIT module for DC motor control, PESC '88 Record; 1988. p 703–708.
[30] Nakamura Y, Tadano H, Takigawa M, Igarashi I, Nishizawa J. Very high speed static induction thyristor. IEEE Trans Ind Appl 1986;IA-22:1000–1006.
[31] Nishizawa J, Tamamushi T. Recent development and future potential of the power statice induction (SI) devices, PECSD'88; 1988. p21–24.
[32] Xu, Z, Li Y, Huang AQ. Performance characterization of 1-kA/4.5-kV symmetrical emitter turn-off thyristor (ETO), Conference Record of the 2000 IEEE Industry Applications Conference; vol. 5; 2000. p 2880–2884.
[33] Li Y, Huang AQ, Lee FC. Introducing the emitter turn-off thyristor (ETO). Proceedings of the IEEE Industrial Applications Society; 1998. p 860–864.
[34] Li H-HP. Bidirectional lateral insulated gate bipolar transistor having increased voltage blocking capability. US Patent 5977569 A 1999 02 Nov.
[35] Weber A, Galster N, Tsyplakov E, (*ABB*). *"A New Generation of Asymmetric and Reverse Conducting GTOs and their Snubber Diodes"*. PCIM Europe; 1997.
[36] Satoh K, Yamamoto M, Morishita K, Yamguchi Y, Iwamoto H. (*Mitsubishi*). *High Power Symmetrical GCT for Current Source Inverter*. ISPSD; 1999.
[37] Weber A, Dalibor T, Kern P, Oedegard B, Waldmeyer J, Carroll E. *Reverse Blocking IGCTs for Current Source Inverters*, (*ABB*). PCIM Europe: ; 2000.
[38] Zeller H-R, (*ABB*). *High Power Components: From the State of the Art to Future Trends*. PCIM Europe; 1998.
[39] Chai FK, Odekirk B, Maxwell E, Caballero M, Fields T, Mallinger M, Sdrulla D. A SiC static induction transistor (SIT) technology for pulsed RF power amplifiers. Proceedings of the 23rd International Symposium on Power Semiconductor Devices & IC's; 2011. p 300–303.
[40] Hille F, Niedernostheide FJ, Ruething H, Schulze HJ. Reverse conducting IGBT with vertical carrier lifetime adjustment. US patent 7557386 B2. 2009 July 7.
[41] Rashid MH. *Power Electronics Circuits Devices and Applications (Livro)*. Englewood Cliffs: Prentice Hall, Inc.; 1988.
[42] Steimer, PK, Gruning, HE, Werninger J, Carroll E, Klaka S, Linder S. IGCT-a new emerging technology for high power, low cost inverters. Thirty-Second IAS Annual Meeting, Conference Record of the 1997 IEEE, Vol. 2; 1997 Oct 5–9 Oct. p 1592–1599.
[43] Ryu S-H, Capell C, Jonas C, Cheng L, O'Loughlin M, Burk A, Agarwal A, Palmour J, Hefner A. Ultra high voltage (>12 kV), high performance 4H-SiC. IGBTs Power Semiconductor Devices and ICs (ISPSD), 2012 24th International Symposium; 2012. p 257–260.
[44] Zhao B, Qin H, Wen J, Yan Y. Characteristics, applications and challenges of SiC power devices for future power electronic system. Power Electronics and Motion Control Conference (IPEMC), 2012 7th International; 2012. p 23–29.
[45] Wang G, Huang A, Li C. ZVS range extension of 10A 15kV SiC MOSFET based 20kW dual active half bridge (DHB) DC-DC converter. Energy Conversion Congress and Exposition (ECCE), 2012 IEEE; 2012. p 1533–1539.

[46] Lorenz L. Power semiconductor devices and smart power IC's - the enabling technology for future high efficient power conversion systems. Proceedings of IPEMC; 2009. p 193–201.
[47] Shigekane H, Fujihira T, Sasagawa K et al. Macro-trend and a future expectation of innovations in power electronics and power devices. Proceedings of IPEMC; 2009. p 35–39.
[48] Klaka S, Frecker M, Grüning H. *The Integrated Gate-Commutated Thyristor: A New High-Efficiency, High-Power Switch for Series or Snubberless Operation*. Nürnberg: PCIM; 1997.
[49] Arsov GL, Slobodan M. The sixth decade of the thyristor. Electronics 2010;14(1):3–7.
[50] Riordan M, Hoddeson L. The origins of the p-n junction. IEEE Spect 1997;34(6):46.
[51] Baliga BJ. How the super-transistor works. Scientific American Magazine, Special Issue on "The Solid State Century", pp. 34–41, 1998.
[52] Bedford BD, Hoft RG. *"Principles of Inverter Circuits"*. New York: John Wiley; 1964.
[53] Gentry FE, Gutzwiller FW, Holonyak N, Von Zastrow EE. *Semiconductor Controlled Rectifiers*. Englewood Cliffs: Prentice Hall; 1964.
[54] Łukasiak L, Jakubowski A. History of semiconductors. J Telecommun Inf Technol 2010;1:3–9.
[55] Baliga J. *Fundamentals of Power Electronics Devices*. Springer Science; 2008.
[56] Moll JL, Tanenbaum M, Goldey JM, Holonyak , N Jr.. p-n-p-n transistor switches. Proceedings of IRE; 1956; vol. 44. p 1174–1182. New way to change ac to dc, Business Work; 1957 Dec 28. p :114–:116.
[57] Nishizawa J, Terasaki T, Shibata J. Field effect transistor versus analog transistor (static induction transistor). Res. Inst. Electrical Comm., Tohoku University, Tech. Rep. RIEC TR-36; 1973; also IEEE Trans Electron Devices 1975; ED-22(4):185–197.
[58] Houston DE, Khrishna S, Piccone D, Finke RJ, Sun YS. Field controlled thyristor (FCT)—a new electronic component. IEDM Tech Digest 1975:379–382.
[59] Burgess MPD. *Semiconductor Research and Development at General Electric*. US Semiconductor Manufactures; 2008 and 2011.
[60] Kahng D, Atalla MM, DRC, Pittsburg (1960); US patent 3,102, 230. 1963.
[61] Russell JP et al. The COMFET: a new high conductance MOS gated device. IEEE ED Lett 1983;EDL 4(3):63–65.
[62] Temple VAK. MOS GTO—a new class of power devices. IEEE Trans Elect Devices 1986;33:1609–1618.
[63] Stoisiek M, Strack H. Turn-off thyristor with MOS-controlled emitter shorts. IEEE IEDM Tech Digest 1885:158–161.
[64] US patent 3,239,728. Semiconductor switch, Richard W. Aldrich and Nick Holonyak, assignors to General Electric Company, Filed July 17, 1962, Date of Patent: 1965 Nov 5.
[65] US patent 3,303,360. Semiconductor switch. Finis E. Gentry, assignor to General Electric Company Original application. 1963 Sept 3, Ser. No. 306,147, now Patent no. 3,265,909.
[66] (Insulated Gated Controlled Semiconductor) United States Patent 4,551,643 Power switching circuitry. Inventors: John P. Russell, Pennmgton; Alvin M. Goodman, Filed 10/1983, Date of Patent: 1985 Nov 5.
[67] Thomas KM, Backlund B, Toker O, Thorvaldsson B. *The Bidirectional Control Thyristor*. ABB; 1999.
[68] Application Note 5SYA 2006-03, Bi-directionally controlled thyristors ABB, 2013. Storm, F. Herbert, Introduction to turn-off silicon-controlled rectifiers.
[69] Baliga BJ, Adler MS, Love RP, Gray PV, Zommer ND. The insulated gate transistor: A new three-terminal MOS-controlled bipolar power device. IEEE Trans Electron Devices 1984;31(6):821–828.
[70] Adler MS, Owyang KW, Jayant Baliga B, Kokosa RA. The evolution of power device technology. IEEE Trans Electron Devices 1984;11:1570–1591.
[71] Kokosa RA, Tuft BR. A high-voltage, high-temperature reverse conducting thyristor. IEEE Trans Electron Devices 1970;17(9):667–672.
[72] Shockley W. The theory of p-n junctions in semiconductor; and p-n junction transistors. Bell Syst Tech J 1949;28:435–489.
[73] Teal GK. Some recent developments in silicon and germanium materials and devices. National IRE Conference; 1954.

[74] Piccone DE, De Doncker RW, Barrow JA, Tobin WH. The MTO thyristor—a new high power bipolar MOS thyristor. IEEE Industry Applications Society 31 Annual Meeting; 1996 Oct 6–10. p 1472–1473.

[75] Gruning H, Odegbrd B, Rees J. High-power hard-driven GTO module for 4.5kV/3kA snubberless operation; 1996. . PCIM 96.

[76] Jakubowski A, Lukasiak L. CMOS evolution. Development limits. Mater Sci-Pol 2008;26(1):5–20.

CHAPTER 3

POWER ELECTRONICS CONVERTERS PROCESSING ac VOLTAGE AND POWER BLOCKS GEOMETRY

3.1 INTRODUCTION

This chapter provides a brief review of the main power electronics converters that process ac voltage and an introduction to power blocks geometry (PBG), to understand the power converters described later in this book. In fact, this chapter has a compilation of the topologies studied throughout this book. The fundamentals of PBG and its correlation to the development of power electronics converters are presented in a general way; details regarding each converter are presented in the following chapters. This approach is called geometry because it deals with interconnection of geometric shapes following specifics axioms and conjectures.

The main goal of the PBG is to provide an alternative and systematic approach for the presentation of the power converters used to process ac voltage. This strategy employs formal methods based on a simple geometrical representation with established rules. A universe with axioms and conjectures is defined in order to establish a formation law. Through this method, which is called PBG, power electronics configurations observed in the technical literature, especially the traditional ones, can be obtained by following the proposed rules with a high level of abstraction. The PBG method is a general approach that can be employed for the creation and development of dc–ac and ac–dc–ac converters from two levels to n levels. It is demonstrated that different types of converters processing ac voltage can be easily built by using PBG.

The converters studied in this book have at least one stage that synthesizes ac voltage (v_o), as observed in Fig. 3.1(a). This kind of converter should be able to create ac waveforms from a dc voltage, with at least two levels, as in Fig. 3.1(b). Figure 3.1(c) shows a pulse width modulation (PWM) waveform for the output converter, also with two levels, but with reduced harmonic distortion compared to that in

Advanced Power Electronics Converters: PWM Converters Processing AC Voltages,
Forty Fifth Edition. Euzeli Cipriano dos Santos Jr. and Edison Roberto Cabral da Silva.

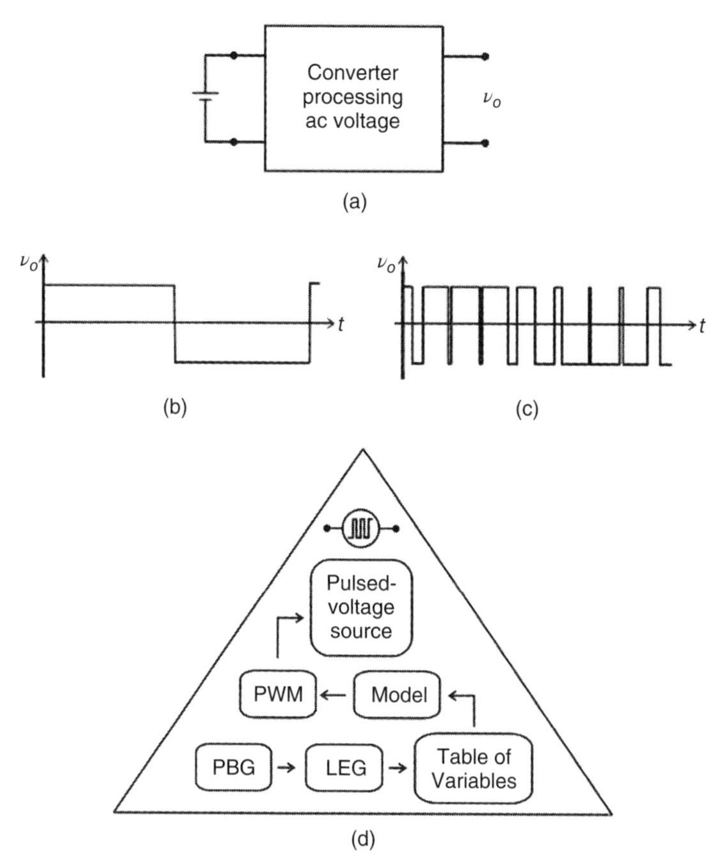

Figure 3.1 (a) Power converter able to synthesize ac voltage. (b) ac Output voltage with two levels. (c) PWM ac output voltage with two levels. (d) Approach employed to determine the pulsed voltage source.

Fig. 3.1(b). The focus of this book is topologies dealing with PWM voltages with two or more levels obtained from a dc stage. Indeed, the block in Fig. 3.1(a) that processes a PWM ac voltage can also be referred to as a pulsed voltage source. Figure 3.1(d), in turn, shows the approach employed throughout this book to obtain the pulsed voltage sources systematically. First, the PBG is considered to define what a single leg looks like (LEG in Fig. 3.1(d) – circuit constituted by a set of switches able to generate a waveform with two or more levels). Then the table of variables will be employed to define both the model of the converter and the PWM strategy. Such a leg can be defined from the PBG, considering different figures of merit, for example, number of levels, quantity of semiconductors and passive elements desired.

This chapter is organized as follows: Section 3.2 introduces the concept of PBG with the concepts, axioms, and postulates defining a systematic way to present power electronics converters; Section 3.3 describes each power block (PB) used in the PBG; Sections 3.4 and 3.5 describe the application of the PBG for multilevel and back-to-back configurations, respectively. Finally, Section 3.6 summarizes the entire chapter.

3.2 PRINCIPLES OF POWER BLOCKS GEOMETRY (PBG)

By definition, engineering is the application of science to practical uses, and one of the most important approaches employed in engineering is the act of abstracting, that is, simplifying a complex system in a simple statement or simple equation. An example of abstraction is in analog and digital electronics where

OpAmp and Inverter Logic symbols are considered abstract elements that describe the behavior of electrical variables. There is a high level of abstraction from the physical elements to the symbols. Such a level of abstraction is crucial to the development of important devices such as computers and satellites, as the designer can focus on the functionality of each element (high level consideration) instead of, for instance, either the interactions of atoms or operation of internal transistors (low level consideration).

Also, the presentation of power electronics converters can involve some level of abstraction, with the power switches (e.g., insulated-gate-bipolar transistor (IGBT) and metal oxide semiconductor field-effect-transistors (MOSFETs)) being considered as a simple and ideal (loss-free) two-state switch, on (1) or off (0). In this sense, the goal of the PBG is to bring a higher level of abstraction to the concept of converters by furnishing a set of concepts, axioms, and postulates, as presented in the rest of this chapter.

Propositions, as considered in general statements of geometry relationships, are basically divided into two types: (i) those assumed to be true, and (ii) those proven to be true. The former are called (in Euclid's terminology) axioms if they deal with quantities and postulates if they deal with geometrical figures, while the latter are called theorems. In this way, to axiomatize a particular system is to show that its influences can be defined by a short set of sentences. Hence, it is possible to use some concepts, axioms, and postulates to define a formation law for many power electronics converters that process ac voltage as follows.

Concepts

Concept 1. PB is a basic unit of the power electronics configurations. There are six different PBs, which are able to represent a large number of topologies by just changing their arrangement.

Concept 2. Shape is either a slot or tooth with triangular or square geometry. Each block is constituted by either three or four shapes. Triangular and square shapes are used to represent specific types of variables, that is, dc or ac. On the other hand, slot and tooth are employed in the PBG to bring up the intuitive aspect of mechanical connection among blocks.

Concept 3. PB–dc (PB which processes dc variables) is a PB with at least one triangular shape. If a PB presents one or more triangular shapes, it is considered a unit that processes dc voltage.

Concept 4. PB–ac (PB which processes ac variables) is a PB without any triangular shape. If a PB presents only square shapes, it is considered a unit that processes ac voltage.

Axioms

Axiom 1. Every PB is constituted of at least three shapes.

Axiom 2. Every mechanical connection between PBs is an electrical connection.

Axiom 3. A dashed line denotes ac variable and a solid line denotes dc variable.

Postulates

Postulate 1. A triangular shape connects only with another triangular shape.

Postulate 2. A square shape connects only with another square shape.

Postulate 3. The triangular shape is associated with dc variables.

Postulate 4. The square shapes could be associated with either dc or ac variables.

Regarding these concepts, the axioms and postulates described above, it is possible to construct a large number of configurations proposed in the technical literature that process ac voltage.

Normally, power switches are considered the basic elements or cells for construction of power electronics converters. Figure 3.2 shows different topologies conceived for different applications, but with the same number of controlled power switches (i.e., four power switches).

While Fig. 3.2(a) depicts an H-bridge configuration, Fig. 3.2(b) and 3.2(c) show a leg for the three-level NPC and flying capacitor topologies, respectively; finally Fig. 3.2(d) shows an ac–dc–ac single-phase half-bridge (at both sides) configuration. It is worth mentioning that the configurations in Fig. 3.2 also process ac voltage

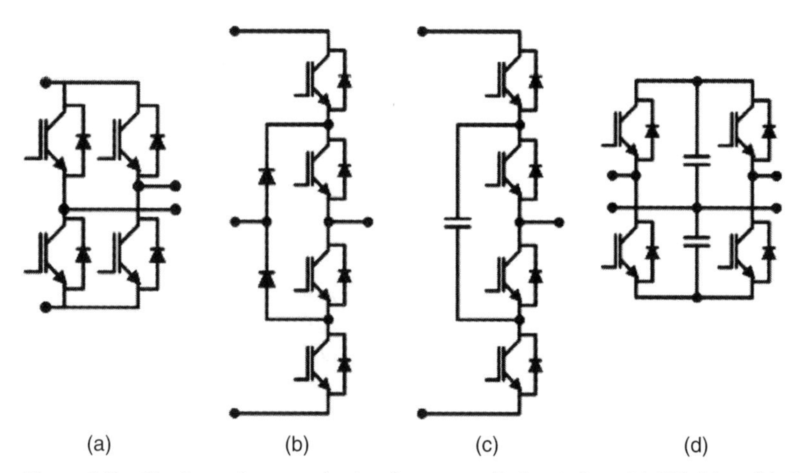

Figure 3.2 Configurations employing four controlled switches: (a) H-bridge, (b) three-level NPC, (c) three-level flying capacitor, and (d) ac–dc–ac single-phase half-bridge.

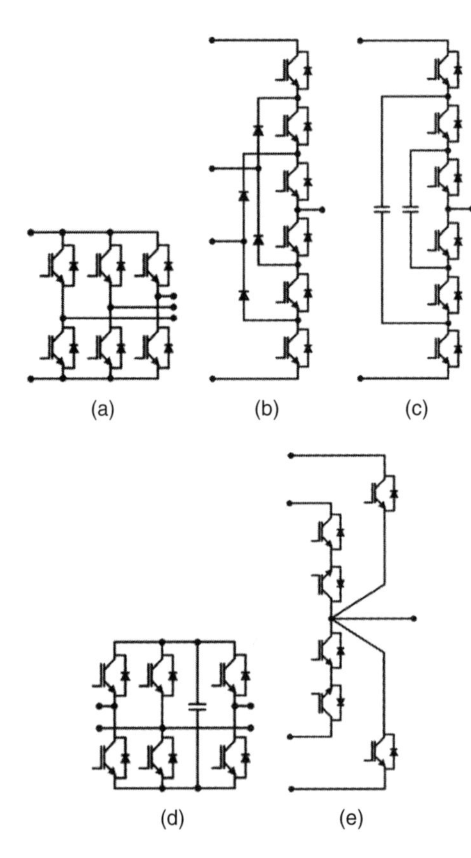

Figure 3.3 Configurations employing six controlled switches: (a) three-phase inverter, (b) four-level NPC, (c) four-level flying capacitor, (d) ac–dc–ac single-phase shared leg, and (e) nested configuration.

through a dc stage and have the same number of switches. The conventional way to present power converters with the switches as basic cells assumes the configurations in Fig. 3.2 as distinct categories. As a consequence, students are unable to understand how those configurations were created. A similar situation can be observed in Fig. 3.3.

Also, in this figure, an isolated presentation hides the understanding of creation of the topologies. In this case the configurations have six controlled power switches and Fig. 3.3(e) shows a different type of topology (as compared to those presented in Fig. 3.2) named nested converter.

To overcome such a difficulty, the conception of topologies by using the principles of the PBG brings up a systematic presentation with a simple and familiar concept of connecting blocks. Such an approach guarantees a formal outline needed for a generalized and didactic presentation. It is worth emphasizing that with PBG it is possible to create a large number of topologies from two- to n-level, just following simple rules. The PBs used in the PBG is shown in Fig. 3.4. Figure 3.4(a) and 3.4(b) shows three PB-dc and three PB-ac, respectively. See Concepts 3 and 4.

Each PB constitutes a set of power switches, which guarantees specific characteristics for the blocks. Figure 3.5 shows what is obtained inside each PB. During the process of creation of the configurations the PBs are considered in their

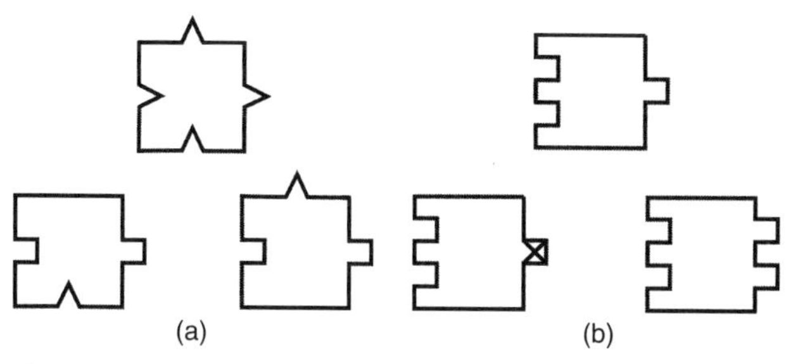

Figure 3.4 Basic units (power blocks – PB) used in the power blocks geometry (PBG): (a) PB-dc and (b) PB-ac.

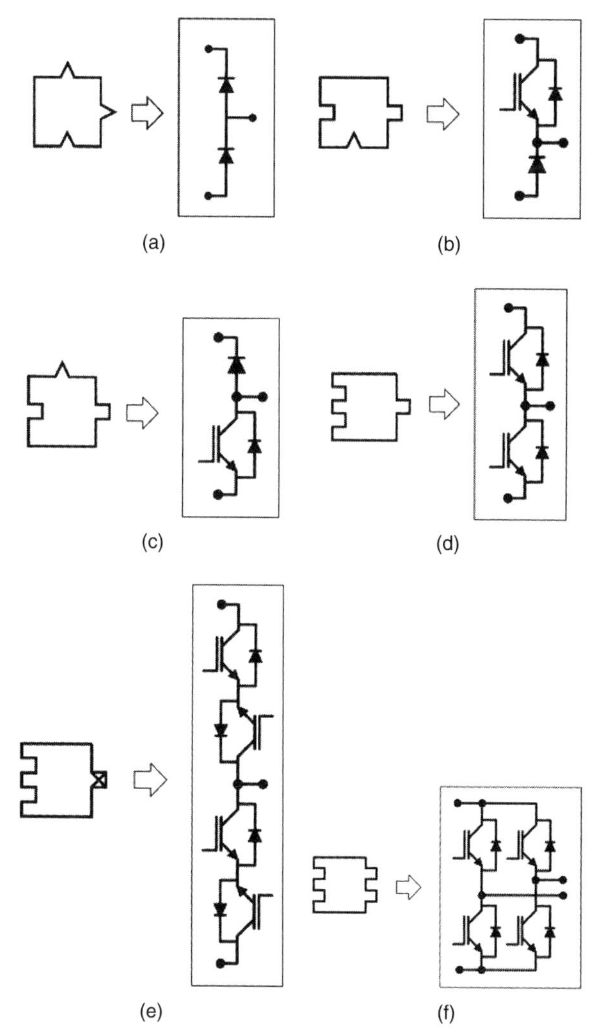

Figure 3.5 Basic units (power blocks – PBs) used in the power blocks geometry (PBG): (a) first PB-dc, (b) second PB-dc, (c) third PB-dc, (d) first PB-ac, (e) second PB-ac, and (f) third PB-ac.

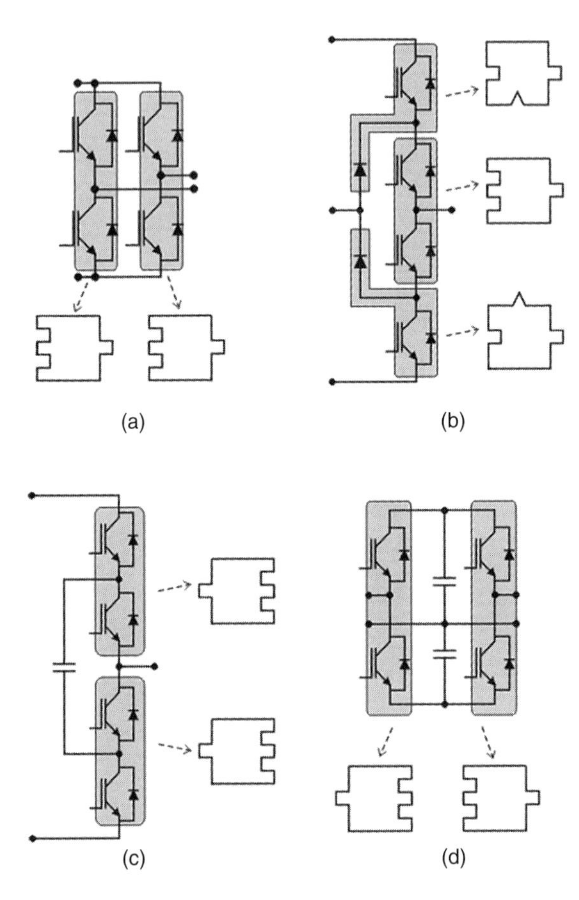

Figure 3.6 Configurations in Fig. 3.2 obtained from the PBs.

higher level (i.e., with strong abstraction from the details inside the blocks). The connections among the PBs, considering the postulates mentioned previously, will be employed for construction of the classical and nonconventional configurations, as considered in the next sections. In fact, the topologies showed in Figs 3.2 and 3.3 are obtained from the PBs following the rules presented in Section 3.2, as observed in Fig. 3.6.

3.3 DESCRIPTION OF POWER BLOCKS

The first PB-dc [see Fig. 3.5(a)] is constituted only by diodes, which means that there is no generation of gating signals.

On the other hand, there is at least one controlled power switch inside all other PBs, which requires some understanding. Figure 3.7 shows the variables associated with the second PB-dc, as well as the table with their related variables, called Table of Variables. In this table, q is the state of the controlled switch ($q = 1$ means closed switch, while $q = 0$ means open switch), v_o, V_p, and V_n are the output voltage,

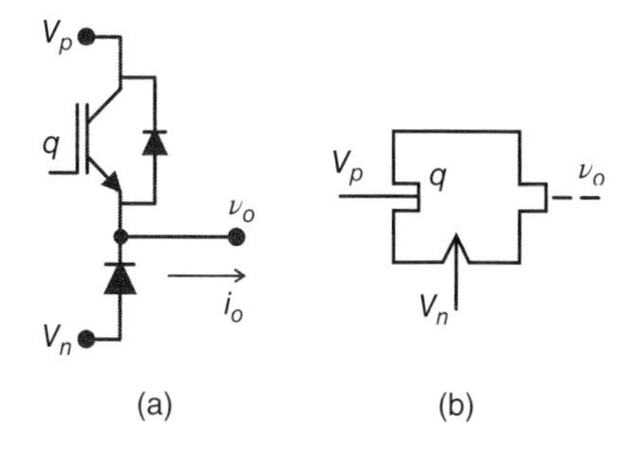

row	q	i_o	v_o
1	1	> 0	V_p
2	1	< 0	V_p
3	0	> 0	V_n
4	0	< 0	V_p

Figure 3.7 Variables associated with the second PB-dc and Table of Variables.

positive terminal of the input voltage, and negative terminal of the input voltage, respectively.

In principle, considering Postulate 4, it is possible to generate an ac output voltage, since a square tooth is observed at the output of this block. Since the desirable scenario indicates that the output voltage could be defined by the switching state independent of the load current, there is no full control associated with the first PB-dc.

Hence, taking into account the possibilities shown in the Table of Variables (see Fig. 3.7), some restrictions related to the controllability of the output voltage (v_o) are observed. When the output current is negative (rows 2 and 4) the output voltage is always V_p, independent of q, which means that there is no control of the output voltage in this case (i.e., $i_o < 0$). Then, considering the PB-dc operating alone, as depicted in Fig. 3.7, the full controlled cell is obtained only when $i_o > 0$. Equation (3.1) describes the behavior of the output voltage for the second PB-dc when $i_o > 0$.

$$v_o = qV_p + (1 - q)V_n \tag{3.1}$$

A similar restriction is observed in the third PB-dc, when $i_o > 0$, which means that the full controlled output voltage is obtained when $i_o < 0$, as observed in Fig. 3.8. In this case it turns out that

$$v_o = qV_n + (1 - q)V_p \tag{3.2}$$

Although such current restriction observed in Figs 3.7 and 3.8 can prohibit the application of those blocks to process ac voltage, their associations will guarantee the use of these blocks to generate multilevel circuits able to process ac voltage.

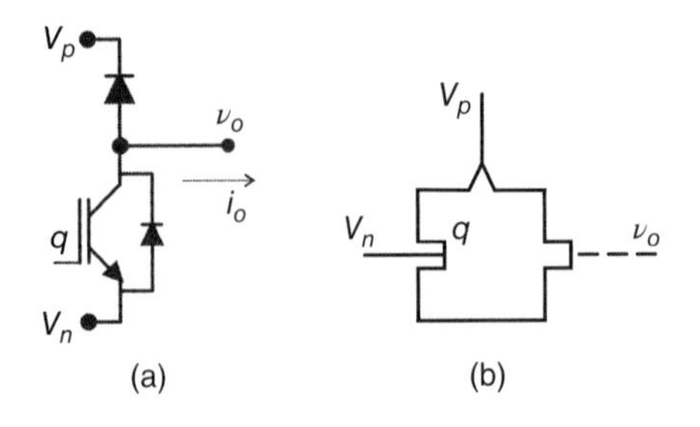

row	q	i_o	v_o
1	1	> 0	V_n
2	1	< 0	V_n
3	0	> 0	V_n
4	0	< 0	V_p

Figure 3.8 Variables associated with the third PB-dc and Table of Variables.

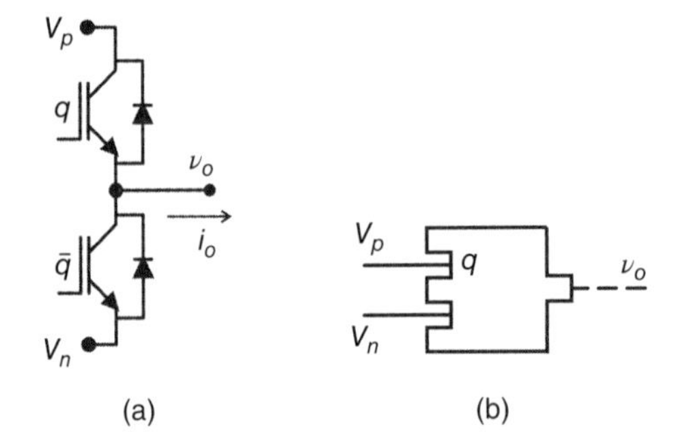

row	q	i_o	v_o
1	1	> 0	V_p
2	1	< 0	V_p
3	0	> 0	V_n
4	0	< 0	V_n

Figure 3.9 Variables associated with the first PB-ac and Table of Variables.

On the other hand, the current restriction does not appear in the PBs that process ac voltage (PB-ac). Figure 3.9 shows the variables associated with the first PB-ac, as well as its Table of Variables. Since there are two controlled switches, there is no restriction in terms of the controllability of the output voltage, even in the case of

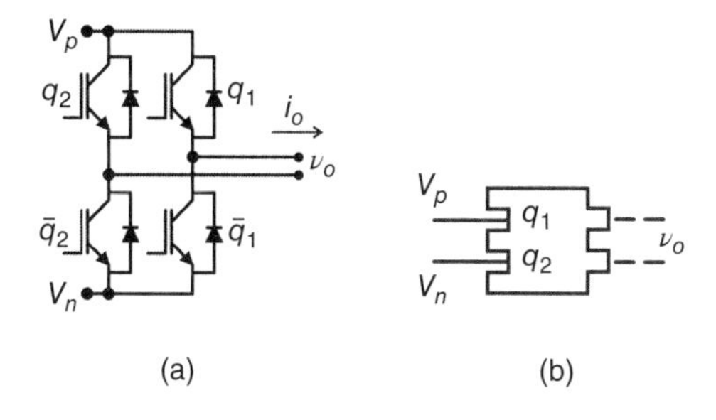

row	q_1	q_2	i_o	v_o
1	0	0	< 0	0
2	0	0	> 0	0
3	0	1	< 0	$V_n - V_p$
4	0	1	> 0	$V_n - V_p$
5	1	0	< 0	$V_p - V_n$
6	1	0	> 0	$V_p - V_n$
7	1	1	< 0	0
8	1	1	> 0	0

Figure 3.10 Variables associated with the first PB-ac and Table of Variables.

positive and negative currents. In this case, equation (3.1) is valid for any value of i_o, and $\overline{q}$ is complementary to q (i.e., $\overline{q} = 1 - q$) to avoid a short-circuit between V_p and V_n. The second PB-ac [see Fig. 3.5(e)] is in fact a particular case of the first instance with the ability to control positive and negative currents due to the antiseries connection of the switches.

The third PB-ac [see Fig. 3.5(f)] has four controlled switches q_1 and q_2 and their complementary switches $\overline{q}_1$ and $\overline{q}_2$. Like the first PB-ac, the output voltage is a function of only the state of switches (q_1 and q_2). But unlike the case in Fig. 3.9, Fig. 3.10 shows that the output voltage has three different values, or three levels ($V_n - V_p$, 0, $V_p - V_n$), and in this case the output voltage is given by

$$v_o = (q_1 - q_2)V_p + (q_2 - q_1)V_n \tag{3.3}$$

The process of building power electronics converters processing ac voltage by using the PBG does not take into account the switching states, just the geometry of the blocks and the connections among them, which make possible the process of topologies creation. The Table of Variables presented in Figs 3.7–3.10 and equations (3.1)–(3.3) will be employed for the generation of the PWM waveforms, discussed in the following chapters. The next section of this chapter will deal only with the creation of the configurations by using the axioms and postulates.

Example 3.1

Assume a dc motor drive system requires a converter with controlled voltage at the output converter side and demanding positive currents only. (a) Choose one of the PBs with minimum components able to deal with these requirements. (b) If the motor average current is given by $I_o = 50\,\text{A}$ and the dc-link voltage is given by $V_{pn} = 200\,\text{V}$ (with $V_{pn} = V_p - V_n$) specify the switches employed in this converter in terms of voltage and current values.

Solution

(a) Since the output has a positive dc current and there is a need for a controlled output voltage, the PB with fewer components will be the first PB-dc [i.e., configuration constituted by a controlled switch q and a diode – see Fig. 3.7(a)].

(b) Considering the operation of this PB: (1) when $q = 1$ the output current (I_o) will be going through the controlled switch, the diode current will be zero, the controlled switch voltage will be zero and the reserve diode voltage will be V_{pn}; (2) when $q = 0$ the output current (I_o) will be going through the diode, the controlled switch current will be zero, the controlled switch voltage will be V_{pn} and the diode voltage will be zero. The specification of the voltage and current for both devices (controlled switch and diode) are 200 V and 50 A.

Example 3.2

A battery formation apparatus is constituted by circuits that aim to generate a specific voltage waveform in order to guarantee the charge and discharge process until the battery is considered formed. Sometimes the discharge process is obtained by burning the stored energy with resistors (i.e., inefficient process). To overcome this problem, it is possible to employ power electronics solutions to charge and discharge the battery by sending energy back to the primary source of energy with higher efficiency, as observed below. Choose one of the PBs with minimum components that can deal with these requirements.

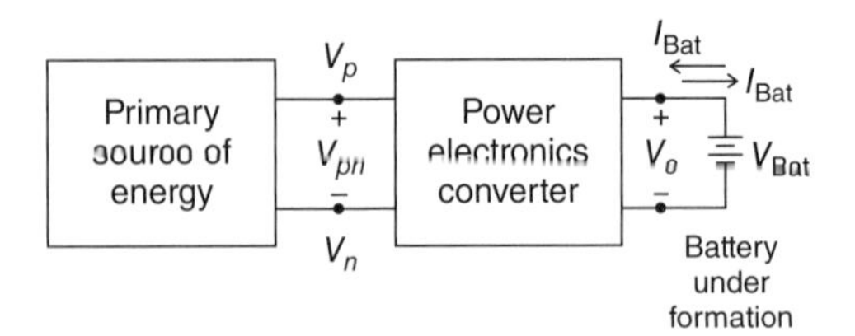

Solution

The power electronics converter for this particular application must generate controlled output voltage either for positive or negative current, which means that both PB-ac (half-bridge or H-bridge configurations) could be used.

3.4 APPLICATION OF PBG IN MULTILEVEL CONFIGURATIONS

Multilevel converters were first conceived for high voltage and high power applications, beginning with the neutral-point-clamped (NPC) inverter. Since that time, many configurations have been proposed to establish the highly desirable characteristics for high power applications such as low blocking voltage by switching devices and reduced waveform distortion. The number of levels can be increased as far as necessary, limited by both the complexity of the control and the price of the converter itself, due to the increased number of semiconductor devices. To figure out the advantages of waveforms with increased number of levels, Fig. 3.11 shows the total harmonic distortion (THD) versus modulation index (or amplitude-modulation ratio) for two-, three-, and four-level waveforms. As defined in the previous chapter,

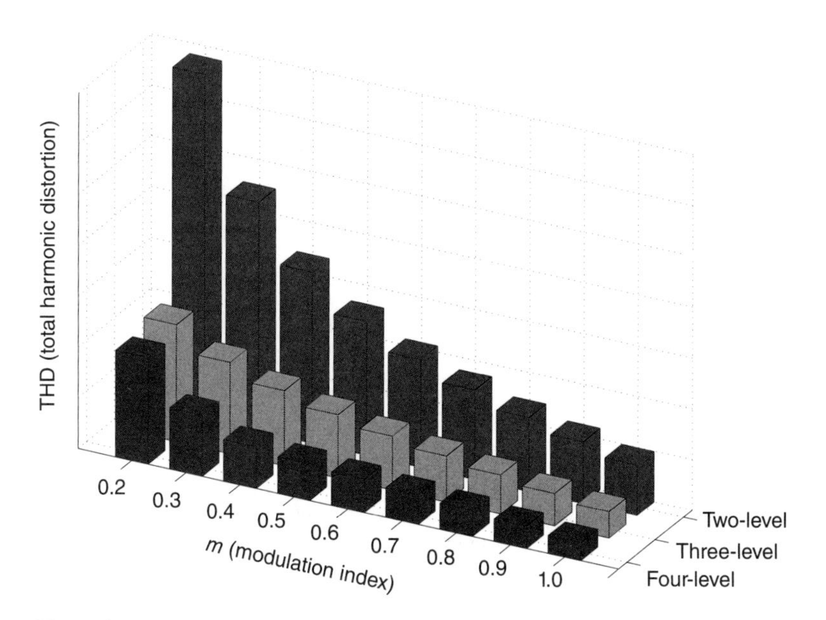

Figure 3.11 THD versus modulation index for voltages with two, three, and four levels.

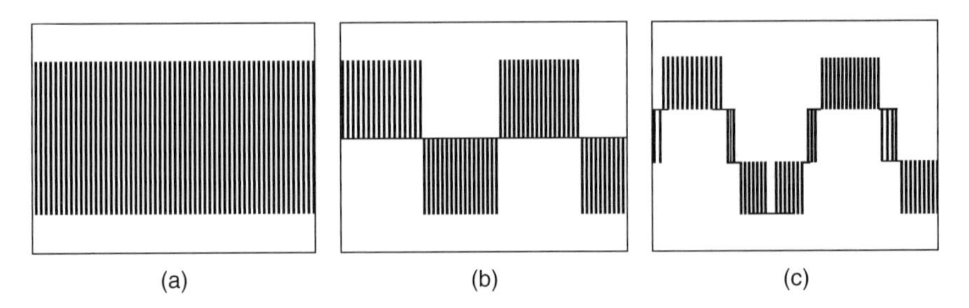

Figure 3.12 Output converter voltage with two-, three-, and four-level.

the modulation index (m_a) is a relationship between the amplitude of the sinusoidal waveform (also called modulating or control voltage – V_m) and triangular waveform (also called carrier voltage – V_t), that is, $m_a = V_m/V_t$. It can be seen from Fig. 3.11 that the increase in the number of levels results in THD reduction for a given value of m_a. Figure 3.12 depicts the waveforms produced by converters with two-, three- and four-levels, respectively. From this figure it is easy to understand the behavior shown in Fig. 3.11 with better THD for a higher number of levels – notice that the waveforms in Fig. 3.12 become closer to a sinusoidal voltage.

The most employed multilevel configurations, with industry interest, are diode-clamped (or NPC), cascade, and flying capacitor multilevel inverters. Those multilevel configurations will be considered throughout this book as the conventional ones, while different multilevel topologies presented in the technical literature will be treated as nonconventional solutions. The nonconventional converters are important in power electronics basically because they may provide an alternative solution to improve a specific aspect of the conversion system. Furthermore, such nontraditional power electronics converters play an important role in teaching students due to the extra effort required in modeling and dealing with specific requirements.

3.4.1 Neutral-Point-Clamped Configuration

Considering the PBG, the constructions of the NPC configurations can be presented in a systematic way, as described briefly in this subsection. Chapter 4, however, will provide details of this configuration. With two PBs-dc cells [see Fig. 3.5(b) and 3.5(c)] and the first PB-ac cell [see Fig. 3.5(d)], and connecting them following the Postulates 1 and 2 presented previously, it is possible to generate the three-level NPC configuration, as observed in Fig. 3.13(a). This is a three-level converter because the blocks (and consequently the switches) are connected in such a way that one of the three inputs can be selected to appear at the output converter side.

Other levels can be obtained following the same strategy, that is, connecting a triangular slot (or tooth) with a triangular tooth (or slot) (i.e., following Postulate 1); and connecting a square slot (or tooth) to a square tooth (or slot) (i.e., following Postulate 2).

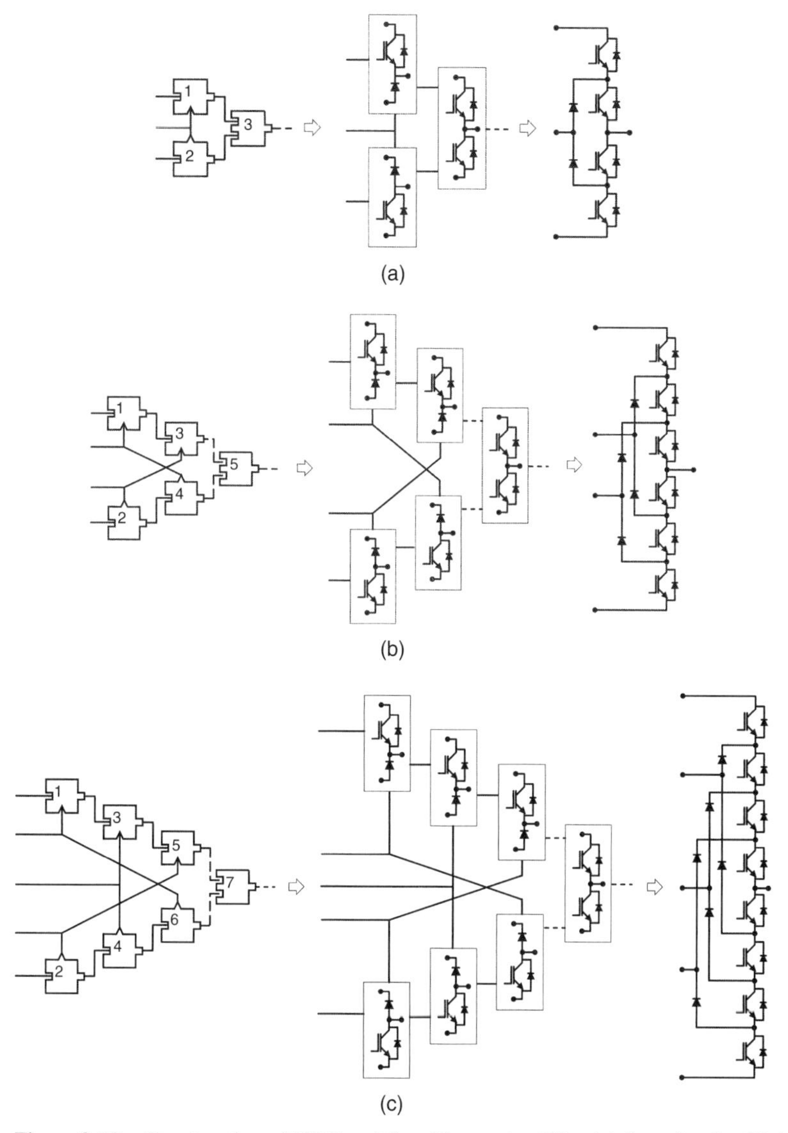

Figure 3.13 Construction of NPC multilevel legs using PBs: (a) three-levels, (b) four-levels, and (c) five-levels.

From the four-level NPC configuration onward there are always two or more possibilities of connections among the triangle slots and teeth. For instance, considering the configuration in Fig. 3.13(b), the triangle slot of PB 1 is connected to the triangle tooth of PB 4, while the slot of PB 3 is connected to the tooth of PB 2. There are other ways to connect the blocks differently from that shown in Fig. 3.13. These different ways and the PWM approach are considered in Chapter 4.

Example 3.3

Write the Table of Variables for the three-level NPC configuration; see Fig. 3.13(a). Prove that the switching devices can be represented by a binary variable, with $q_j = 1$ used when the switch is on and $q_j = 0$ when the switch is off (with $j = 1, 2, 3$, and 4), as highlighted below.

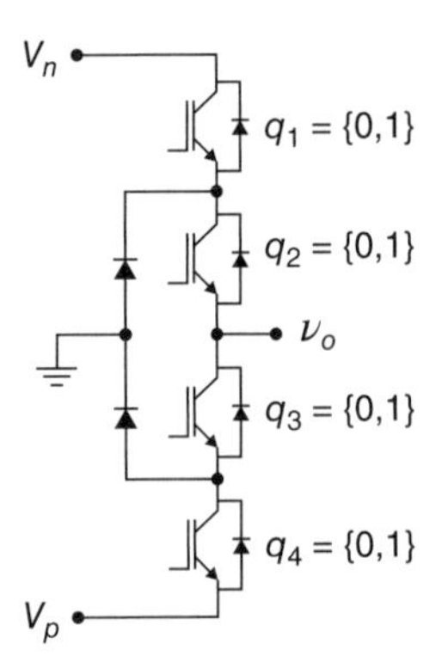

Solution

Since there are four switches with two states each, there are sixteen ($2^2 = 16$) possibilities for the total switching states, as shown below:

row	q_1	q_2	q_3	q_4	v_o
1	0	0	0	0	*Undefined*
2	0	0	0	1	*Undefined*
3	0	0	1	0	*Undefined*
4	0	0	1	1	V_n
5	0	1	0	0	*Undefined*
6	0	1	0	1	*Undefined*
7	0	1	1	0	0
8	0	1	1	1	*Short-circuit*
9	1	0	0	0	*Undefined*
10	1	0	0	1	*Undefined*
11	1	0	1	0	*Undefined*
12	1	0	1	1	0
13	1	1	0	0	V_p
14	1	1	0	1	V_p
15	1	1	1	0	*Short-circuit*
16	1	1	1	1	*Short-circuit*

Notice that many of the states obtained in this Table are prohibited since they are either undefined or short-circuit, and so must be avoided. In this context, undefined means that the output voltage is not defined exclusively by the switching states. Without taking into account the prohibited states, this converter can generate three different voltages at its output (i.e., $V_n, 0, V_p$) for either positive or negative currents.

Example 3.4

Many of the switching states in Example 3.3 cannot be applied since they are prohibited states. Since there are only three inputs and those inputs should be selected to appear at the output converter side (v_o), determine: (a) the minimum number of independent switching states to select any input voltage at the output converter side (b) how those switching states can be associated with the converter in Fig. 3.13(a) with four switches to generate a three-level output voltage.

Solution

(a) As there are three inputs and the requirement is to use the switching states to select each of these inputs at the output, the minimum number of independent switching states is 2, since $2^2 = 4$.

(b) As observed in Fig. 3.13(a), the three-level NPC topology requires four switches to guarantee an output with three levels. On the other hand, with two independent switching states (as concluded previously) it is possible to select three inputs available ($V_n, 0, V_p$), defining the complementary switching states as $\overline{q}_1$ and $\overline{q}_2$, where $\overline{q}_1 = 1 - q_1$ and $\overline{q}_2 = 1 - q_2$. Below is presented the three-level NPC topology, using only two independent switches, as well as its Table of Variables. The converter as presented below is able to select any input by using two independent switches.

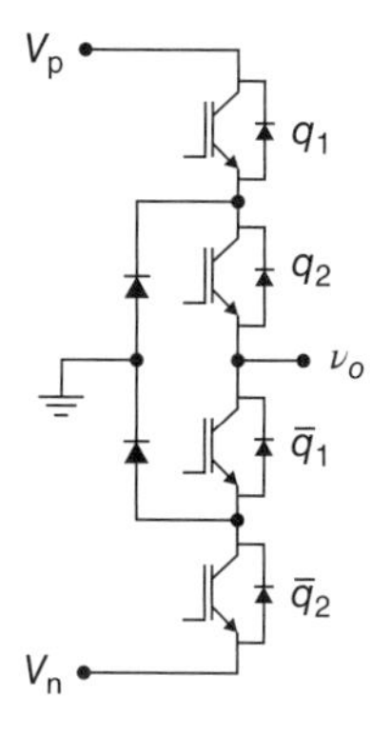

State	q_1	q_2	v_o
1	0	0	V_n
2	0	1	0
3	1	0	*Undefined*
4	1	1	V_p

3.4.2 Cascade Configuration

As the number of levels increases, the practical control limits observed in the NPC multilevel converter restrict its use, due to capacitor voltage control issues. Such operation limits are presented in Chapter 4. Another topology that deals with generation of multilevel voltages without dramatically increasing the control complexity is the cascade configuration, since it normally uses isolated dc sources. A high power motor drive system with high level reliability is one of the applications where this kind of converter is popular.

The construction of the cascade multilevel configurations are observed in Fig. 3.14, with the direct use of the third PB-ac (see Fig. 3.5) following Postulate 2 with the connection of the teeth in a series arrangement. Dashed lines at the output of these converters mean that they will deal with ac variables, while solid lines at the input side of the converter mean dc variables. Details of the cascade configuration are furnished in Chapter 5.

Example 3.5

For the converter presented below [as in Fig. 3.14(a)], write the Table of Variables for the case in which the dc-link voltage of the converters are given by $V_{\mathrm{dc1}} = V_{\mathrm{dc}}$ and $V_{\mathrm{dc2}} = 2V_{\mathrm{dc}}$ and then determine how many levels will be generated at the output converter side with $\overline{q}_j = 1 - q_j$ and $j = 1, 2, 3, 4$.

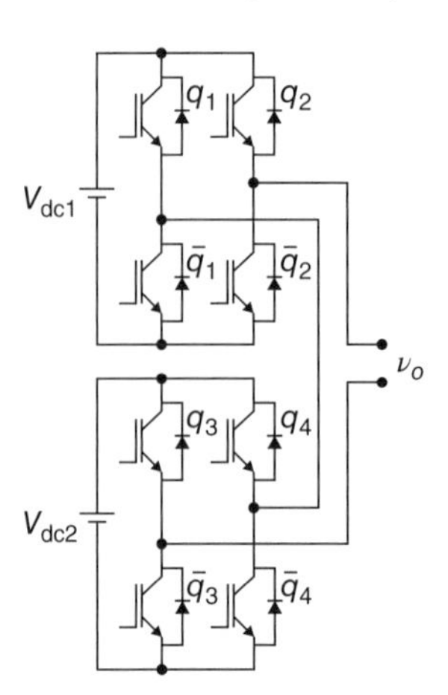

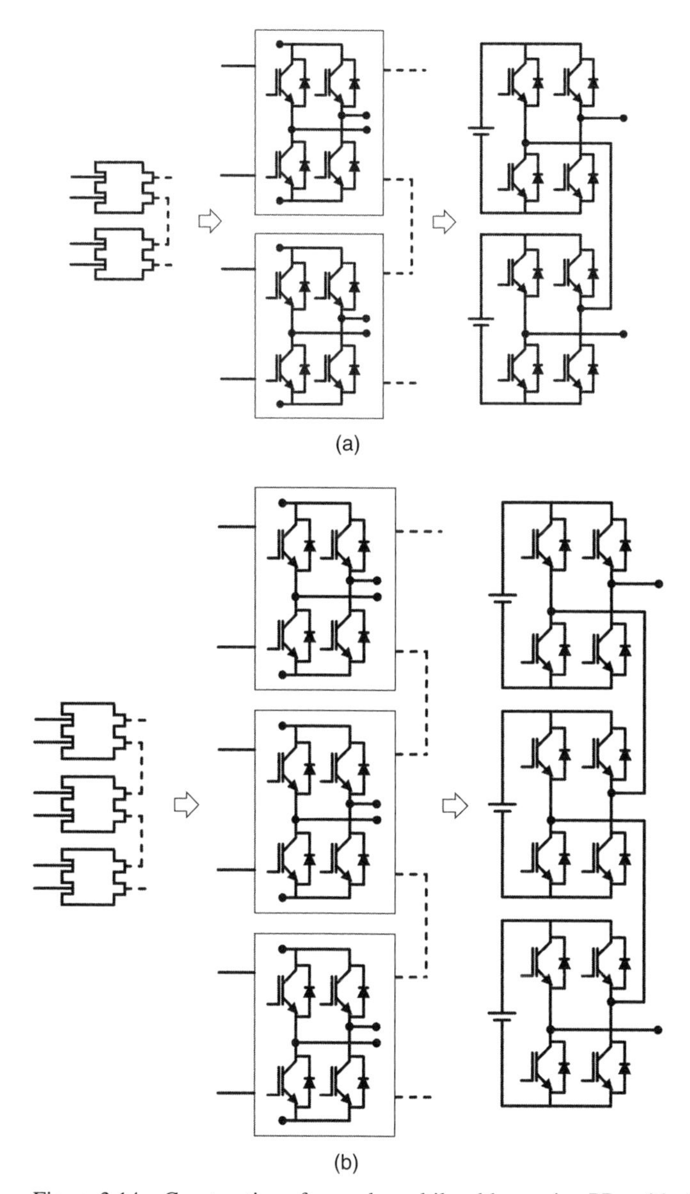

Figure 3.14 Construction of cascade multilevel legs using PBs with: (a) two power blocks and (b) three power blocks.

Solution

The Table of Variables for this converter is presented below. The output voltage will have seven levels given by 0, $\pm V_{dc}$, $\pm 2V_{dc}$ and $\pm 3V_{dc}$.

row	q_1	q_2	q_3	q_4	v_o
1	0	0	0	0	0
2	0	0	0	1	$2V_{dc}$
3	0	0	1	0	$-2V_{dc}$
4	0	0	1	1	0
5	0	1	0	0	V_{dc}
6	0	1	0	1	$3V_{dc}$
7	0	1	1	0	$-V_{dc}$
8	0	1	1	1	V_{dc}
9	1	0	0	0	$-V_{dc}$
10	1	0	0	1	V_{dc}
11	1	0	1	0	$-3V_{dc}$
12	1	0	1	1	$-V_{dc}$
13	1	1	0	0	0
14	1	1	0	1	$2V_{dc}$
15	1	1	1	0	$-2V_{dc}$
16	1	1	1	1	0

Example 3.6

Considering two cells of the first PB-ac, as presented in Fig. 3.5(d) and Postulates 2 and 4, draw a different configuration from the ones presented so far with three inputs and one output and write the possible output voltages.

Solution

The postulates applied for two cells of the first PB-ac as presented in Fig. 3.5(d) allow a possible solution as depicted below.

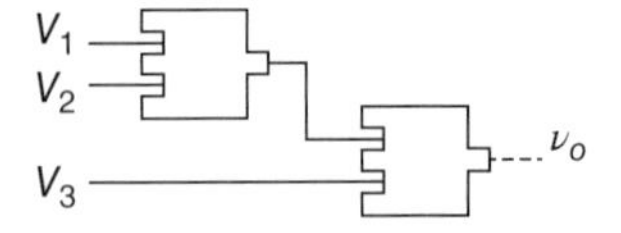

After the process of defining the topology, it is possible to write the output voltages, taking into account the switches inside each block, as shown below.

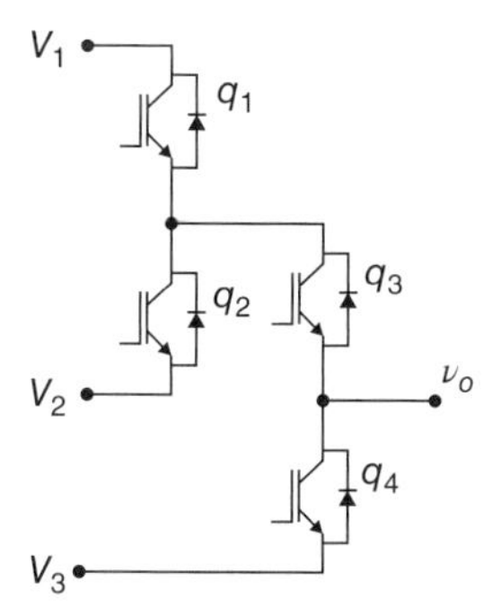

Here, the output voltages are given respectively by

- $v_o = V_1$ when $q_1 = q_3 = 1$ and $q_2 = q_4 = 0$
- $v_o = V_2$ when $q_2 = q_3 = 1$ and $q_1 = q_4 = 0$
- $v_o = V_3$ when $q_4 = 1$ and $q_1 = q_2 = q_3 = 0$

3.4.3 Flying Capacitor Configuration

There are some drawbacks associated with both multilevel topologies (NPC and cascade) presented earlier. One of the disadvantages of the NPC topology is the need to control dc-link capacitor voltages. On the other hand, an obvious drawback for a cascade topology is the high number of dc voltage sources required.

The flying capacitor converter is another topology that can generate multilevel voltages. Although needing additional flying capacitors (FCs), the capacitance values are smaller than the dc-link capacitor and there are two switching states to control the charging and discharging of FC. Unlike the NPC topology, in which just one switching state is available to obtain zero voltage at the output converter side, the FC converter has two states to generate zero voltage at the output.

In practical implementation of the NPC topology there is only one dc voltage available; so two dc-link capacitors are normally used to guarantee three voltages at the input converter side, as in Fig. 3.15(a). Figure 3.15(b) shows the Table of Variables when two capacitors are used to generate three levels. For a symmetrical three-level output voltage [as in Fig. 3.12(b)] each capacitor voltage should be the same, for example, $V_{C1} = V_{C2} = V_{pn}/2$. A control scheme should be added to guarantee this requirement, but as shown in Fig. 3.15(b), there is a missing switching state (State 3) $q_1 = 1$ and $q_2 = 0$, which reduces the degree of freedom for controlling capacitor voltage. This problem is minimized for the capacitor flying topologies since there are two switching states to generate zero voltage at the output.

The PBG approach can also be employed to systematically obtain the flying capacitor converters. For example, Fig. 3.16(a) brings up the three-level flying

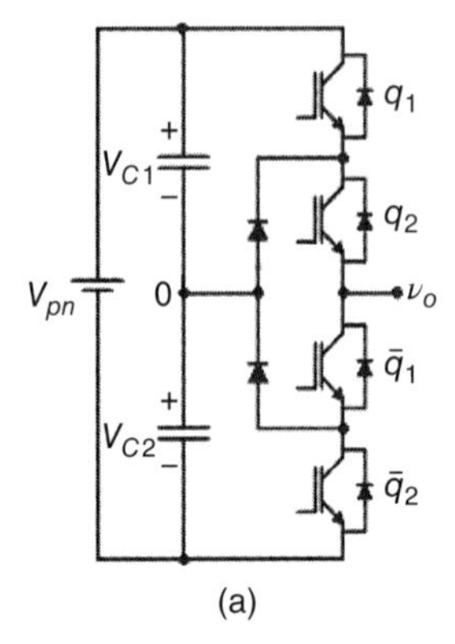

State	q_1	q_2	v_o
1	0	0	$-V_{C2}$
2	0	1	0
3	1	0	
4	1	1	V_{C1}

(a) (b)

Figure 3.15 (a) NPC topology with a single dc voltage source and two dc-link capacitors. (b) Table of Variables.

capacitor topology under PBs representation. The converters with four and five levels are depicted in Fig. 3.16(b) and 3.16(c), respectively. Details about this circuit are given in Chapter 6.

Example 3.7

Write the Table of Variables for the converter given below, which is based on the three-level flying capacitor topology, with a dc voltage source instead of flying capacitor. Prove also that the switching devices can be represented by a binary variable, that is, $q_j = 1$ is used when the switch is on and $q_j = 0$ when the switch is off (with $j = 1, 2, 3,$ and 4).

Solution

Since there are four switches with two states each, there exist sixteen possibilities for the total switching states, as shown below:

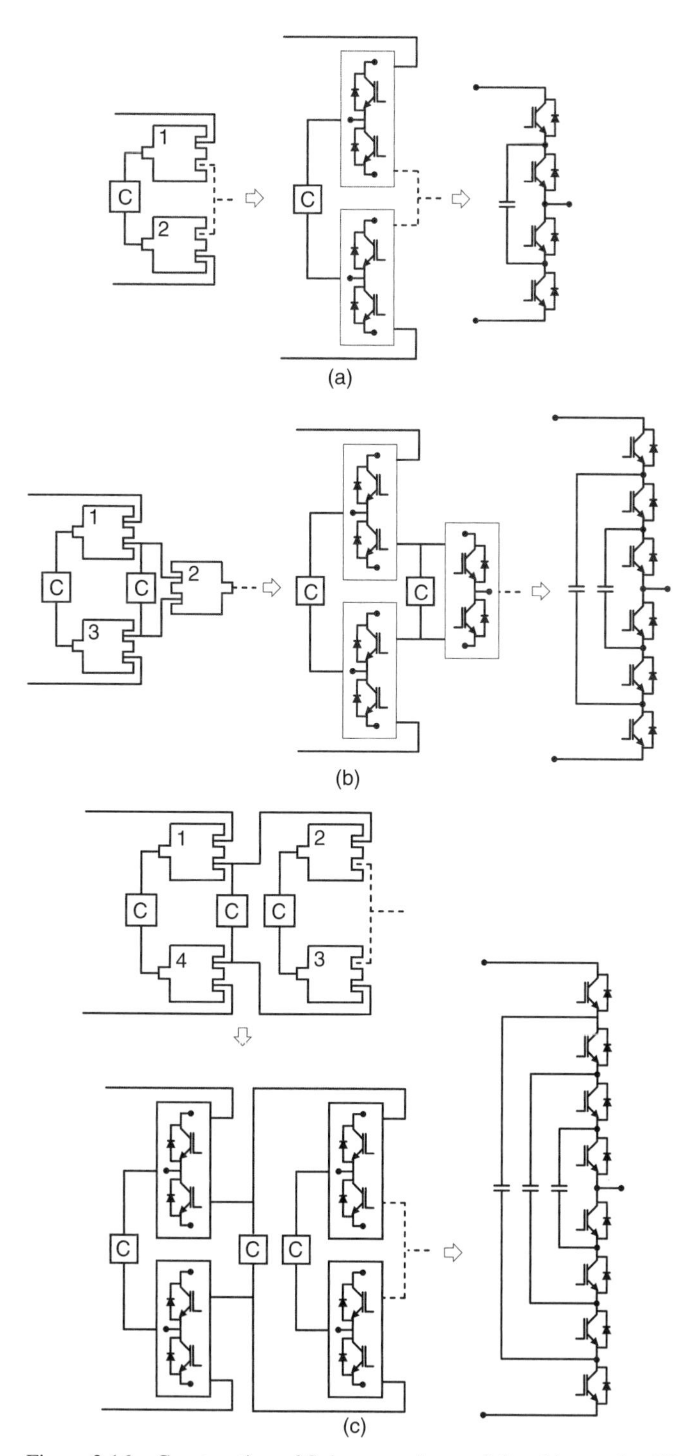

Figure 3.16 Construction of flying capacitor multilevel legs using PBs: (a) three-level, (b) four-level, and (c) five-level.

row	q_1	q_2	q_3	q_4	v_o
1	0	0	0	0	*Undefined*
2	0	0	0	1	*Undefined*
3	0	0	1	0	*Undefined*
4	0	0	1	1	$-V_{dc}$
5	0	1	0	0	*Undefined*
6	0	1	0	1	0
7	0	1	1	0	*Short-circuit*
8	0	1	1	1	*Short-circuit*
9	1	0	0	0	*Undefined*
10	1	0	0	1	*Undefined*
11	1	0	1	0	0
12	1	0	1	1	$-V_{dc}$
13	1	1	0	0	V_{dc}
14	1	1	0	1	V_{dc}
15	1	1	1	0	*Short-circuit*
16	1	1	1	1	*Short-circuit*

Notice that many of the states obtained in this table are prohibited, that is, either undefined or short-circuit, and so must be avoided. Without taking into account the prohibited states, this converter can generate three different voltages at its output (i.e., $-V_{dc}, 0, V_{dc}$) for either positive or negative currents.

Example 3.8

(a) Considering Example 3.7, what is the minimum number of independent switching states that still allows three levels at the output converter side? b) Draw the power converter with this minimum number of independent switches to guarantee a three-level output voltage.

Solution

(a) If there are three input voltages and the requirement is to use the switching states to select each input, the minimum number of independent switching states is 2, since $2^2 = 4$.

(b) As observed in Fig. 3.16(a), the three-level flying capacitor topology requires four switches to guarantee an output with three levels. Moreover, as shown earlier, with two independent switching states (q_1 and q_2) it is possible to select the three inputs available ($V_{dc}, 0, -V_{dc}$), which ends up with the definition of the complementary switching states $\overline{q}_1$ and $\overline{q}_2$ ($\overline{q}_1 = 1 - q_1$ and $\overline{q}_2 = 1 - q_2$). Below is presented the three-level flying capacitor configuration, using only two independent switching states and the complementary switching states, as well as the Table of Variables. Comparing this Table with that in Fig. 3.15, there is no prohibited state for this topology, and this degree of freedom is used to control the voltage of the flying capacitor.

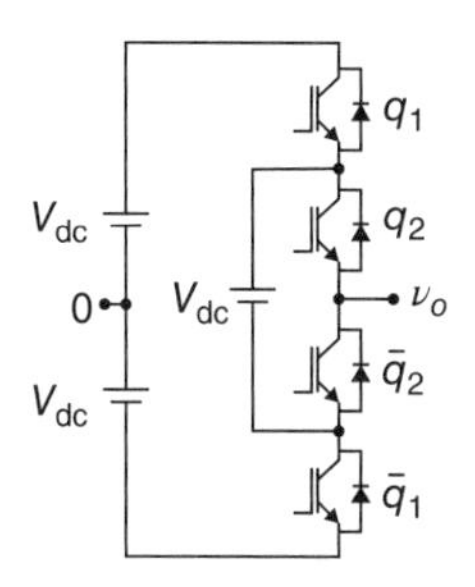

State	q_1	q_2	v_o
1	0	0	$-V_{dc}$
2	0	1	0
3	1	0	0
4	1	1	V_{dc}

3.4.4 Other Multilevel Configurations

Although the configuration obtained in Example 3.6 can generate an output voltage with three levels, it is clearly a nonconventional topology, since it is not identified as NPC, cascade, or flying capacitor converter. This section presents briefly some nonconventional converters and Chapter 7 provides the necessary details.

Figure 3.17(a) shows a possible four-level topology by using the first and second PBs-ac. Since this is a four-level converter, one of its counterparts is the four-level NPC configuration shown in Fig. 3.13(b). Considering just the number of semiconductor devices employed for both circuits, it favors the nonconventional one in Fig. 3.17(a) due to the elimination of four diodes.

In addition to the drawbacks of the NPC converters mentioned earlier, such topologies present unequal loss distribution among the semiconductor power devices. The active NPC (ANPC) achieves a better loss distribution and wider semiconductor devices utilization. ANPC configurations can also be presented following the PBG approach (basically considering Postulates 2 and 4), as depicted in Fig. 3.17(b) for three-level and in Fig. 3.17(c) for four-level output voltage.

Many of the nonconventional configurations proposed in the technical literature can be obtained by following the axioms and postulates of the principles of the PBG, as in Fig. 3.18.

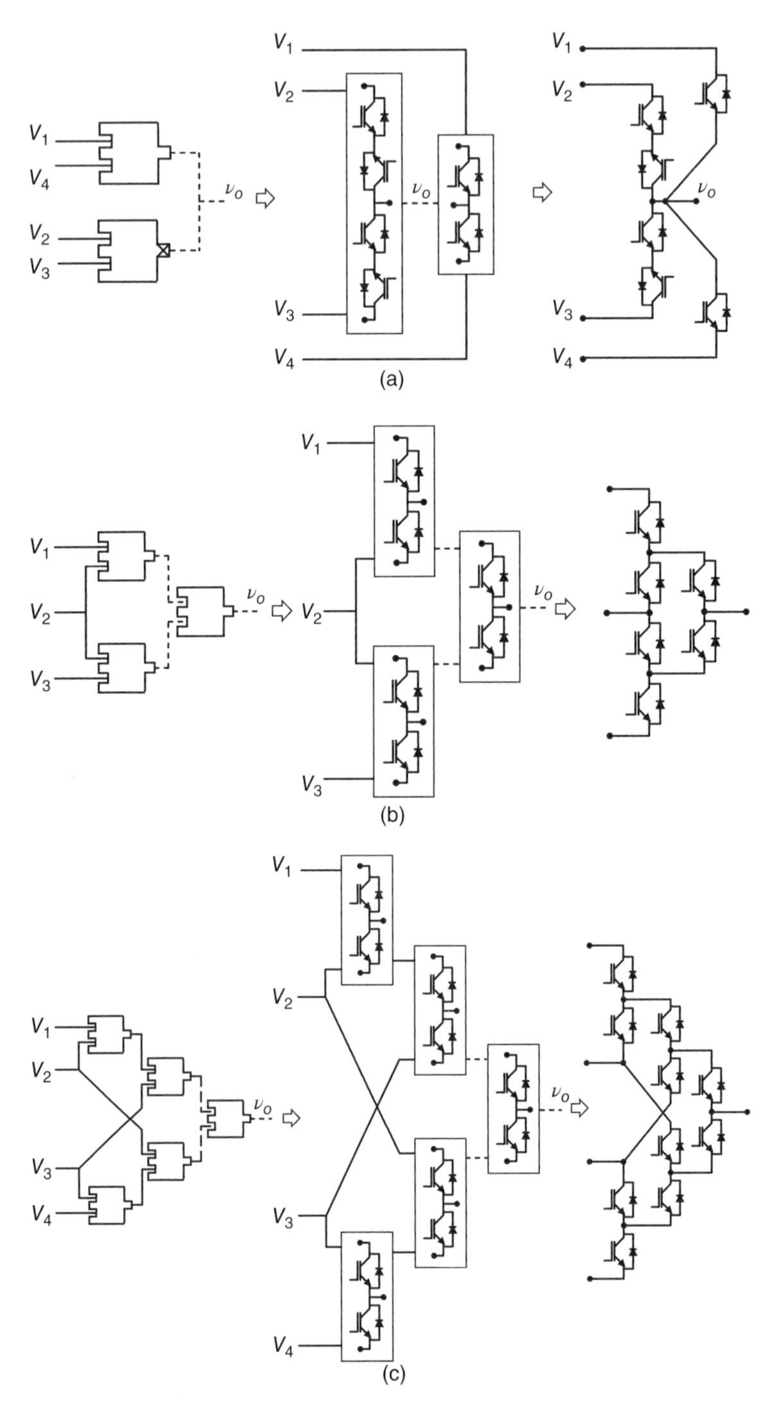

Figure 3.17 Construction of nonconventional converters. (a) Nested multilevel converter. (b) Three-level ANPC multilevel legs using PB. (c) Four-level ANPC multilevel legs using PB.

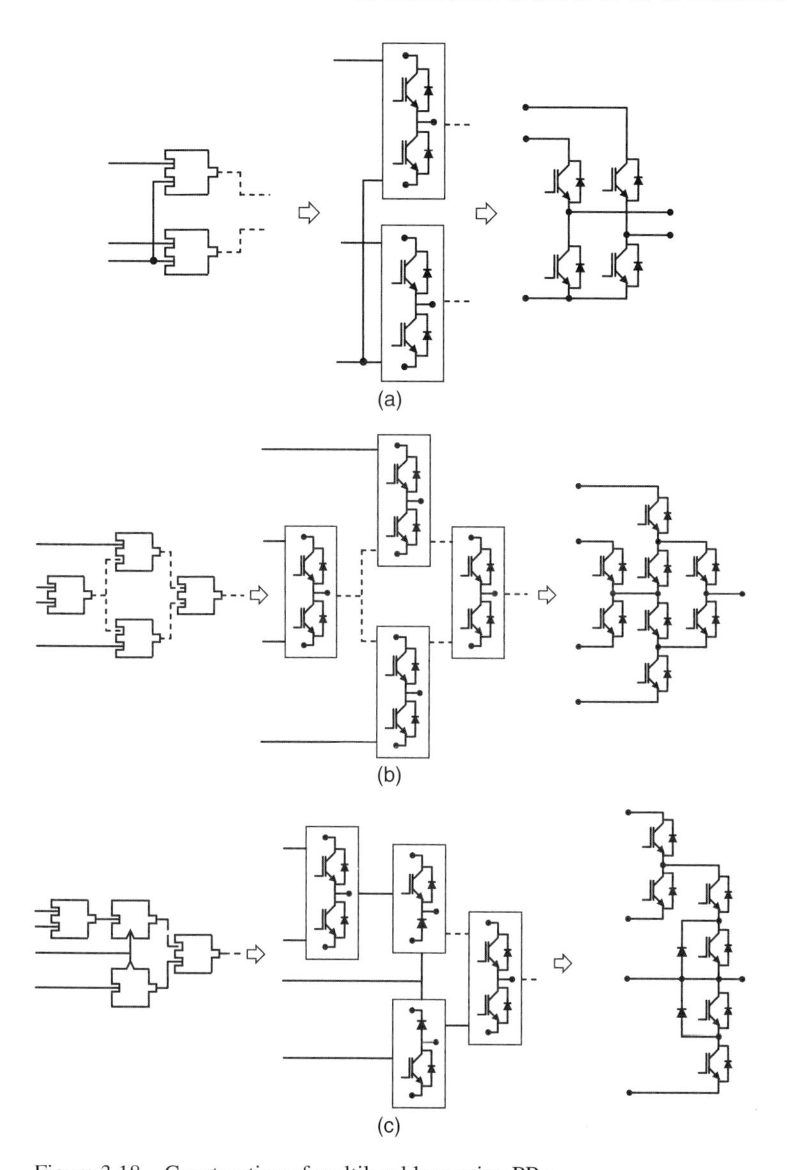

Figure 3.18 Construction of multilevel legs using PBs.

3.5 APPLICATION OF PBG IN ac–dc–ac CONFIGURATIONS

Another important class of configuration that processes ac voltage is the one with two conversion stages, that is, with both rectification (ac–dc conversion) and inversion (dc–ac conversion) stages connected through a dc-link capacitor link. Such topologies are normally called ac–dc–ac configurations (or back-to-back configurations) and are usually described as three-phase to three-phase (3ph–3ph), single-phase to single-phase (1ph–1ph), and single-phase to three-phase (1ph–3ph) conversions. Although each power conversion system (3ph–3ph, 1ph–1ph or 1ph–3ph) presents many possibilities of implementation in terms of the topology structure employed, the ac–dc–ac configurations can be formulated considering the principles of the PBG.

Many applications require this kind of converter: (i) three-phase motor drive system with active power factor correction [see Fig. 3.19(a)], (ii) universal active power filter with a combination of series and shunt active filters [see Fig. 3.19(b)], and (iii) wind generation system [see Fig. 3.19(c)].

3.5.1 Three-Phase to Three-Phase Configurations

A three-phase back-to-back converter feeds a three-phase load from a three-phase grid employing six legs (twelve switches) with one leg connected to each phase of the system, as shown in Fig. 3.20(a). Considering the PBs and the geometry involving them, the dashed line means that all input and output variables are of ac type, while the solid line represents the dc variables of the dc-link voltage.

Some other topologies can be derived from the conventional one in Fig. 3.20(a), especially when it is necessary to reduce the number of semiconductor devices. In this way, both dc-link midpoint and shared-leg connections have been employed as an approach to reduce the power switches. It means that at least one of the legs, constituted by two power switches, is no longer used for the solutions presented in

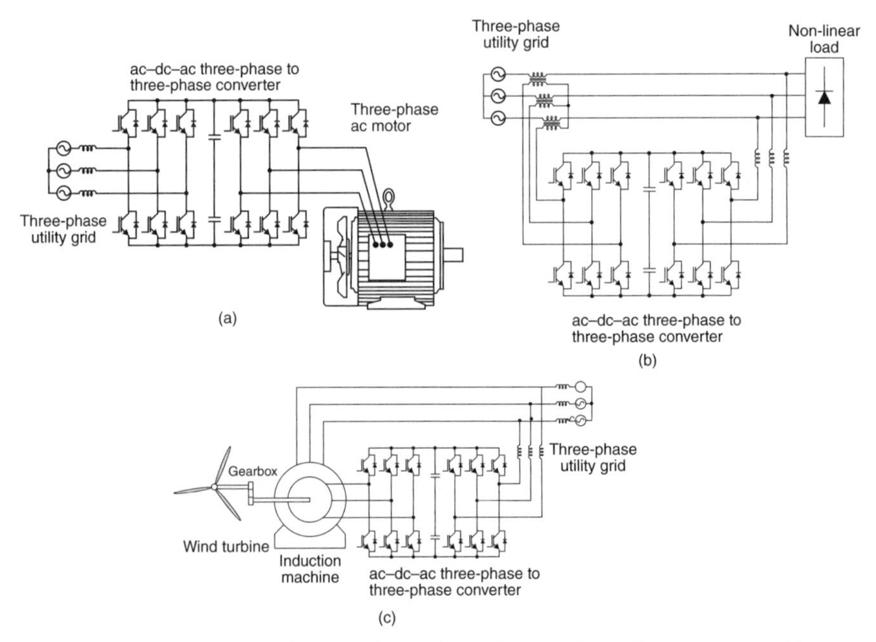

Figure 3.19 Applications of the ac–dc–ac three-phase to three-phase converter: (a) ac motor drive system, (b) active power filter, and (c) double-fed induction generation system used in wind turbines.

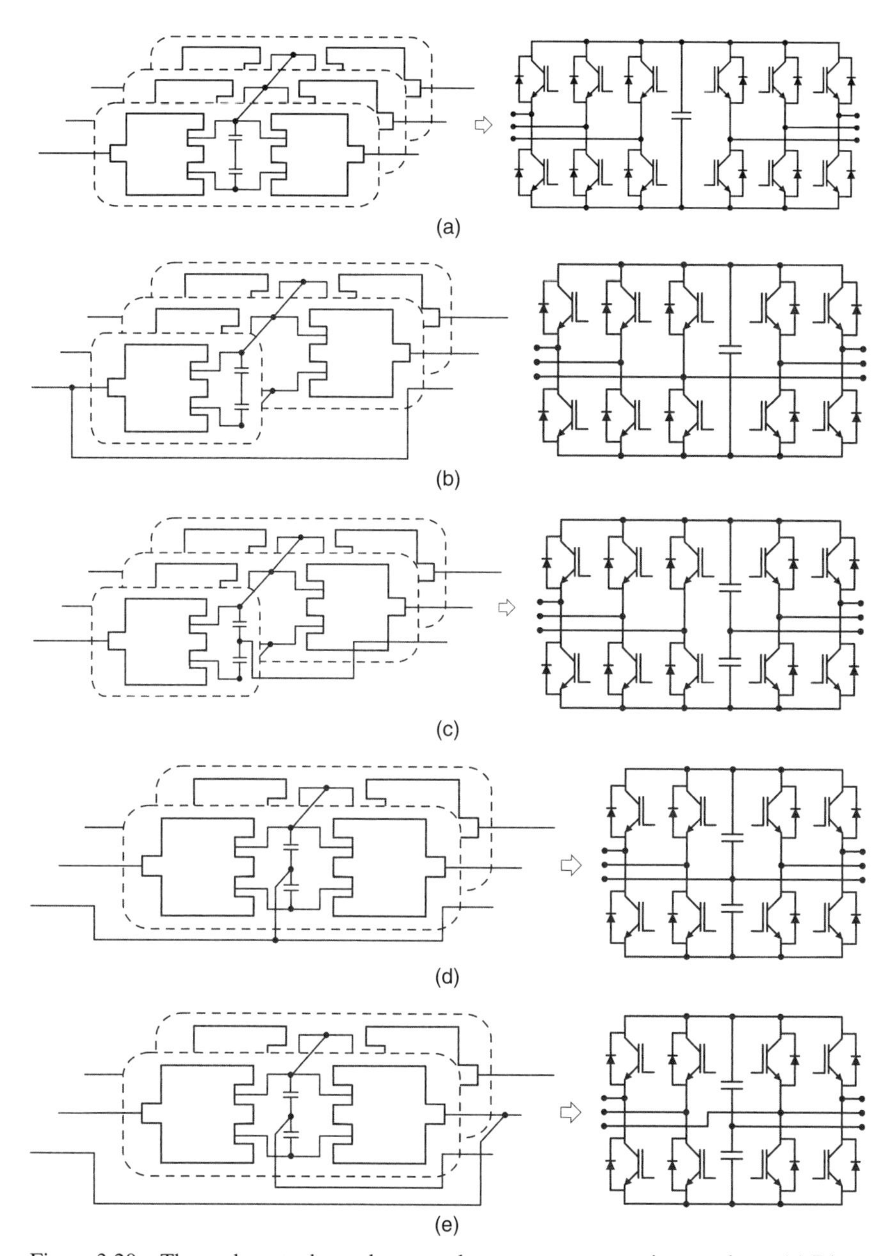

Figure 3.20 Three-phase to three-phase ac–dc–ac power conversion topology: (a) Direct solution with six legs, (b) one shared leg, (c) one dc-link midpoint connection, (d) two dc-link midpoint connections, and (e) shared-leg and dc-link midpoint connections.

Fig. 3.20(b)–3.20(e). However, regarding cost reduction, this approach make sense if what is saved by the number of power semiconductor devices is not lost in higher device ratings and higher values of electrolytic capacitor.

Figure 3.20(b) shows a solution with a shared leg between input-output converter sides. Figure 3.20(c) presents a full-bridge configuration at the input side (left side), and a half-bridge configuration at the output side (right side).

Although using the same number of power switches, the configuration in Fig. 3.20(e) has advantages compared to that in Fig. 3.20(d). Notice from Fig. 3.20(e) that one of the converter sides presents full-bridge characteristics (no connection with the dc-link midpoint capacitors). On the other hand, in Fig. 3.20(d), both converter sides are half-bridge. The topologies presented in Fig. 3.20 are discussed in detail in Chapter 11.

Exercise 3.1

The common principle behind the topologies proposed in Fig. 3.20 is the inherent redundancy of the three-phase systems. For a set of balanced three-phase voltages, the relationship among the voltages is given by $v_1 + v_2 + v_3 = 0$, which means that if two voltages are defined, the third one will be indirectly defined (they are linearly dependent). Considering the three-phase load connected to a voltage source with only two phases [$v_A = V_A \cos(\omega t + \Phi_A)$ and $v_B = V_B \cos(\omega t + \Phi_B)$], as depicted below, define the amplitude of the voltages (V_A and V_B) and the values of phases (Φ_A and Φ_B) to guarantee a balanced three-phase load voltage given by $v_1 = V\cos(\omega t)$, $v_2 = V\cos(\omega t + 120)$, and $v_3 = V\cos(\omega t - 120)$.

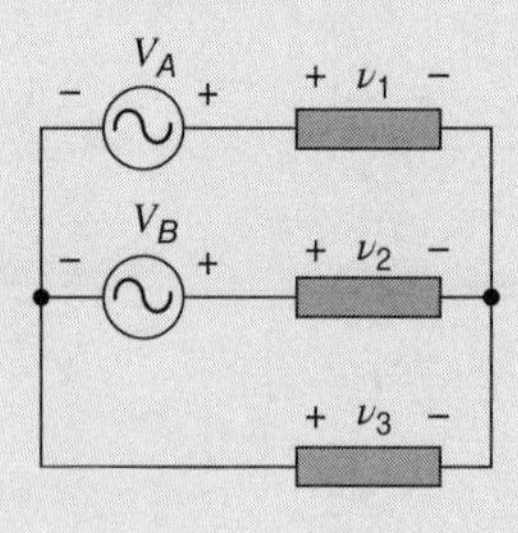

In addition to reducing the number of power switches to reduce costs, it is also possible to increase the number of semiconductor devices, as depicted in Fig. 3.21 paralleling back-to-back converters, to reduce the amount of power processed by each conversion stage and improve the quality of the waveform generated by converters.

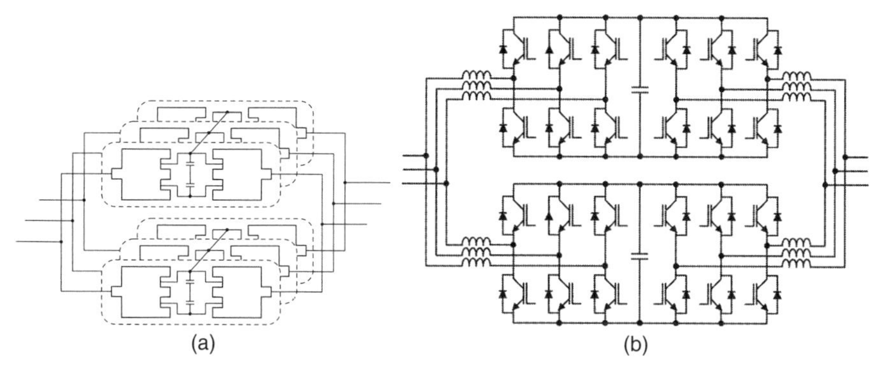

Figure 3.21 Paralleling back-to-back converters: (a) PB representation and (b) representation with switches.

3.5.2 Single-Phase to Single-Phase Configurations

As in the three-phase to three-phase ac–dc–ac configurations, it is possible to establish a systematic presentation for generation of single-phase to single-phase ac–dc–ac configurations, as in Fig. 3.22. Note that the approach of using shared-leg and dc-link capacitor midpoint are still used to guarantee the component count reduction. The single-phase back-to-back converters, including their model, PWM strategy, and analysis are addressed in Chapter 10.

3.6 SUMMARY

This chapter shows that the proposed PBG strategy could be used to establish a systematic method for the presentation of many of the power electronics converters that process ac voltage through a dc stage. A compilation of the main configurations was done under the PBG point of view. Instead of considering power switches as the basic elements in the process of converter construction, the PBs are employed as a higher level component, with descriptions of the details inside the blocks. The connection among the PBs, using the postulates and axioms, furnishes a simple and familiar concept of geometrical blocks. Such a strategy was employed for construction of classical and nonconventional configurations.

As a result of the PBG approach, different configurations are now presented as a part of the same set of converters (i.e., topologies dealing with ac voltage with at least one dc stage). Besides bringing a systematic presentation, the PBG technique will play an important role in the process of topology construction. Even in the case in which practical applications are not directly addressed for a specific configuration obtained with PBG, this configuration still will be considered in this book since it provides challenges in terms of converter understanding itself and creativity in developing its model and control strategies.

Each configuration presented in this chapter is deeply studied in the following chapters, with the development of comprehensive model and PWM strategies. Further details can be found in References from 1 to 21.

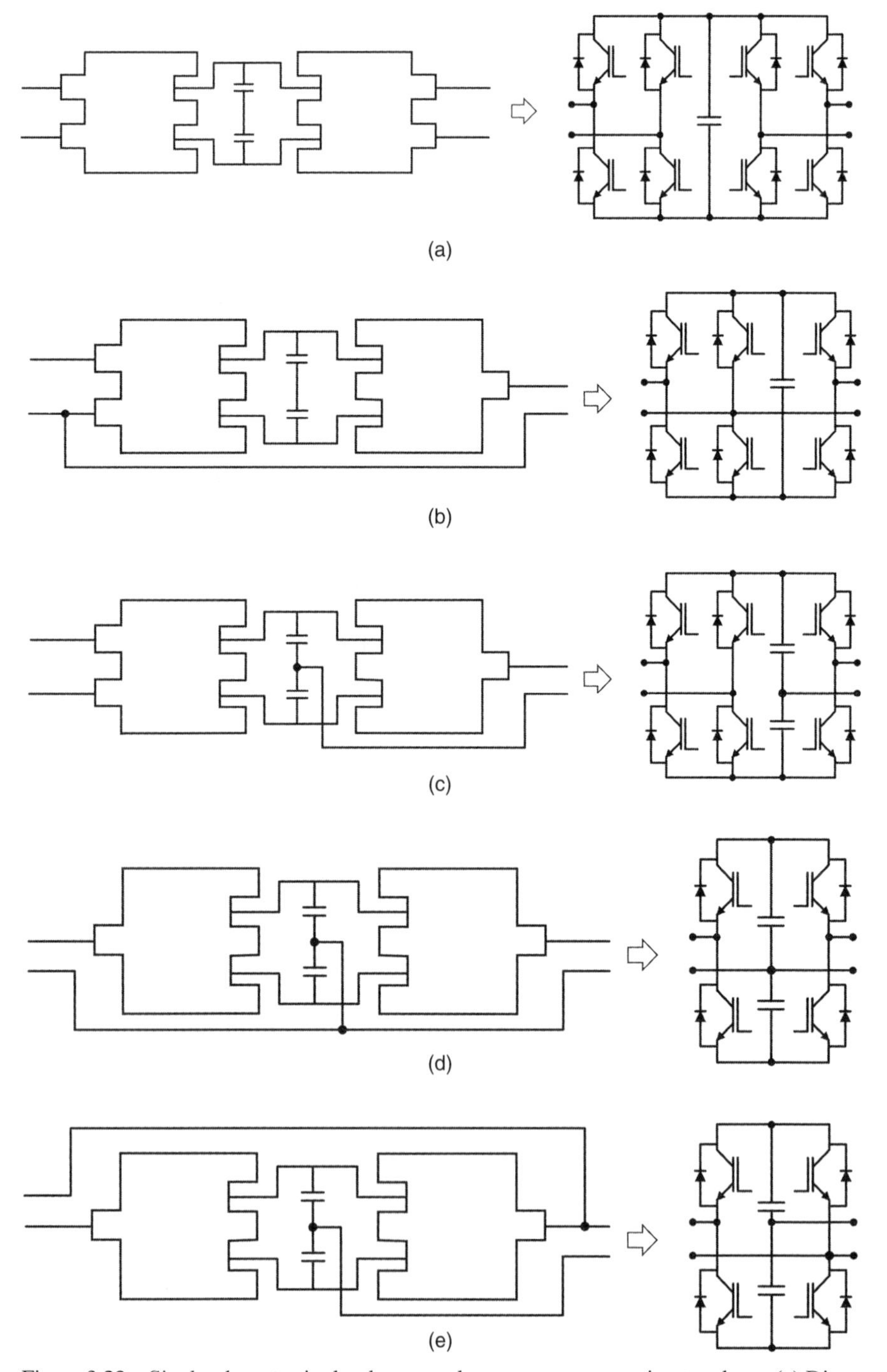

Figure 3.22 Single-phase to single-phase ac–dc–ac power conversion topology: (a) Direct solution with four legs, (b) one shared leg, (c) one dc-link midpoint connection, (d) two dc-link midpoint connections, and (e) shared-leg and dc-link midpoint connections.

REFERENCES

[1] Escaud B, Marty P. Introduction à l'étude des structures des convertisseurs statiques. Electron Ind 1983;56:65–71.

[2] Marty and Escaud B. La commutation dans les convertisseurs statiques, Ibid., No. 60; 1983. p 69–73.

[3] M. V. Miranda, A. M. N. Lima, Formal verification and controller redesign of power electronic converters, Proc IEEE Int Symp Ind Electron, 2004. v. 1. p. 1–6.

[4] E. C. dos Santos Jr, E. R. C. da Silva. Power block geometry applied to the building of power electronics converters, IEEE Trans Educ, v. 2, pp. 191–198, 2013.

[5] Babaei E, Hosseini SH. New multilevel converter topology with minimum number of gate driver circuits, SPEEDAM 08; 2008. p 792–797.

[6] Baker RH. Bridge converter circuit. US patent 4270163. May 1981 (filed in Aug. 79).

[7] Nabae A, Takahashi I, Akagi H. A new neutral-point-clamped PWM inverter. Trans Ind Appl 1981;IA-17(5):518–523.

[8] Meynard TA, Foch H. Multilevel conversion: high voltage chopper and voltage source inverters. Proceedings of IEEE PESC'92; 1992. p 397–403.

[9] Chen A, Hu L, He X. A novel type of combined multilevel converter topologies. IEEE Proceedings of IECON'04; 2004. p 2290–2294.

[10] Andrejak JM Lescure M. High voltage converters promising technological developments. Proc. Rec. EPE Conf.; 1987. p 1.159–1.162.

[11] Choi NS, Cho JG, Cho GH. A general circuit topology of multilevel inverter. Proc Rec IEEE PESC; 1991. p 96–103.

[12] Carpita M, Tenconi S. A novel multilevel structure for voltage source inverter. Proc. Rec. EPE Conf.; 1991. p. 1.90–1.94.

[13] Hammond PW. Medium voltage PWM drive and method. US patent 5,625,545. 1997.

[14] Peng FZ. A generalized multilevel inverter topology with self voltage balancing. IEEE Trans Ind Appl 2001;37(2):611–618.

[15] Brückner T, Bernet S, Güldner H. The active NPC converter and its loss-balancing control, IEEE Trans Ind Electron, 52, No. 3, 2005, pp. 855–868.

[16] T. Bruckner, S. Bernet, P.K. Steimer, Feedforward loss control of three-level active NPC converters, IEEE Trans Ind Appl, 43, no.6, pp.1588–1596, 2007.

[17] Lin B-R, Yang T-Y. Single-phase switching mode multilevel rectifier with a high power factor. IEE Proc-Electr Power Appl 2005;152(3):447–454.

[18] Bhagwat PM, Stefanovic VR. Generalized structure of a multilevel PWM inverter. IEEE Trans Ind Appl 1983;IA-19(6):1057–1069.

[19] dos Santos Jr EC, Muniz JH, da Silva ERC, Jacobina CB. Nested multilevel configurations. IEEE Energy Conversion Congress and Exposition, ECCE 2012; 2012. p 324–329.

[20] Perantzakis GS, Xepapas FH, Manias SN, A new four-level PWM inverter topology for high power applications - effect of switching strategies on power losses distribution, PESC'04; 2004. p 4398–4404.

[21] Tang T, Han J, Zhou L, Yao P, Tan X. Novel hybrid cascade asymmetrical converter based on asymmetrical converter. Proceedings of IEEE ISIE; 2007 Jun. p 1004–1008.

CHAPTER 4

NEUTRAL-POINT-CLAMPED CONFIGURATION

4.1 INTRODUCTION

In this chapter, the neutral-point-clamped (NPC) configuration, also known as diode-clamped configuration, is considered under the power blocks geometry (PBG) point of view. The introduction and fundamental concepts of NPC configurations from three-level to n-level, as well as a detailed description of these circuits, including its model, pulse width modulation (PWM), and control strategies, are addressed in this chapter.

The three-level NPC configuration is one of the most common topologies for medium and high voltage applications, especially in motor drive systems, because of reduced dv/dt, reduced total harmonic distortion (THD), and especially its ability to deal with high voltage demand without connecting switches in series. However, it is also possible to employ this kind of topology in lower power applications, as in photovoltaic energy systems where other factors such as voltage with low distortion and reduced leakage current are important factors.

This chapter is organized as follows: in Section 4.2 the three-level NPC topology is once again obtained from the PBs and also treated in the level of power switches. Its model is obtained by deducing an equation to describe the voltage behavior at the ac side. Such an equation is achieved from the Table of Variables, which also plays an important role in the PWM development. In Section 4.3 the simplest PWM approach is detailed for the half-bridge single-phase three-level converter. The concepts presented previously are expanded to full-bridge single-phase and three-phase converters in Sections 4.4 and 4.5, respectively. Section 4.6 brings up nonconventional converters demonstrating how those topologies can be easily constructed by using PBG. The inherent problem of the unbalanced capacitor voltages in NPC topologies is addressed in Section 4.7. NPC configurations with number of levels higher than three are presented from Section 4.8–4.11. Section 4.12 provides a summary of the chapter.

Advanced Power Electronics Converters: PWM Converters Processing AC Voltages,
Forty Fifth Edition. Euzeli Cipriano dos Santos Jr. and Edison Roberto Cabral da Silva.

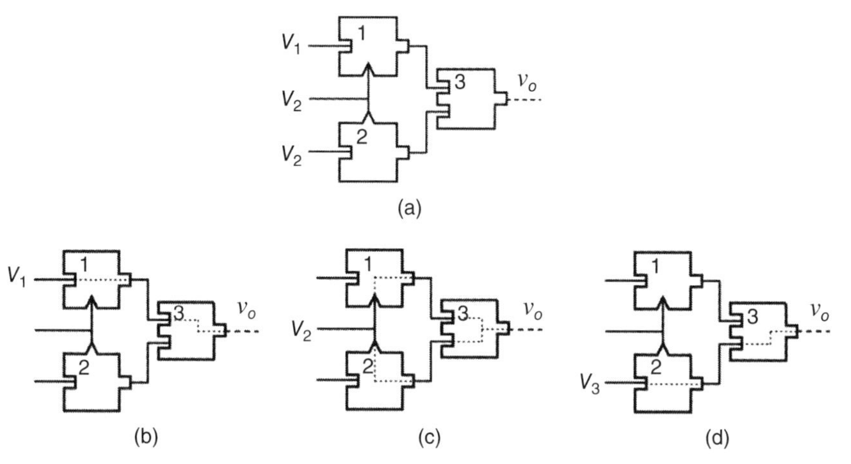

Figure 4.1 (a) Three-level (V_1, V_2, V_3) NPC configuration with PBs. Activation of the PBs to obtain the level: (b) $v_o = V_1$. (c) $v_o = V_2$. (d) $v_o = V_3$.

4.2 THREE-LEVEL CONFIGURATION

As already seen, the power blocks (PBs), can be connected following a systematic approach to guarantee the creation of NPC multilevel configurations. The first and direct arrangement of the blocks is repeated in Fig. 4.1(a), with triangular shape connected to triangular shape and square shapes connected to each other. Figure 4.1(b)–4.1(d) show the ways used to obtain each specific level (V_1, V_2, V_3) at the output converter side. Figure 4.1(b) illustrates that it is necessary to activate PBs 1 and 3 to obtain $v_o = V_1$, while $v_o = V_3$ is obtained when PBs 2 and 3 are activated [see Fig. 4.1(d)]. On the other hand, the voltage V_2 appears at the output side (v_o), when just PB 3 is fully activated, as observed in Fig. 4.1(c). To generate an output waveform with three levels, it is necessary to activate the PBs in a sequence as observed in Fig. 4.2.

Since v_o can assume just one input voltage V_1, V_2, or V_3 at any time, it is possible to write an equation to describe the output voltage v_o as a function of input voltages and two independent binary variables (q_1 and q_2). Since there are three inputs to be selected, the minimum number of binary variables employed to select those inputs is two. These three inputs must appear at the output converter side through the use of two binary variables, which means that one possible combination to achieve this goal is given in Table 4.1.

Defining complementary variables ($\overline{q}_1$ and $\overline{q}_2$) as $\overline{q}_1 = 1 - q_1$ and $\overline{q}_2 = 1 - q_2$, the output voltage can be defined as follows:

$$v_o = \overline{q}_1\overline{q}_2 V_3 + \overline{q}_1 q_2 V_2 + q_1 q_2 V_1 \tag{4.1}$$

Developing (4.1) we get

$$v_o = (1 - q_1)(1 - q_2)V_3 + (1 - q_1)q_2 V_2 + q_1 q_2 V_1 \tag{4.2}$$

$$v_o = (1 - q_2 - q_1)V_3 + q_2 V_2 + q_1 q_2 (V_3 - V_2 + V_1) \tag{4.3}$$

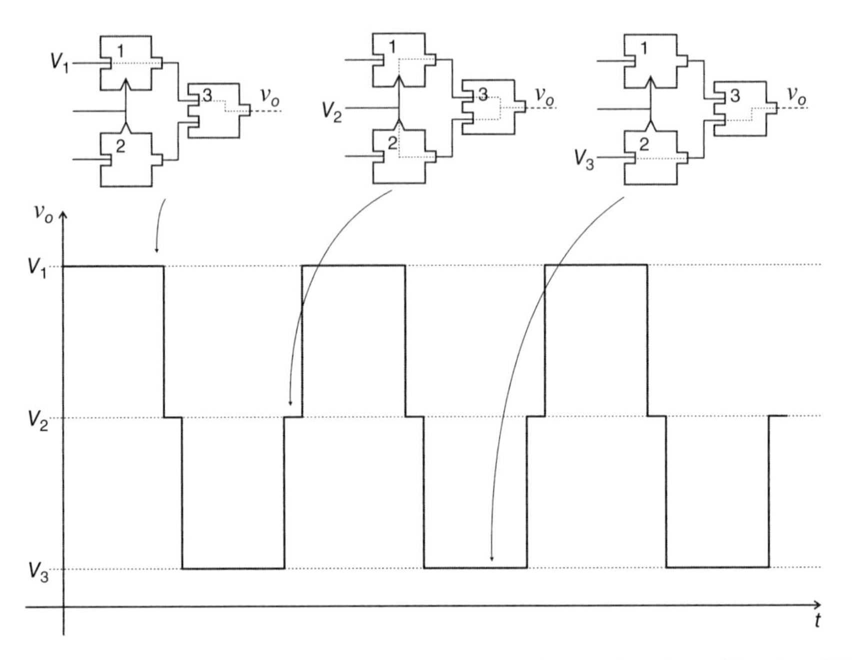

Figure 4.2 Generation of a three-level output voltage for generic values of V_1, V_2 and V_3.

TABLE 4.1 Table of Variables for the Converter in Fig. 4.1(a)

State	q_1	q_2	v_o
1	0	0	V_3
2	0	1	V_2
3	1	0	
4	1	1	V_1

Normally the waveform presented in Fig. 4.2 must be symmetrical, which means that $V_1 = V_{dc}$, $V_2 = 0$ and $V_3 = -V_{dc}$, and then it turns out that

$$v_o = (q_1 + q_2 - 1)V_{dc} \tag{4.4}$$

It is worth mentioning that the binary variables considered in Table 4.1 are related to the switching states inside the blocks in Fig. 4.1. Figure 4.3 shows the same topology presented in Fig. 4.1 along with the binary variables describing the state of the switches for both positive and negative output currents (i_o).

Notice that, unlike PB-dc operating alone (see Figs 3.7 and 3.8), the NPC topology is able to generate the desired output voltage for both positive and negative currents, which means that each voltage value V_1, V_2, and V_3 can be obtained at the output converter side independent of the current values. In other words any

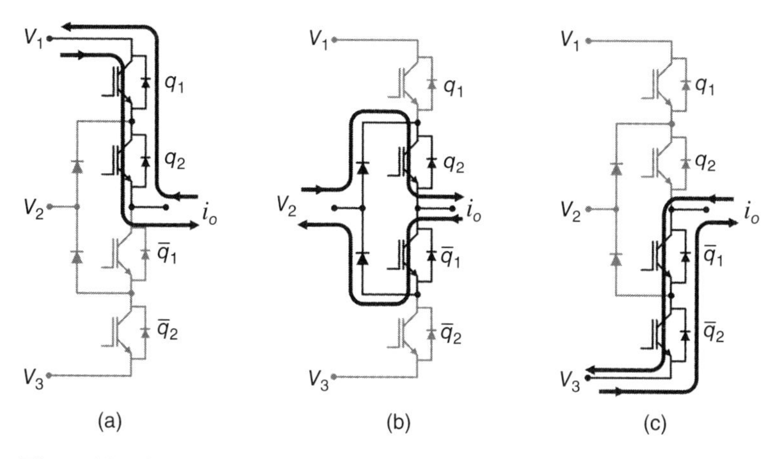

Figure 4.3 Generation of levels: (a) V_1, (b) V_2, and (c) V_3 independent of the output current.

desired output voltage is obtained for both cases $i_o > 0$ and $i_o < 0$, as observed in Fig. 4.3.

Example 4.1

Considering the Tables of Variables presented below, place each switch (q_1, q_2, $\overline{q}_1$, and $\overline{q}_2$) in the correct position in the circuit below to guarantee the generation of the levels as demanded in the following tables.

State	q_1	q_2	v_o
1	0	0	
2	0	1	
3	1	0	
4	1	1	

State	q_1	q_2	v_o
1	0	0	
2	0	1	
3	1	0	
4	1	1	

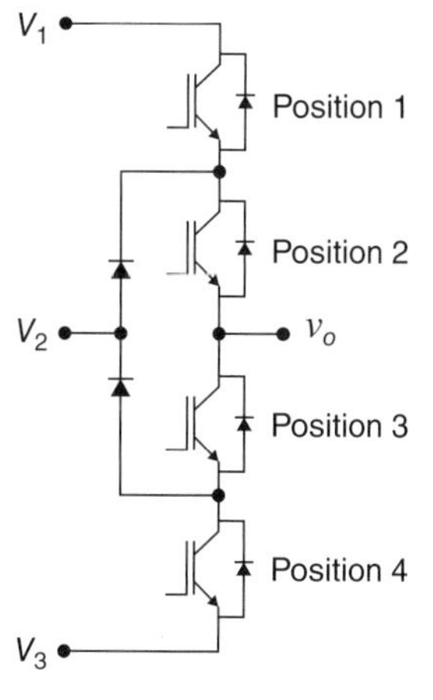

Solution

To guarantee an output voltage v_o as demanded in the first Table of Variables, the switches q_1, q_2, $\overline{q}_1$, and $\overline{q}_2$ should be placed, respectively, in the positions 3, 4, 1, and 2 as below. There is no possible combination for the second table.

State	q_1	q_2	v_o
1	0	0	V_1
2	0	1	–
3	1	0	V_2
4	1	1	V_3

State	q_1	q_2	v_o
1	0	0	V_1
2	0	1	V_2
3	1	0	V_3
4	1	1	–

Example 4.2

Build the Table of Variables for the circuits presented below, with respect to the position of each switch.

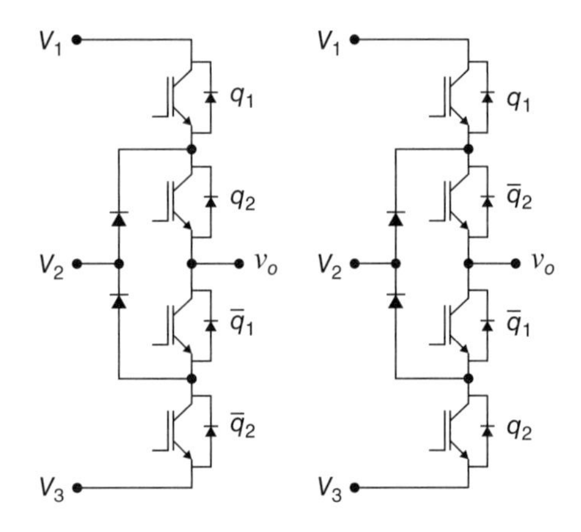

Solution

The Table of Variables for each circuit is given, respectively, by

State	q_1	q_2	v_o
1	0	0	V_3
2	0	1	*undefined*
3	1	0	*undefined*
4	1	1	V_3

State	q_1	q_2	v_o
1	0	0	V_2
2	0	1	*undefined*
3	1	0	*undefined*
4	1	1	*undefined*

Exercise 4.1

Considering the tables presented in Example 4.2, show through equivalent circuits and assuming positive and negative currents, why some of the states are undefined.

Notice that for the two arrangements in Example 4.2 there are more than one undefined states, which means that there is no way to obtain three levels at the output converter side, as seen in Fig. 4.3. Such a characteristic is not desirable for a power converter processing ac voltage.

4.3 PWM IMPLEMENTATION (HALF-BRIDGE TOPOLOGY)

If the aim is to generate an ac voltage at the output converter side with both amplitude and frequency controllable, a PWM strategy is applied to generate the switching gating signals. From a converter processing ac voltage, ideally an output sinusoidal voltage is expected. However, due to the characteristics of the NPC converter presented in Fig. 4.2, the number of levels is limited to three. Hence, the aim of the PWM strategy is to guarantee that the three-level voltage at the output of the converter emulates the desirable sinusoidal voltage. The average value of the square output waveform must be equal to a sinusoidal reference (inside the switching period) as presented in the figure below.

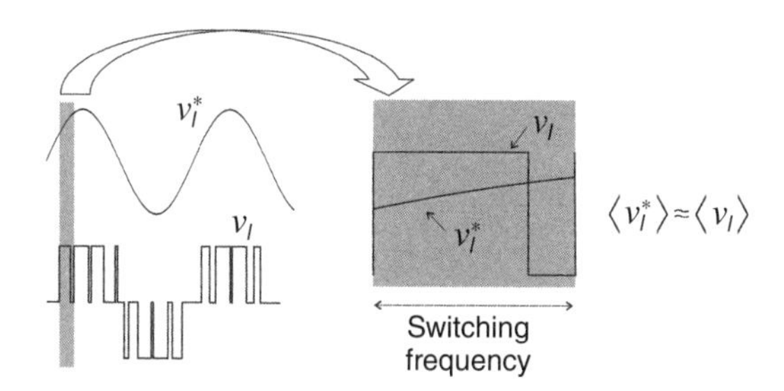

Normally, the switching frequency is kept high (compared to the modulating signal) to increase the accuracy of $v_l \approx v_l^*$. In fact many works in the technical literature have proposed solutions in terms of modulation approaches applied for NPC converters to improve different aspects of the converter. In this chapter, the simplest way to generate the PWM waveforms is considered. However, an optimized PWM approach is derived in Chapter 8.

Figure 4.4 shows a simple analog solution, in which the states of the four switches are determined by a comparison among a sinusoidal waveform (v_{sin}^*) and two high frequency triangular waveforms (v_{t1}^* and v_{t2}^*). Such PWM can be expanded for NPC converters with higher levels, which uses $N-1$ (N is the number of levels) triangular carrier signals with just one modulating signal. The sinusoidal waveform is given by $v_{\text{sin}}^* = V_l \sin(\omega_l t)$, where V_l is the amplitude and ω_l is the angular frequency desired to the load. The block mentioned as *driver* must guarantee the levels of voltage and current in the switch's gate circuit needed to operate the power devices adequately as well as guarantee galvanic isolation.

Notice from Fig. 4.4(c) and 4.4(d) that when the switches q_1 and $\overline{q}_1$ are changing their states (during the positive half cycle of v_{sin}^*), switch q_2 is on and switch $\overline{q}_2$

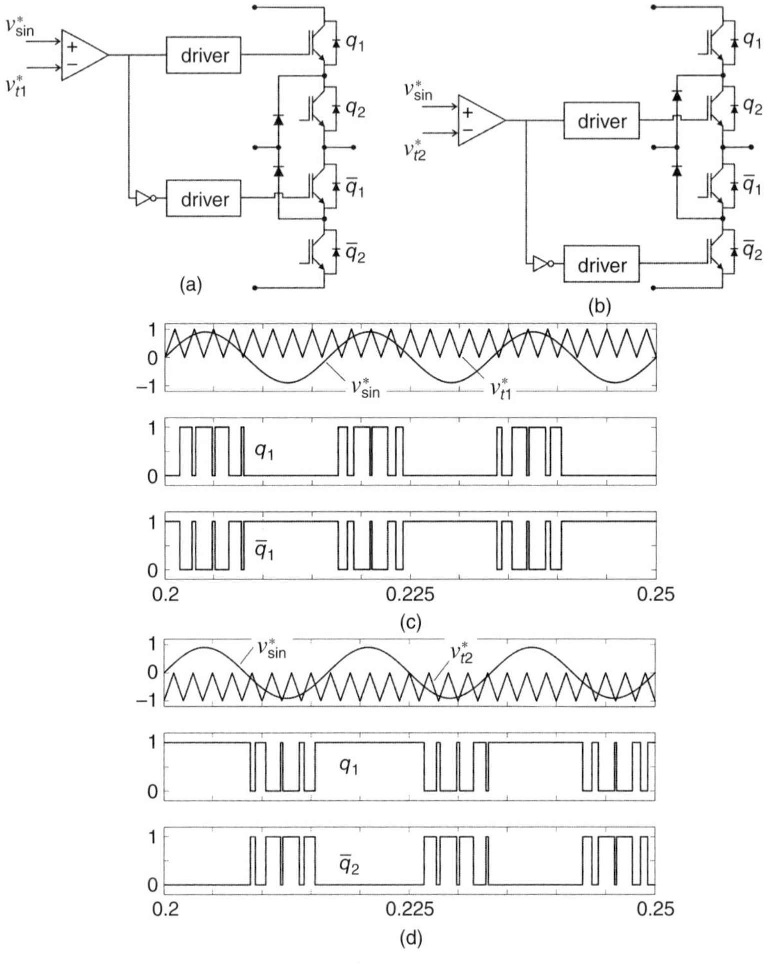

Figure 4.4 (a) PWM generation circuit for switches: (a) q_1 and $\overline{q}_1$; and (b) q_2 and $\overline{q}_2$. (c) Variables associated with switches q_1 and $\overline{q}_1$ (from top to bottom) – reference voltages, gating signal of switch q_1 and gating signal of switch $\overline{q}_1$. (d) Variables associated with switches q_2 and $\overline{q}_2$ (from top to bottom) reference voltages, gating signal of switch q_2 and gating signal of switch $\overline{q}_2$.

is consequently off for the entire half cycle, while in the negative half cycle of v^*_{sin}, when switches q_2 and $\overline{q}_2$ are changing their states, switch q_1 is off and switch $\overline{q}_1$ is on.

The modulation presented in Fig. 4.4 describes the states in Table 4.1, when q_2 is on and q_1 can be either off (State 2 in Table 4.1) or on (State 4 in Table 4.1). In the same way, when q_1 is off, q_2 can be either off (State 4.1 in Table 4.1) or on (State 2 in Table 4.1). Also, this modulation approach avoids the undesired State 3.

Figure 4.5(a) shows the half-bridge three-level configuration, in which the load is connected between the dc-link midpoint and the center point of the leg.

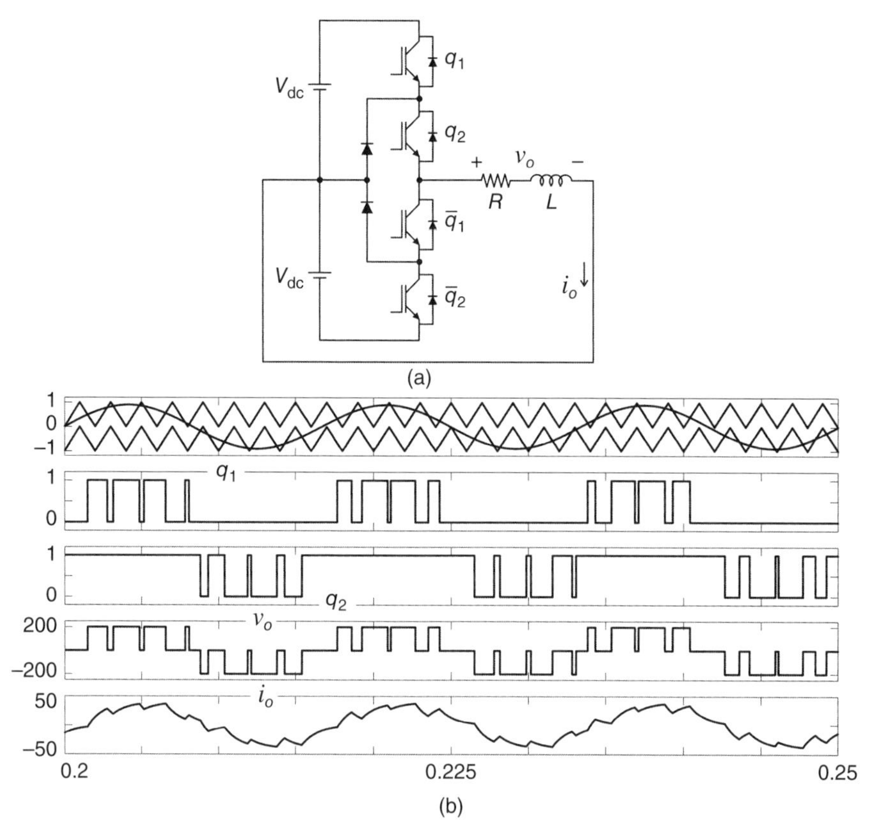

Figure 4.5 (a) Three-level NPC half-bridge converter. (b) (from top to bottom) v^*_{sin}, v^*_{t1}, v^*_{t2}; gating signal of switch q_1; gating signal of switch q_2; load voltage and load current.

Figure 4.5(b) depicts (from top to bottom) PWM reference waveforms, gating signal of switches q_1 and q_2, load voltage, and load current, respectively. The parameters employed in these simulation tests were: $V_{dc} = 200$ V, $R = 5\Omega$ and $L = 5$mH, and switching frequency equal to 500Hz. For the sake of comparison, the same parameters are used for the other simulation results in this chapter.

Sometimes in practical applications just one dc source is available, which means that both dc sources shown in Fig. 4.5(a) are changed by capacitors connected to a single dc source. In this sense, it is necessary to guarantee the control voltage for both capacitors. This specific issue is considered later on in this chapter.

4.4 FULL-BRIDGE TOPOLOGIES

The three-level NPC topology with two legs, that forms a full-bridge single-phase converter, can be obtained from PBG by connecting the PBs given in Fig. 4.1 appropriately, as seen in Fig. 4.6(a). Following this approach it is also possible to build

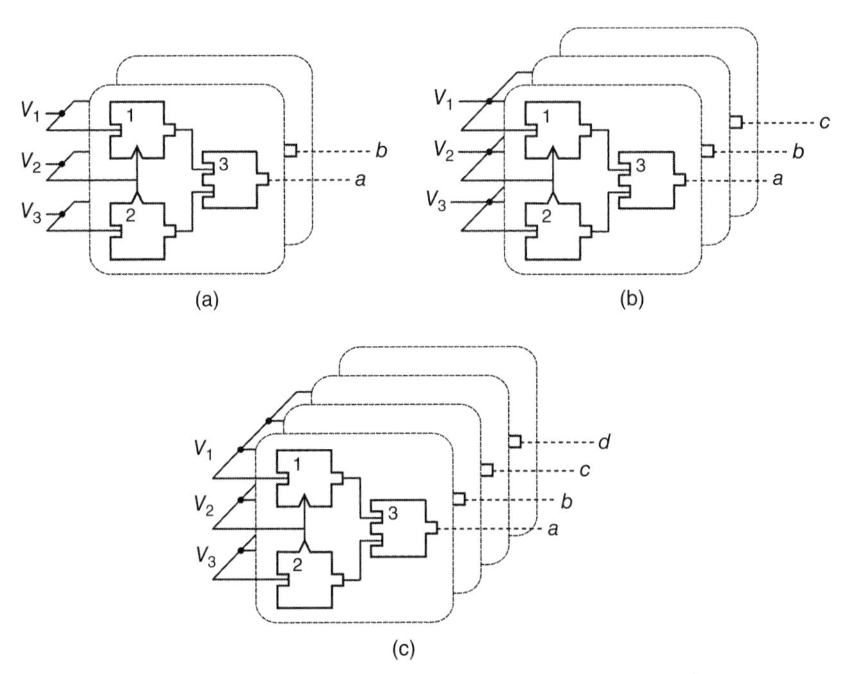

Figure 4.6 NPC three-level topology with: (a) two legs (full-bridge single-phase load), (b) three legs (three-phase load), and (c) four legs (three-phase four-wire load).

configurations with n legs. Figure 4.6(b) shows a three-level NPC topology with three legs for applications demanding a three-phase system (e.g., three-phase motor drive systems or three-phase rectifier converter), while Fig. 4.6(c) shows a three-level NPC topology with four legs to be employed, for instance, in a three-phase four-wire system. These three configurations in Fig. 4.6 are described in the following sections.

Figure 4.7(a) presents the full-bridge single-phase converter [as in Fig. 4.6(a)] highlighting the connections of the switches. It is worth mentioning that even though the full-bridge topology in Fig. 4.7(a) is constituted by three-level legs, the output voltage has five levels.

Figure 4.7(b) shows (from top to bottom) pole voltage (voltage between the midpoint of the leg and the dc-link capacitor midpoint v_{a0}) of the first leg with three levels, pole voltage of the second (v_{b0}) also with three levels, load voltage, and load current, respectively. Comparing Figs 4.5 and 4.7, it is evident that the increased number of levels observed on the load voltage of the full-bridge topology guarantees a ripple reduction on the load current for the full-bridge topology. In this case the output voltage is given by

$$v_o = (q_{1a} + q_{2a} - q_{1b} - q_{2b})V_{\text{dc}} \tag{4.5}$$

The number of levels considered in this chapter for conventional topologies is defined by the pole voltages. For instance, both converters in Figs 4.5 and 4.7 are

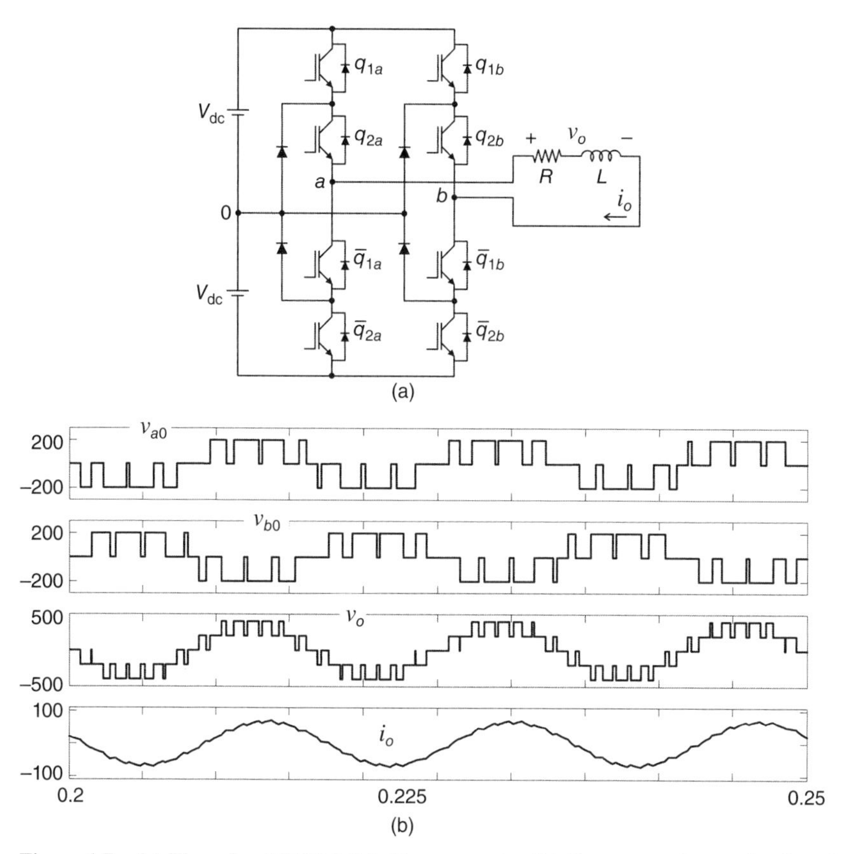

Figure 4.7 (a) Three-level NPC full-bridge converter. (b) (from top to bottom) pole voltage of the first leg (v_{a0}), pole voltage of the second (v_{b0}), load voltage and load current.

considered as single-phase three-level NPC topologies, although they present different number of levels for the load voltages.

Example 4.3

Considering the topology presented in Fig. 4.7(a), find a solution for the PWM generation circuit to guarantee the waveforms shown in Fig. 4.7(b).

Solution

A set with a reference sinusoidal and two triangular waveforms plus two OpAmps for each leg (as seen in Fig. 4.4 for the half-bridge topology) can be employed for the PWM signal generation for the full-bridge topology. Since the control of both legs is independent, those sinusoidal waveforms should have 180° of phase apart to guarantee a maximum output voltage.

Example 4.4

What happens if different phase displacements are employed for both reference sinusoidal waveforms?

Solution

The figure below shows a phase or diagram representation for the pole voltages v_{a0} and v_{b0} (just for the sinusoidal components) as well as for the load voltage $v_o = v_{a0} - v_{b0}$. The output voltage is a function of the phase displacement of the pole voltages, meaning that the maximum output voltages will be obtained when this phase displacement angle is equal to 180°.

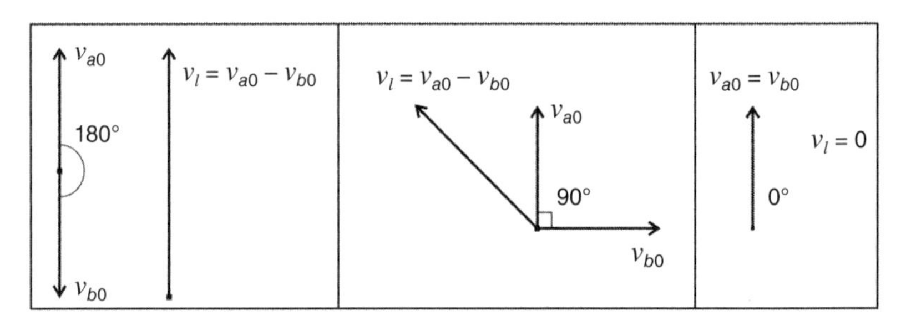

4.5 THREE-PHASE NPC CONVERTER

A three-phase three-level full-bridge NPC configuration is depicted in Fig. 4.8(a) by employing switches. Notice that this circuit was easily obtained by just connecting the PBs as seen in Fig. 4.6(b). Figure 4.8(b), in turn, shows some variables associated with this converter, that is, (from top to bottom): pole voltage, voltage between the load neutral point (NP) (n) and dc-link midpoint connection (0), load phase voltage, and load current, respectively. The pole voltages can be written as follows:

$$v_{a0} = (q_{1a} + q_{2a} - 1)V_{\text{dc}} \tag{4.6}$$

$$v_{b0} = (q_{1b} + q_{2b} - 1)V_{\text{dc}} \tag{4.7}$$

$$v_{c0} = (q_{1c} + q_{2c} - 1)V_{\text{dc}} \tag{4.8}$$

The load phase voltages are given by

$$v_a = v_{a0} - v_{n0} \tag{4.9}$$

$$v_b = v_{b0} - v_{n0} \tag{4.10}$$

$$v_c = v_{c0} - v_{n0} \tag{4.11}$$

Considering a balanced three-phase condition ($v_a + v_b + v_c = 0$), the voltage between the NP of the three-phase load (n) and the dc-link midpoint can be defined

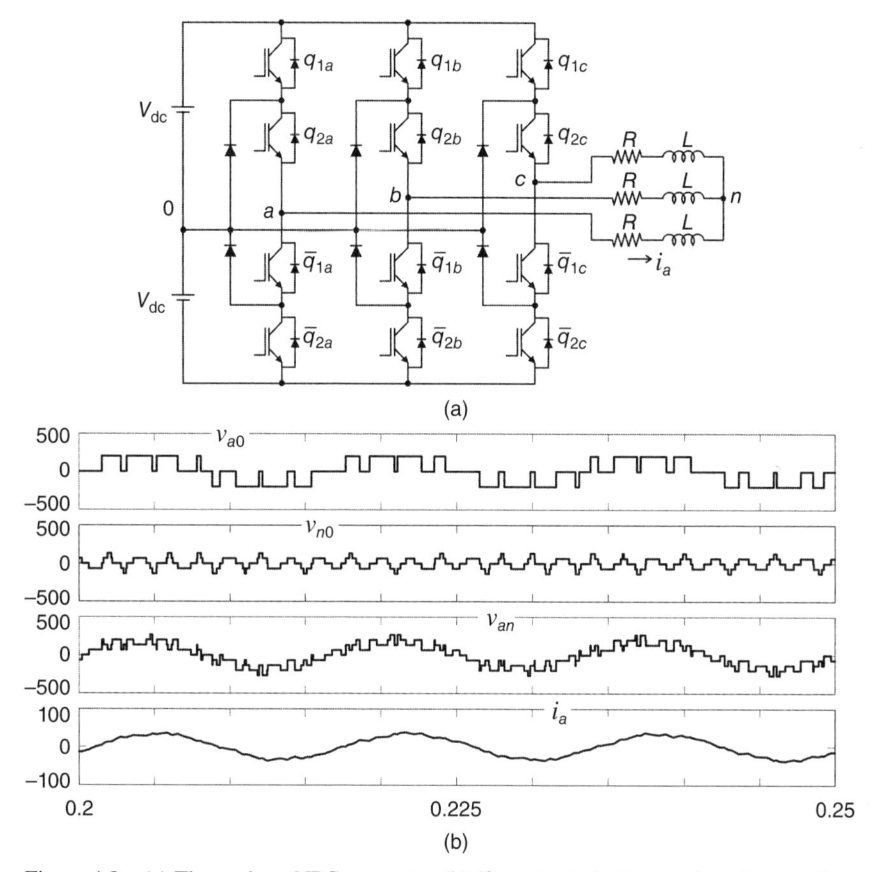

Figure 4.8 (a) Three-phase NPC converter. (b) (from top to bottom) pole voltage, voltage v_{n0}, load voltage, and load current.

analytically by

$$v_{n0} = \frac{1}{3}(v_{a0} + v_{b0} + v_{c0}) \tag{4.12}$$

Substituting (4.10)–(4.12) into (4.13), we have

$$v_{n0} = \frac{V_{dc}}{3}(q_{1a} + q_{2a} + q_{1b} + q_{2b} + q_{1c} + q_{2c}) - V_{dc} \tag{4.13}$$

From (4.13), the voltage for each phase of the three-phase load can be written as follows:

$$v_a = \left(\frac{2}{3}q_{1a} + \frac{2}{3}q_{2a} - \frac{1}{3}q_{1b} - \frac{1}{3}q_{2b} - \frac{1}{3}q_{1c} - \frac{1}{3}q_{2c}\right)V_{dc} \tag{4.14}$$

$$v_b = \left(-\frac{1}{3}q_{1a} - \frac{1}{3}q_{2a} + \frac{2}{3}q_{1b} + \frac{2}{3}q_{2b} - \frac{1}{3}q_{1c} - \frac{1}{3}q_{2c}\right)V_{dc} \tag{4.15}$$

$$v_c = \left(-\frac{1}{3}q_{1a} - \frac{1}{3}q_{2a} - \frac{1}{3}q_{1b} - \frac{1}{3}q_{2b} + \frac{2}{3}q_{1c} + \frac{2}{3}q_{2c}\right)V_{dc} \tag{4.16}$$

In the case of a balanced three-phase load, the load voltages depend only of the state of the switches and of the dc-link voltage as considered in (4.15)–(4.17).

The control circuit for the three-phase converter can be obtained as seen in Fig. 4.4. The difference is the number of modulating signals needed; in this case there are three references: $v^*_{\sin 1} = V_l \sin(\omega t)$, $v^*_{\sin 2} = V_l \sin(\omega t + 120°)$ and $v^*_{\sin 3} = V_l \sin(\omega t + 240°)$, where V_l and ω_l are the desired peak voltage and frequency for the load, respectively.

Exercise 4.2

Create a table relating all states of the switches (64 possibilities) to find the number of levels for v_{n0} and the number of levels for the phase voltage v_a for the converter shown in Fig. 4.8.

Application (High Voltage Motor Drive System)

The three-phase NPC multilevel configuration can be employed in medium voltage (MV) drives to reach a high voltage level without using switching devices in series connection. The switching frequency for MV high power application is normally low as compared to the two-level converter in order to reduce the power losses (it is usually limited to a few hundred hertz) without compromising the quality of the waveforms generated due to the number of levels. Normally, special modulation techniques are needed to produce the minimum harmonic distortion for the motor current for a large range of speed and torque. The main applications are in pumps, fans/exhausts, compressors, mixers, agitators, conveyors, cement mills, steel mills, winders/unwinders, sugar mills, refiners, banburys, and extruders. The figure below shows a common scheme with an NPC topology employed in MV drives.

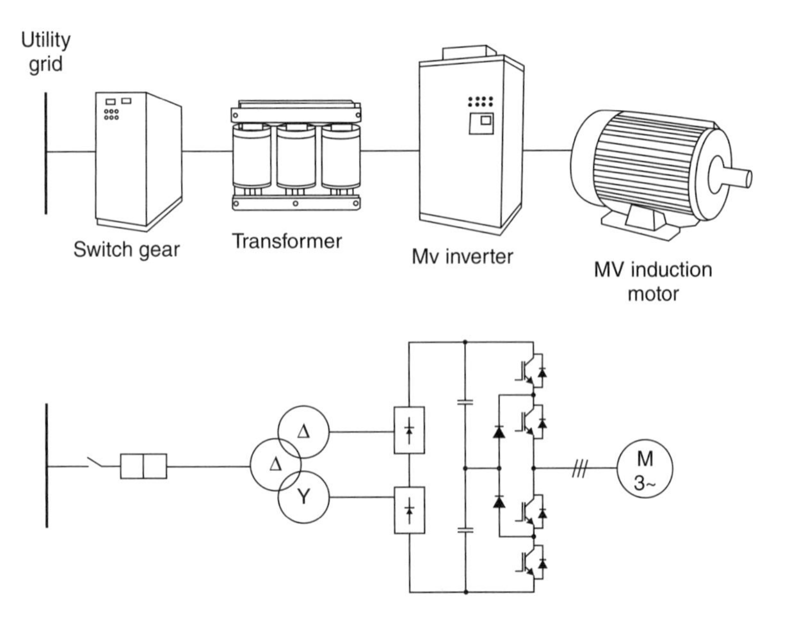

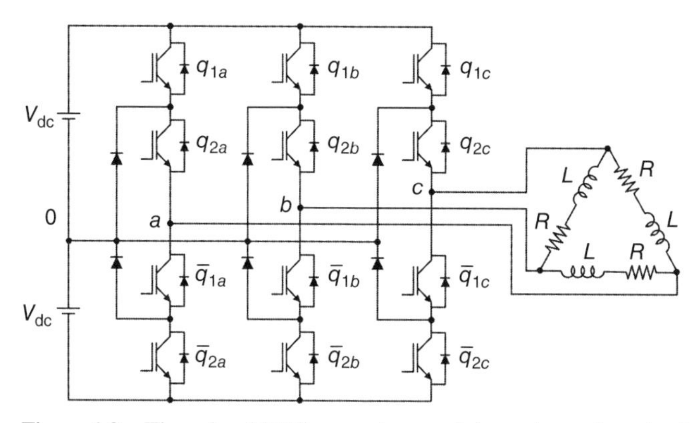

Figure 4.9 Three-level NPC converter supplying a three-phase load connected in Δ arrangement.

Exercise 4.3

An MV three-phase motor has line-to-line and active power given, respectively, by 2.3 kV and 150 kW (with power factor 0.85). Assuming that this motor will be supplied by a three-leg NPC topology, determine the voltage and current ratings of the switches for this configuration.

Figure 4.9 shows the NPC three-phase power converter topology supplying a Δ- connected three-phase load. The main difference between Figs 4.9 and 4.8(a) is related to the values of voltages and current demanded by the load, as well as the number of levels measured on the load voltages. In this case (Δ-connected) the load phase voltage has only five levels, as seen from the equations given below:

$$v_a = (q_{1a} + q_{2a} - q_{1b} - q_{2b})V_{dc} \tag{4.17}$$

$$v_b = (q_{1b} + q_{2b} - q_{1c} - q_{2c})V_{dc} \tag{4.18}$$

$$v_c = (q_{1c} + q_{2c} - q_{1a} - q_{2a})V_{dc} \tag{4.19}$$

4.6 NONCONVENTIONAL ARRANGEMENTS BY USING THREE-LEVEL LEGS

It is possible to combine the NPC leg with two-level legs already presented in this chapter to establish a nonconventional single-phase multilevel converter, as seen in Fig. 4.10. In this case, it is used four PBs with the output load connected between blocks 3 and 4. Since the output of block 3 has three levels (V_1, V_2, and V_3) and the

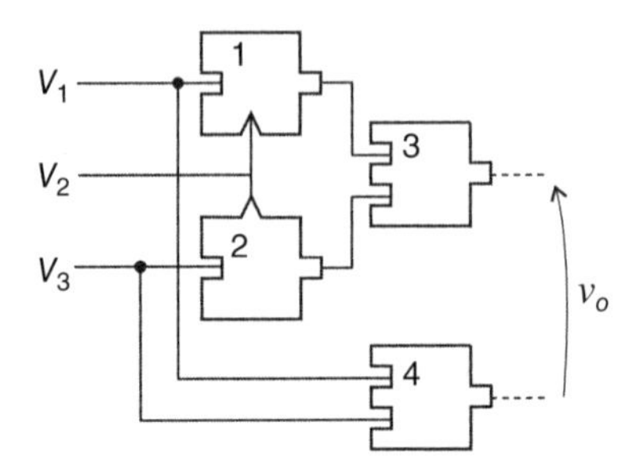

Figure 4.10 Five-level single-phase converter obtained by the combination of NPC and conventional legs – block representation.

TABLE 4.2 Table of Variables for the Converter in Fig. 4.10

State	q_{1a}	q_{2a}	q_{1b}	Output of PB 3	Output of PB 4	v_o
1	0	0	0	$-V_{dc}$	$-V_{dc}$	0
2	0	0	1	$-V_{dc}$	V_{dc}	$-2V_{dc}$
3	0	1	0	0	$-V_{dc}$	V_{dc}
4	0	1	1	0	V_{dc}	$-V_{dc}$
5	1	1	0	V_{dc}	$-V_{dc}$	$2V_{dc}$
6	1	1	1	V_{dc}	V_{dc}	0

output of block 4 has two levels (V_1 and V_3), it leads to an output voltage with five levels (0, $V_1 - V_3$, $V_2 - V_1$, $V_2 - V_3$, $V_3 - V_1$).

For a symmetrical output voltage generation, the input voltages should have the following parameters: $V_1 = V_{dc}$, $V_2 = 0$, and $V_3 = -V_{dc}$, meaning that the output voltage will be with the following values 0, $\pm V_{dc}$, $\pm 2V_{dc}$. For the converter in Fig. 4.10 it is necessary for three independent binary signals (q_{1a}, q_{2a}, and q_{1b}) to synthesize the desired output voltage since there are five levels. Table 4.2 shows the voltage v_o as a function of q_{1a}, q_{2a}, and q_{1b}.

Equations to describe the pole voltages (v_{a0} and v_{b0}) and output voltage v_o as a function of the input voltages and as a function of three independent binary variables (q_{1a}, q_{2a}, and q_{1b}) are given below, respectively:

$$v_{a0} = (q_{1a} + q_{2a} - 1)V_{dc} \tag{4.20}$$

$$v_{b0} = (2q_{1b} - 1)V_{dc} \tag{4.21}$$

$$v_o = (q_{1a} + q_{2a} - 2q_{1b})V_{dc} \tag{4.22}$$

The binary variables considered in both Table 4.2 and in equation (4.22) are in fact the switching states inside the blocks, which means that the topology presented in Fig. 4.10 can be considered in its lower level representation (switches instead of blocks), as in Fig. 4.11.

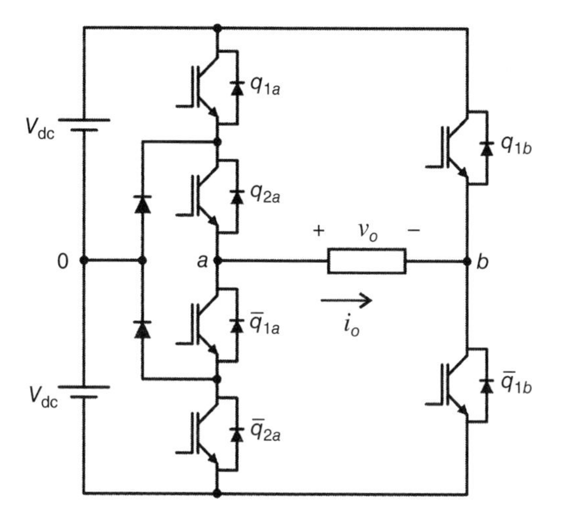

Figure 4.11 Five-level single-phase converter obtained by the combination of NPC and conventional legs with switches.

Exercise 4.4

Assuming all switching states given in Table 4.2, write a Table showing the blocking voltages and current through the switches for the six switches employed in Fig. 4.11.

It is also possible to generate a four-level topology with the same PBs as in Fig. 4.10, but with a different arrangement as presented in Fig. 4.12. Notice that the capacitors in Fig. 4.12(a) are used to guarantee the NP needed by the NPC leg. Figure 4.12(b) shows the same configuration with the switches. In this case, it needs three dc sources with the advantage of using all switches with the same blocking voltage, as discussed in this section.

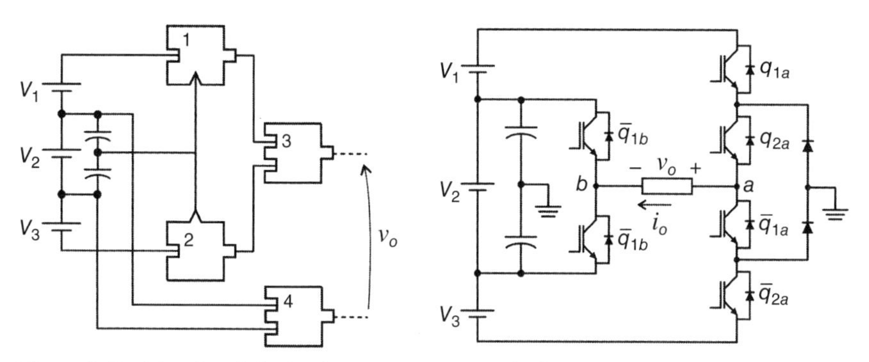

Figure 4.12 Four-level single-phase converter: (a) block representation and (b) circuit with switches.

TABLE 4.3 Table of Variables for the Converter in Fig. 4.12

State	q_{1a}	q_{2a}	q_{1b}	v_o
1	0	0	0	$-V_3$
2	0	0	1	$-V_3-V_2$
3	0	1	0	$-V_2/2$
4	0	1	1	$V_2/2$
5	1	1	0	V_1+V_2
6	1	1	1	V_1

The converter in Fig. 4.12 has six switches (q_{1a}, $\overline{q}_{1a}$, q_{2a}, $\overline{q}_{2a}$, q_{1b}, and $\overline{q}_{1b}$), two clamping diodes, three dc sources, and two dc-link capacitors. Considering all possible switching states available and eliminating the prohibited ones, the output voltage is determined in Table 4.3. To guarantee that all power switches will operate under the same blocking voltage as well as a symmetrical output voltage, it is necessary to establish $V_1 = V_3 = V_{dc}/4$ and $V_2 = V_{dc}/2$.

The modulation strategy for the converter shown in Fig. 4.12 can be determined assuming a combination of the two-level and three-level PWM approaches, which means that, for the two-level leg [PB 4 in Fig. 4.12(a)] only one triangular carrier signal will be employed, while for the three-level leg [PBs 1, 2, and 3 in Fig. 4.12(a)] two triangular carrier signals will be employed, as observed in Fig. 4.13(a). Since the triangular waveform is proportional to the dc-link voltage associated with each leg, the voltages v_{t1}^*, v_{t2}^*, and v_{t3}^* can be obtained as shown in Fig. 4.13(b). In this sense, the two-level leg will deal with one-third of the output voltage while the three-level leg will deal with two-third of the desired output voltage.

The pole voltage for the three-level, two-level legs and output voltage is given respectively by

$$v_{a0} = (q_{1a} + q_{2a} - 1)2V_{dc} \tag{4.23}$$

$$v_{b0} = (2q_{1b} - 1)V_{dc} \tag{4.24}$$

$$v_o = (q_{1a} + q_{2a} - 2q_{1b} - 1)V_{dc} \tag{4.25}$$

Figure 4.14(a) and 4.14(b) shows the output voltage and current for a switching frequency equal to 500 Hz and 1 kHz, respectively.

Although the configuration shown in Fig. 4.12 can generate a four-level output voltage, it needs either three independent dc voltages or an auxiliary circuit to obtain three controlled voltages from a unique dc voltage source, as in Fig. 4.14(c) – top. Figure 4.14(c) – bottom shows the equivalent topological states when the switches in the auxiliary circuit are turned on and off simultaneously. A duty cycle of 50% should be generated to guarantee a desired capacitor voltage in the dc-link of the converters.

Using the three-level NPC topologies and a three-phase load type, it is possible to create a three-phase NPC topology with both two- and three-level characteristics

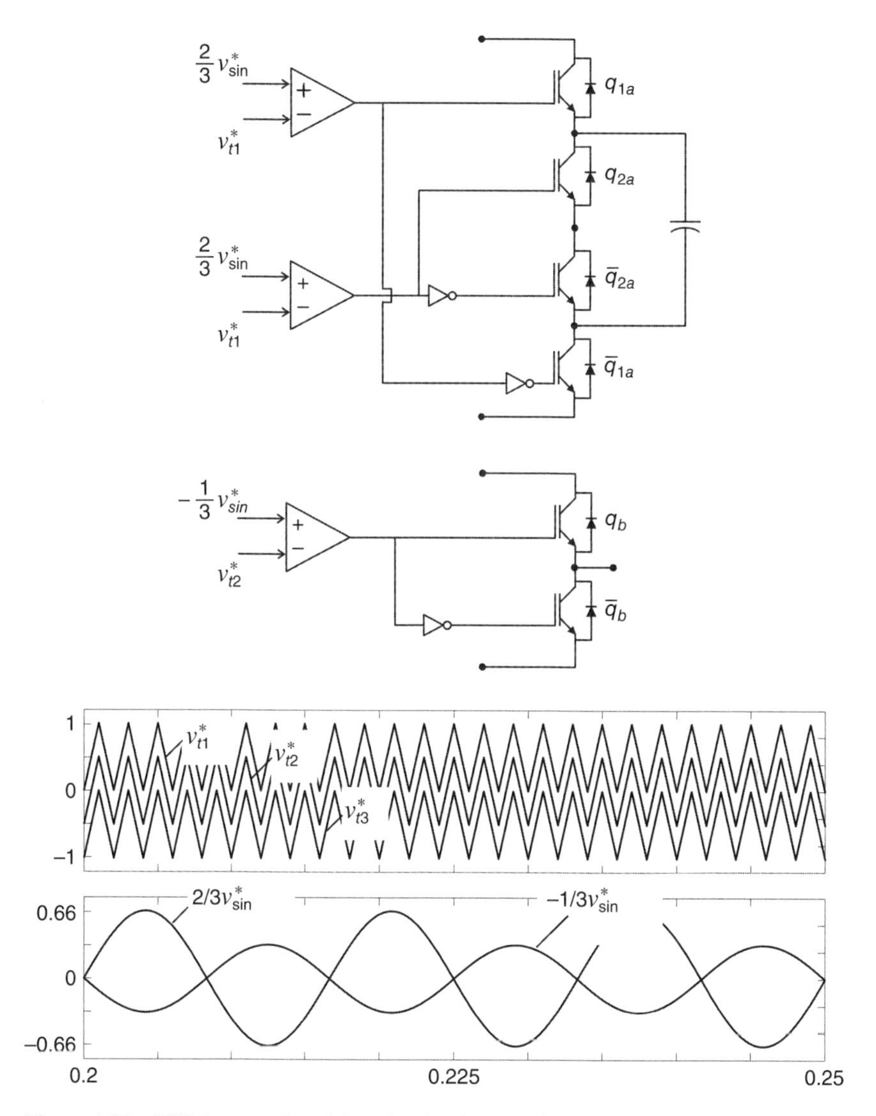

Figure 4.13 PWM generation: (a) analog implementation and (b) waveforms.

for the pole voltage. In this case, the dc-link of the three-phase converter is connected to two PB-dc blocks, as depicted in Fig. 4.15. Figure 4.15(a) shows such hybrid topology with the PBs perspective, while Fig. 4.15(b) depicts this topology by using power switches. For a symmetrical waveform generation, it is necessary to define the input voltages as follows: $V_1 = V_{dc}$, $V_2 = 0$, and $V_3 = -V_{dc}$.

The operation of this converter takes into account the fact that it is possible to obtain different levels per phase, depending on the values for the reference voltages. For instance, Fig. 4.16(a) – top shows the desired phase voltages (v_a^*, v_b^*, and v_c^*) divided by sectors. The first sector is defined by the interval of time in which v_a^* is

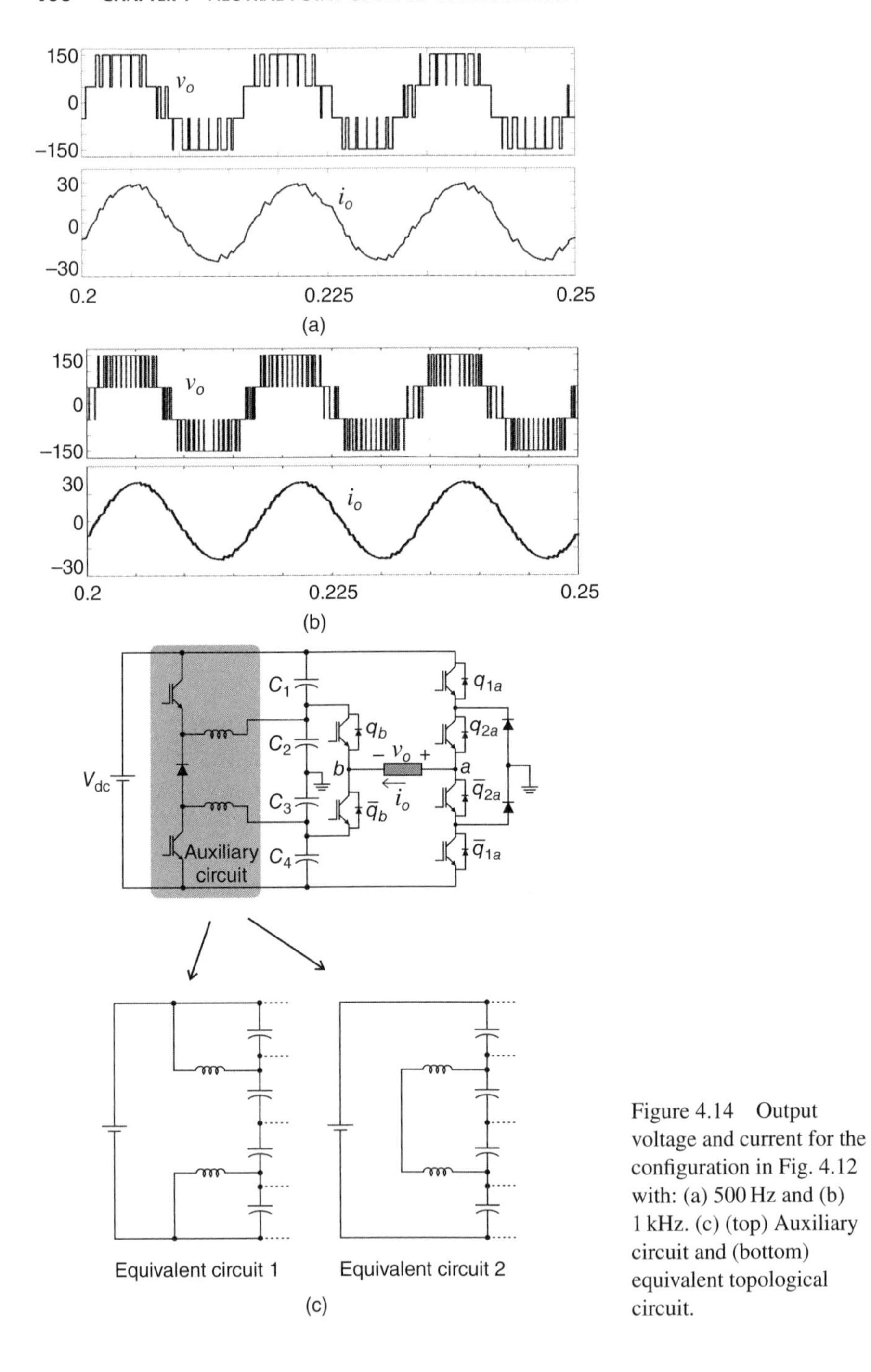

Figure 4.14 Output voltage and current for the configuration in Fig. 4.12 with: (a) 500 Hz and (b) 1 kHz. (c) (top) Auxiliary circuit and (bottom) equivalent topological circuit.

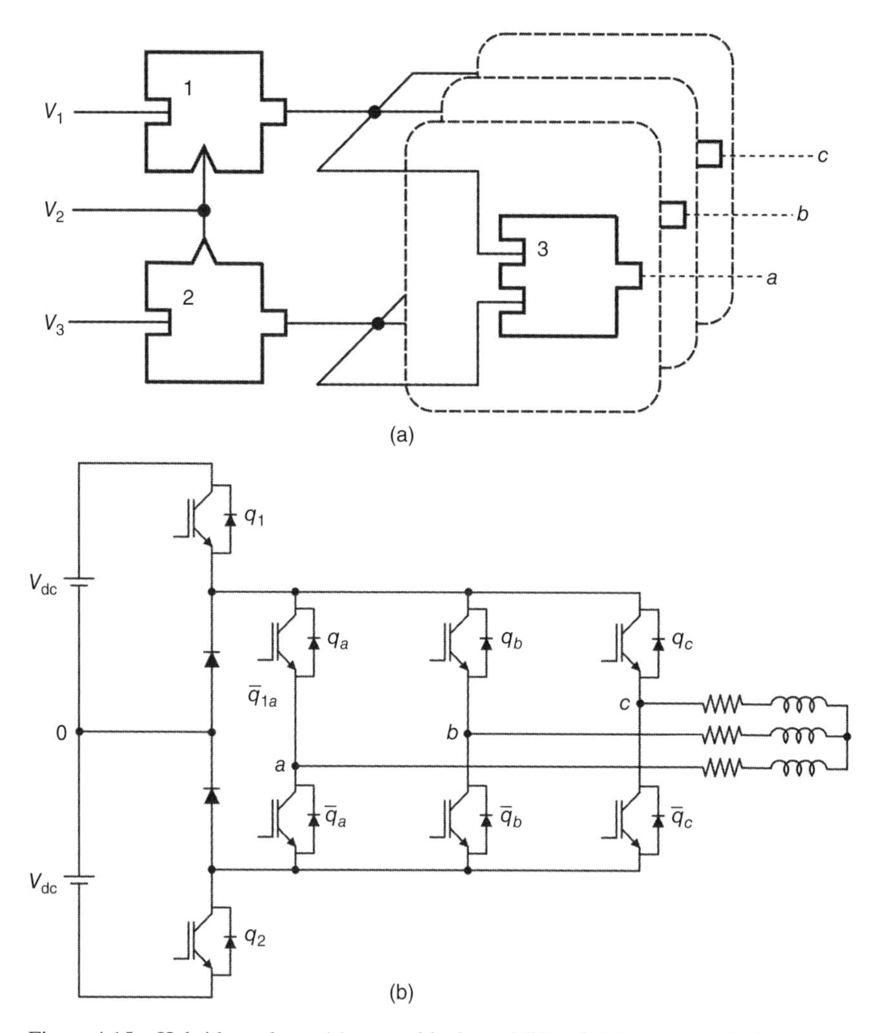

Figure 4.15 Hybrid topology: (a) power blocks and (b) switching representation.

maximum, v_b^* is minimum, and v_c^* is in-between. It means that: (i) phase a can be modulated as a three-level leg (with pole voltage v_{a0} between V_{dc} and 0), since it is expected to generate the maximum voltage; (ii) phase b can be modulated as a three-level leg (with pole voltage v_{b0} between 0 and $-V_{dc}$); and finally (iii) phase c can be modulated as a two-level leg (with the pole voltage between V_{dc} and $-V_{dc}$).

The modulation strategy for the leg dealing with the highest and smallest voltages (three levels) are obtained by comparing the desired waveforms with two triangular waveforms (with level-shift approach) as shown in Fig. 4.4, while the modulation for the leg dealing with middle voltages (two levels) is obtained through a comparison between the reference voltage and one triangular waveform.

Still considering nonconventional configurations, a half-bridge three-phase NPC topology is shown in Fig. 4.17(a).

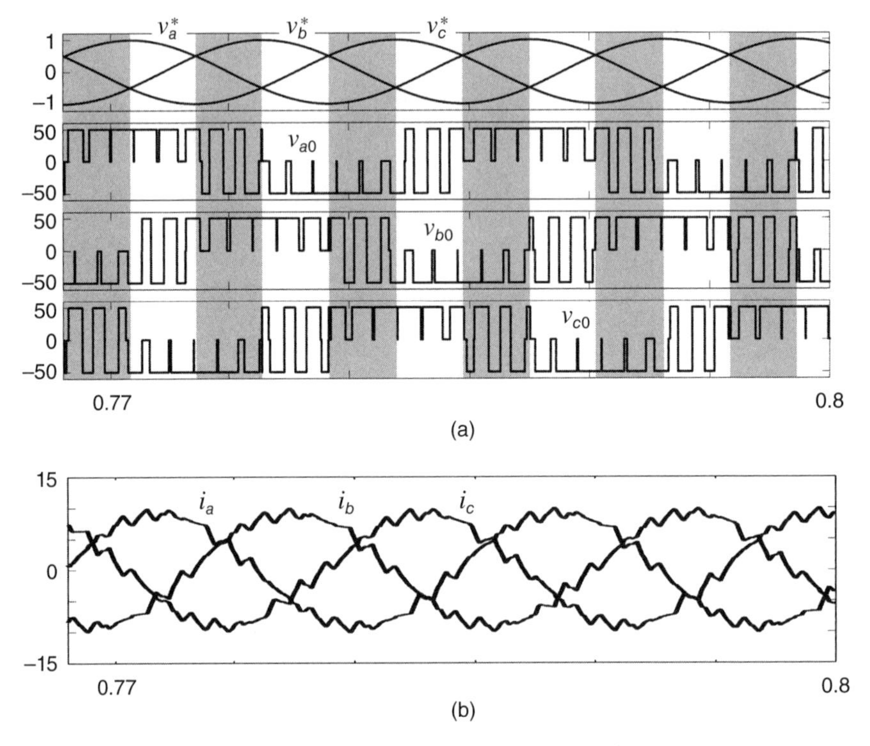

Figure 4.16 (a) Desired reference voltage and pole voltages. (b) Load current.

As in the case of previous nonconventional topologies, it is necessary to adapt the PWM strategy [see Fig. 4.17(b) and 4.17(c)] employed to guarantee the generation of a balanced three-phase load voltage. The challenge in this case is that the three-phase currents of the load must be regulated in a balanced way, even though there are only two legs available to do it. Two load currents are directly controlled by these two legs, and the third one is indirectly controlled, since $i_a + i_b + i_c = 0$. The maximum amplitude of the line-to-line reference voltages ($v_{ac}^* = v_a^* - v_c^*$) and ($v_{bc}^* = v_b^* - v_c^*$) is limited by the amplitude of the triangular waveforms, as in Fig. 4.17(c). In these, the triangular waveforms have a frequency of 1 KHz.

As this topology cannot be considered conventional, there are some restrictions in its practical application. However, it could be employed in fault-tolerant systems, as a backup configuration in the case of a fault of the three-leg conventional topology (Fig. 4.8).

4.7 UNBALANCED CAPACITOR VOLTAGE

If there is only one dc voltage source available, the NP required for the NPC configuration can be obtained with two capacitors in a series connection, as seen in Fig. 4.18(a). When a leg of the converter is connected (or clamped) to the NP, as in

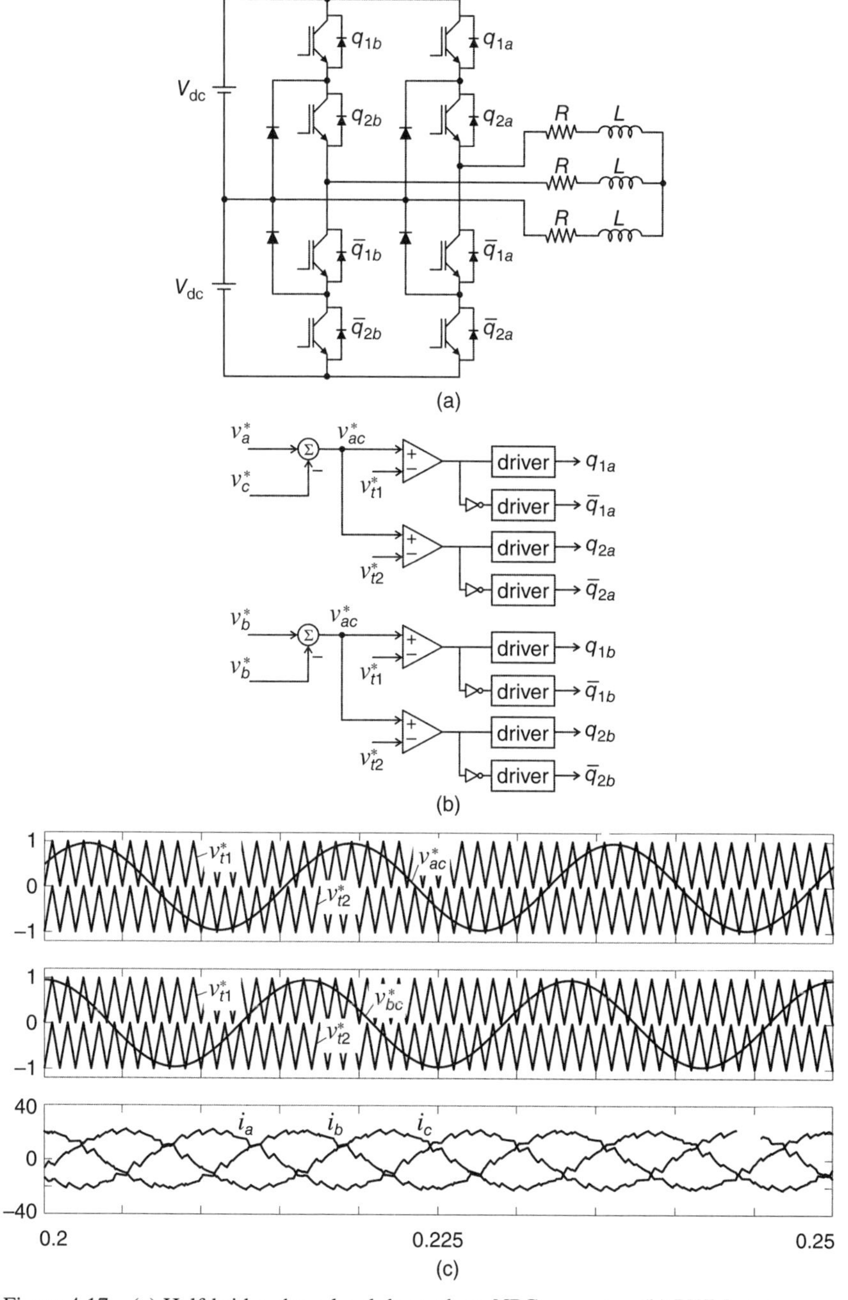

Figure 4.17 (a) Half-bridge three-level three-phase NPC converter. (b) PWM strategy. (c) (from top to bottom) reference waveforms for leg a ($v_a^* - v_c^*$, v_{t1}^*, and v_{t2}^*), reference waveforms for leg b ($v_b^* - v_c^*$, v_{t1}^*, and v_{t2}^*) and load currents.

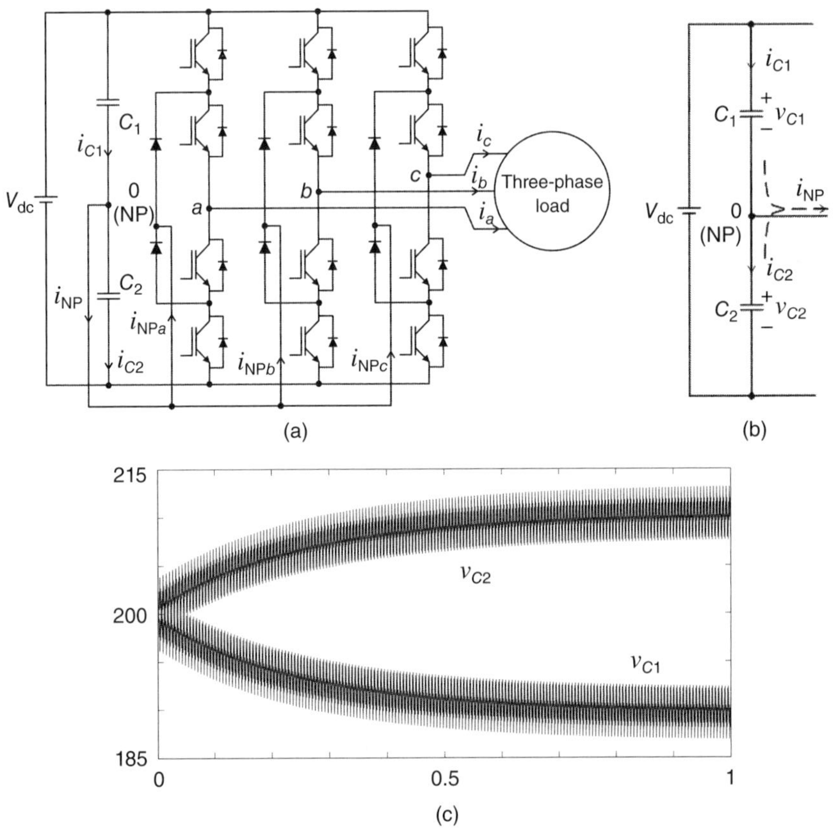

Figure 4.18 (a) NPC three-level topology with one dc voltage source. (b) Circuit showing the currents in the capacitors. (c) Capacitor voltages.

Fig. 4.18(b), its current is injected at this point. The NP current (i_{NP}) can be expressed by the following equation

$$i_{NP} = q_{2a}\overline{q}_{1a}i_a + q_{2b}\overline{q}_{1b}i_b + q_{2c}\overline{q}_{1c}i_c \tag{4.26}$$

Notice that the NP is a floating point. In an open-loop operation and without a specific PWM strategy, there is no guarantee that v_{C1} and v_{C2} will converge to the desired voltage, that is, $V_{dc}/2$. Considering the simple modulation strategy presented in this chapter, the capacitor voltages are unbalanced, as in Fig. 4.18(c), which brings up asymmetry at the output voltage of the converter.

To explain this capacitor voltage behavior it is necessary to observe the variables responsible for charging and discharging C_1 and C_2 highlighting the start-up transient. From equation (4.25) it is seen that each component of the NP current $i_{NPx} = q_{2x}\overline{q}_{1x}i_x$ (with $x = a, b, c$) is a function of load current and switching states.

For example, Fig. 4.19(a) shows two currents associated with phase a, (i.e., i_a and i_{NPa}). Notice that i_{NPa} [the first portion of equation (4.26)] is actually the current i_a multiplied by the state of the switches q_{2a} and $\overline{q}_{1a}$. As expected, such a current has

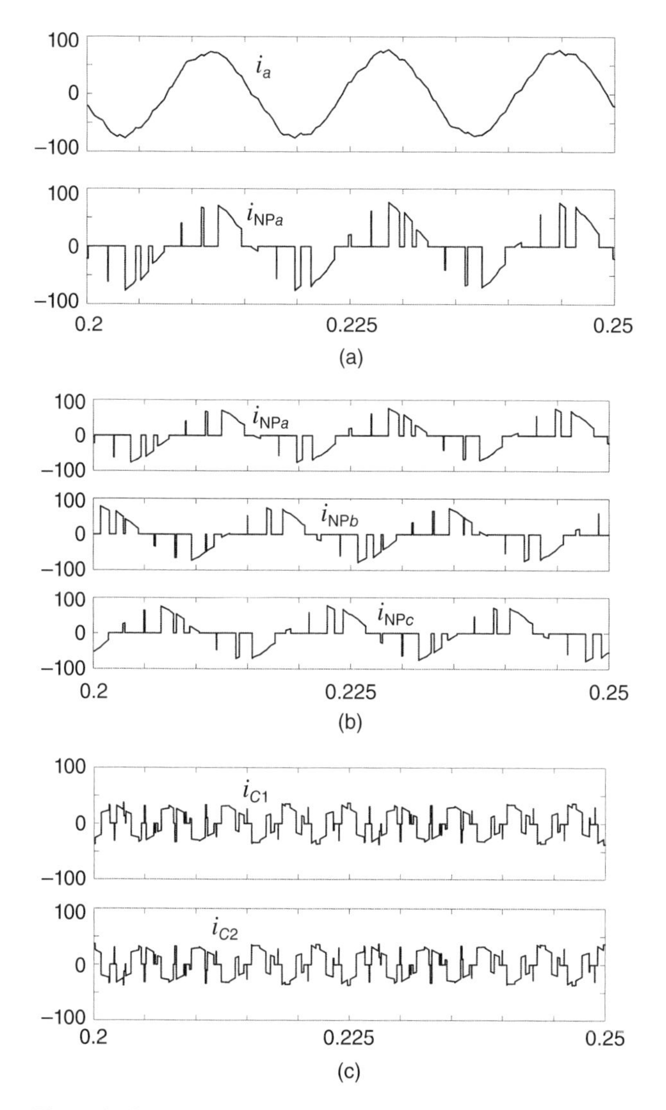

Figure 4.19 (a) i_a and $i_{\mathrm{NP}a}$. (b) $i_{\mathrm{NP}a}$, $i_{\mathrm{NP}b}$, and $i_{\mathrm{NP}c}$. (c) i_{C1} and i_{C2}.

an average value equal to 0. The behavior of the currents $i_{\mathrm{NP}a}$, $i_{\mathrm{NP}b}$, and $i_{\mathrm{NP}c}$ is shown in Fig. 4.19(b).

The capacitor currents i_{C1} and i_{C2} are functions of $i_{\mathrm{NP}a}$, $i_{\mathrm{NP}b}$, and $i_{\mathrm{NP}c}$, as seen in Fig. 4.19(c). The reason why the capacitor voltage v_{C1} goes down and v_{C2} goes up in Fig. 4.18(c) is because i_{C1} starts with a negative value while i_{C2} starts with a positive value, that is, initially C_1 is discharged and C_2 is charged but always maintaining the $v_{C1} + v_{C2}$ constant and equal to V_{dc}. There are different strategies to regulate v_{C1} and v_{C2} by either using feedback controllers or with specific PWM approaches.

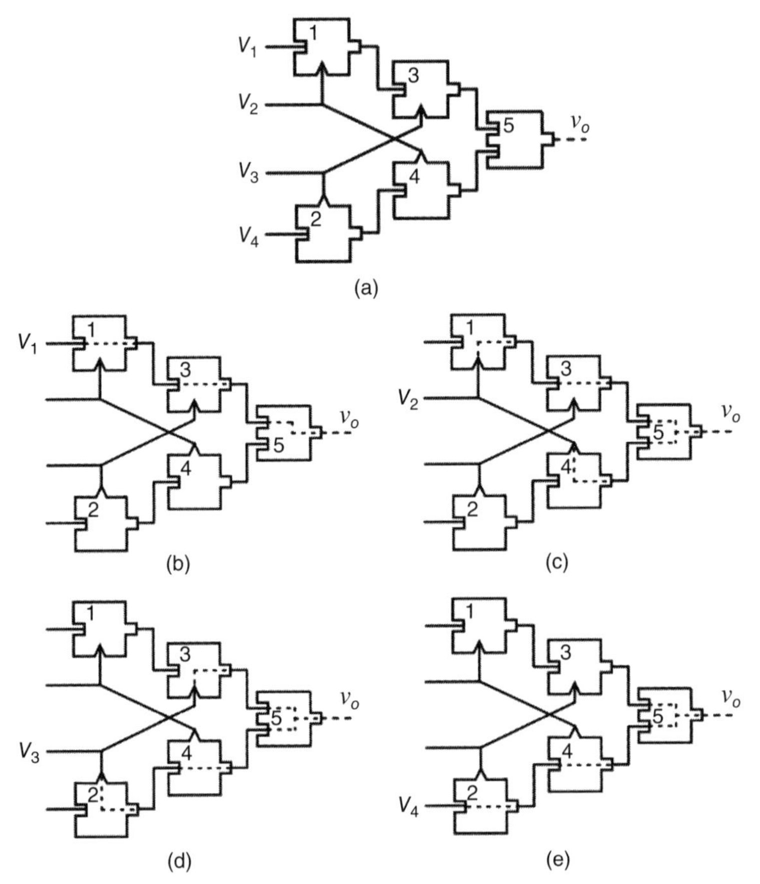

Figure 4.20 (a) Four-level NPC configuration with PBs. (b) $v_o = V_1$. (c) $v_o = V_2$. (d) $v_o = V_3$. (e) $v_o = V_4$.

4.8 FOUR-LEVEL CONFIGURATION

Following the axioms and postulates involving the PBs, presented in the previous chapter, the four-level (NPC) leg can be easily obtained as seen in Fig. 4.20(a). Figure 4.20(b)–4.20(e) shows the ways used to obtain each specific level (V_1, V_2, V_3, and V_4) at the output converter side. Figure 4.20(b) illustrates that it is necessary to activate the PBs 1, 3 and 5 to obtain $v_o = V_1$, while $v_o = V_4$ is obtained when PBs 2, 4, and 5 are activated [see Fig. 4.20(e)]. On the other hand, the intermediate voltages V_2 and V_3 are obtained at the output side when PBs 3 and 5, and PBs 4 and 5 are activated, respectively. In general terms, to generate an output waveform with four levels as in Fig. 4.21 (bottom), it is necessary to activate the PBs in a sequence as observed in Fig. 4.21 (top).

Since v_o is supposed to assume just one input voltage V_1, V_2, V_3, or V_4 at a time, an equation to describe the output voltage v_o as a function of the input voltages can be obtained from the independent binary variables (q_1, q_2, and q_3). As in the

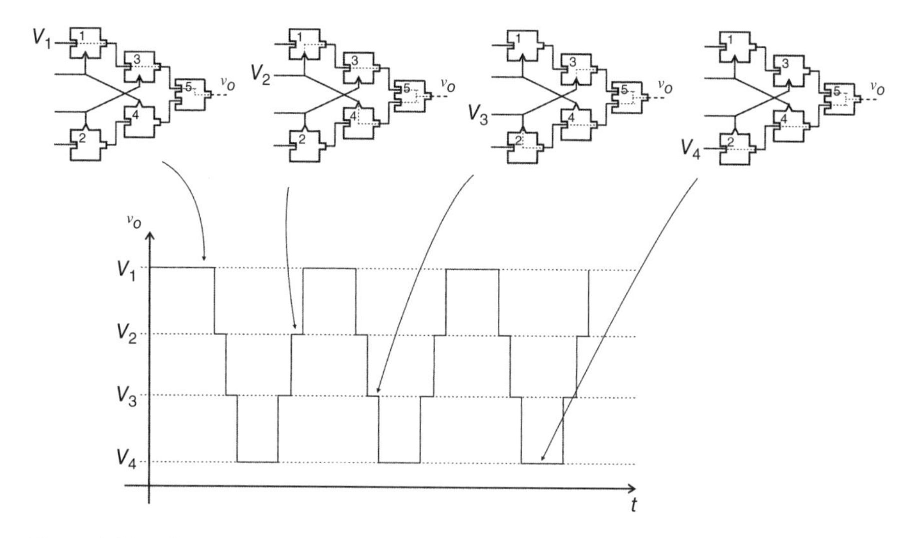

Figure 4.21 Generation of a general four-level output voltage.

TABLE 4.4 Table of Variables for the Converter in Fig. 4.20(a)

State	q_1	q_2	q_3	v_o
1	0	0	0	V_4
2	0	0	1	V_3
3	0	1	0	
4	0	1	1	V_2
5	1	0	0	
6	1	0	1	
7	1	1	0	
8	1	1	1	V_1

three-level NPC topology, the binary variables considered in Table 4.4 are related to the switching states inside the blocks in Fig. 4.20.

In this case, the output pole voltage is given by

$$v_o = \overline{q}_1\overline{q}_2\overline{q}_3 V_4 + \overline{q}_1\overline{q}_2 q_3 V_3 + \overline{q}_1 q_2 q_3 V_2 + q_1 q_2 q_3 V_1 \tag{4.27}$$

where $\overline{q}_1 = 1 - q_1$, $\overline{q}_2 = 1 - q_2$, and $\overline{q}_3 = 1 - q_3$.

Normally the waveform shown in Fig. 4.21 must be symmetrical, which means that $V_1 = V_{dc}$, $V_2 = V_{dc}/3$, $V_3 = -V_{dc}/3$, and $V_4 = -V_{dc}$, and

$$v_o = \left(-\overline{q}_1\overline{q}_2\overline{q}_3 - \frac{\overline{q}_1\overline{q}_2 q_3}{3} + \frac{\overline{q}_1 q_2 q_3}{3} + q_1 q_2 q_3\right) V_{dc} \tag{4.28}$$

If any other combination of output voltages associated with the switching states is considered, instead of the Table of Variables presented in Table 4.4, it will lead to undesired states, as presented shown in Example 4.2.

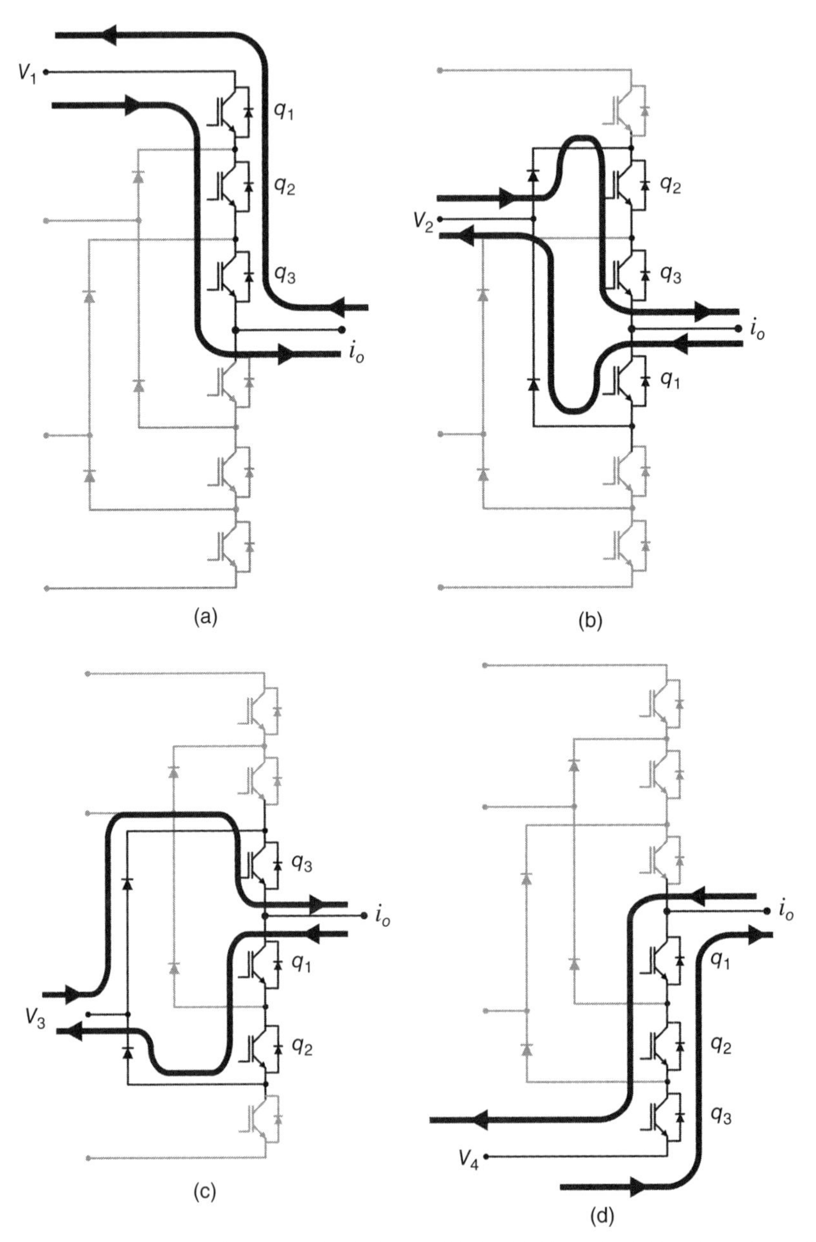

Figure 4.22 Generation of the level: (a) V_1, (b) V_2, (c) V_3, and (d) V_4 independent of the output current.

The four-level NPC configuration has no restriction in terms of the generation of desired output voltage and the values of currents, which means that each desired voltage value V_1, V_2, V_3, and V_4 can be obtained at the output converter side for any values of current ($i_o > 0$ or $i_o < 0$) as in Fig. 4.22.

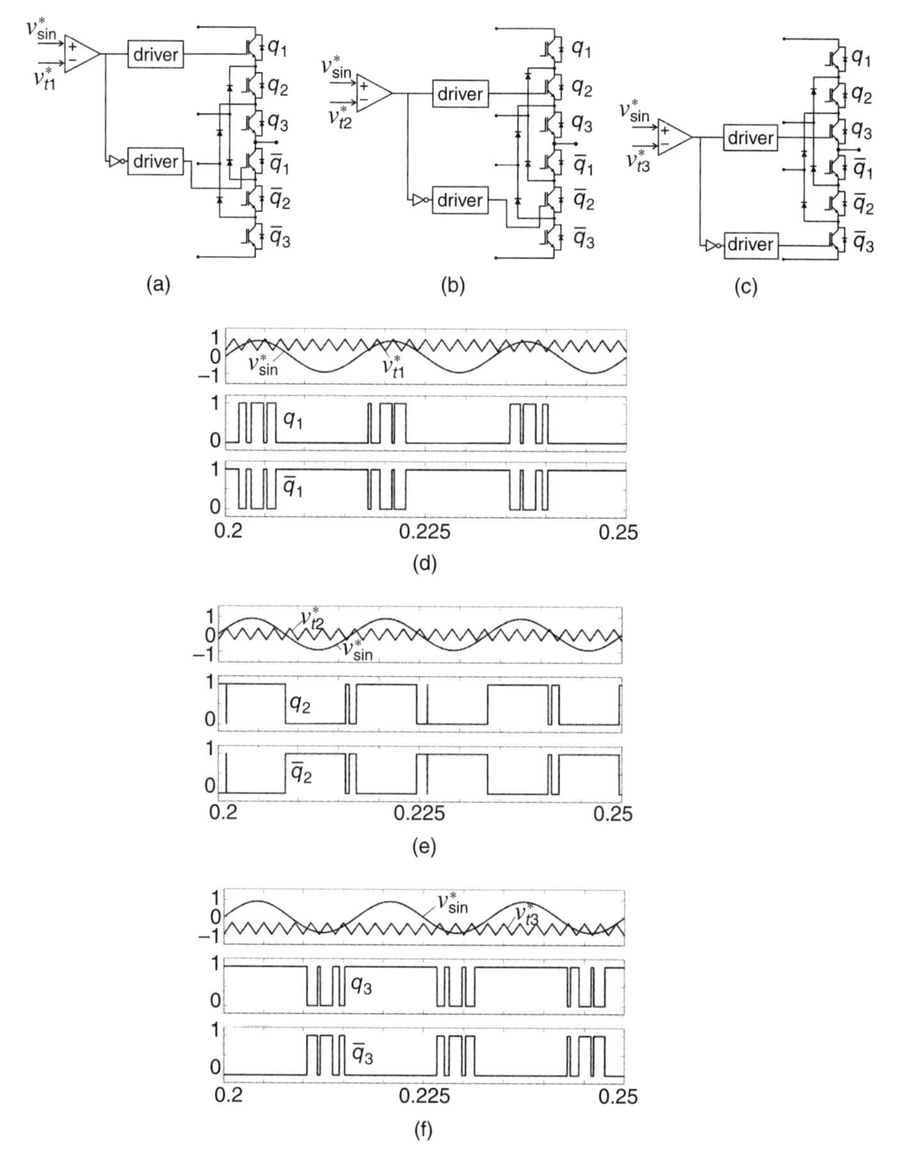

Figure 4.23 Four-level NPC topology – control circuit and signals for the switches.

4.9 PWM IMPLEMENTATION (FOUR-LEVEL CONFIGURATION)

Figure 4.23 shows the analog solution for the PWM generation of one leg. The states of the six switches are determined by a comparison among three high frequency triangular waveforms (v^*_{t1}, v^*_{t2}, and v^*_{t3}) and a sinusoidal waveform (v^*_{sin}).

It can be seen from Fig. 4.23 that these three triangular waveforms (v_{t1}^*, v_{t2}^*, and v_{t3}^*) are intentionally level-shifted to guarantee four levels at the output converter side. During the interval of time in which q_1 and $\overline{q}_1$ are in operation (i.e., turning *on* and *off*) q_2 and q_3 are always *on* and consequently $\overline{q}_2$ and $\overline{q}_3$ are always *off*. This is exactly what is expected from Table 4.4, rows 8 and 4, that is, when the switching sequence is changed from $(q_1,q_2,q_3) = (1,1,1)$ to $(q_1,q_2,q_3) = (0,1,1)$ the output voltage will change from V_1 to V_2. On the other hand, during the interval of time in which q_2 and $\overline{q}_2$ are in operation (turning *on* and *off*) q_1 is always *off* and q_3 is always *on*, which is expected from Table 4.4, rows 2 and 4. In this case, when the switching sequence is changed from $(q_1,q_2,q_3) = (0,0,1)$ to $(q_1,q_2,q_3) = (0,1,1)$ the output voltage will change from V_3 to V_2. Finally, for the interval of time in which q_3 and $\overline{q}_3$ are in operation (turning *on* and *off*) q_1 and q_2 are always *off*, which is also expected from Table 4.4, rows 1 and 2. When the switching sequence is changed from $(q_1,q_2,q_3) = (0,0,0)$ to $(q_1,q_2,q_3) = (0,0,1)$ the output voltage will change from V_4 to V_3.

Figure 4.24 shows the power converter and waveforms for a half-bridge four-level configuration. If only one dc source is available, as in Fig. 4.18, additional

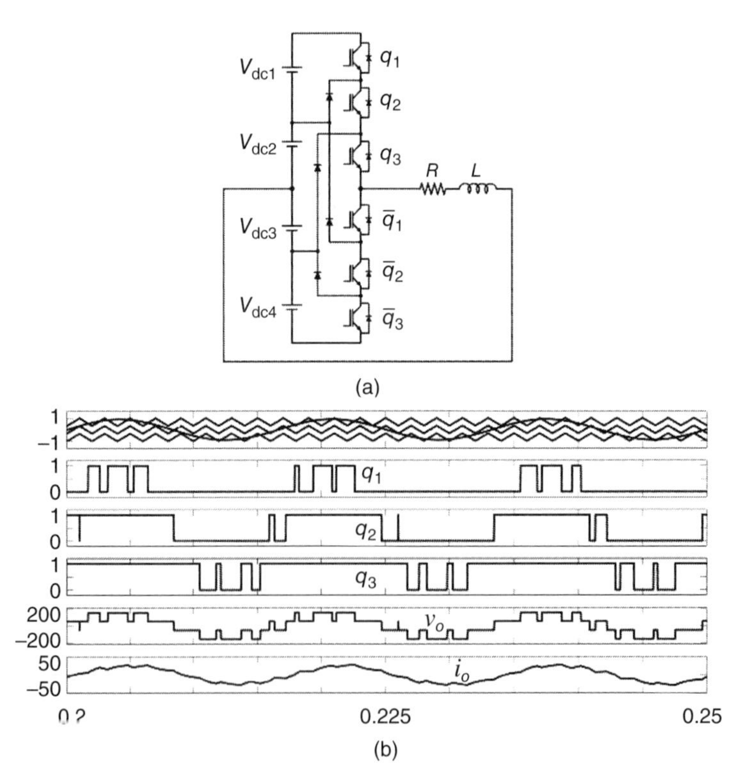

Figure 4.24 (a) Four-level NPC half-bridge converter. (b) (from top to bottom) v_{sin}^*, v_{t1}^*, v_{t2}^*, v_{t3}^*; gating signal of switch q_1; gating signal of switch q_2; gating signal of switch q_3; load voltage and load current.

capacitors can be employed to obtain four voltage levels. However, the challenge to control the capacitor voltages from a single dc source increases with the number of levels of the converters. From four levels onwards, an auxiliary circuit is normally needed to guarantee the control of such capacitors for a large range of load power factor.

It is worth mentioning that the number of triangular waveforms increases proportionally to the number of levels. For a converter with n levels $n-1$ triangular waveforms are needed at the output of the converter. As a consequence of the high number of triangular waveforms, such converters must be designed to operate at high modulation index to guarantee the desired amount of levels at the output converter side. Figure 4.25 shows the influence of the modulation index (m_a) in the converter waveforms. Such a figure highlights the control (top) and output converter (middle and bottom) waveforms for three different values of m_a. When m_a is decreased, the

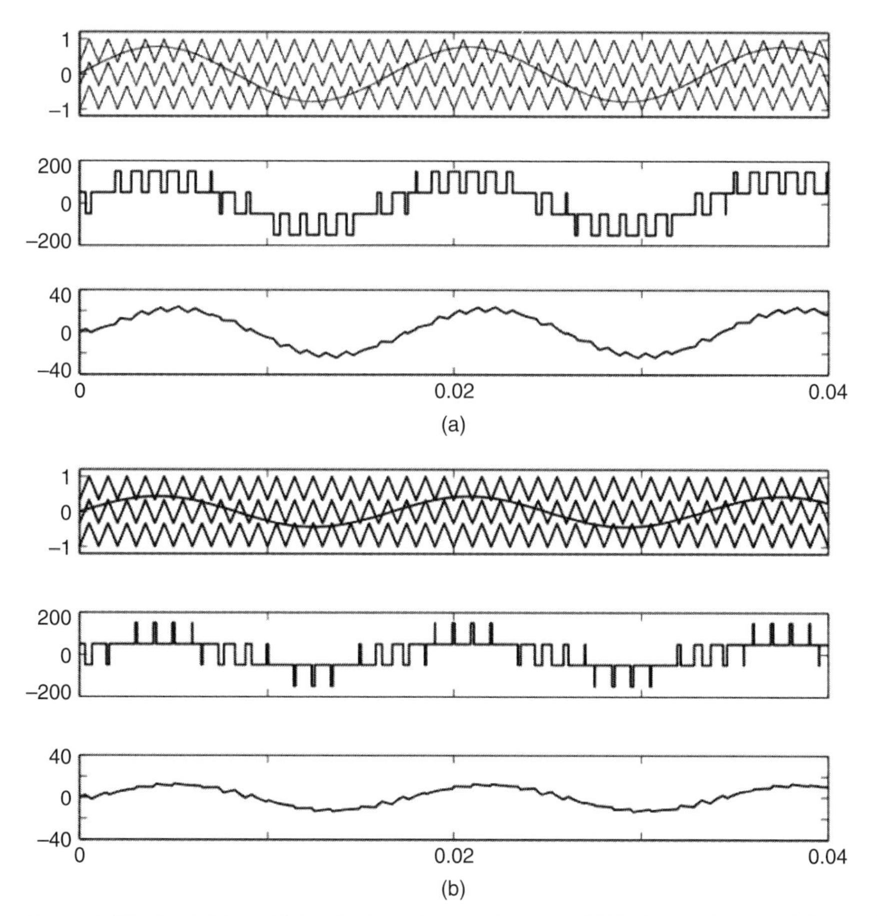

Figure 4.25 (top) Sinusoidal and triangular waveforms, (middle) load voltage and (bottom) load current: (a) $m_a = 0.8$, (b) $m_a = 0.45$, and $m_a = 0.2$.

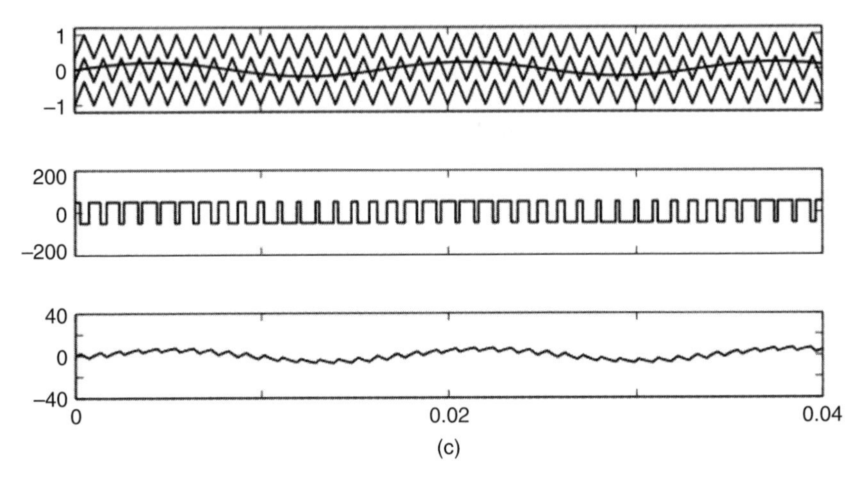

Figure 4.25 (*Continued*)

number of levels reduces as well, as observed in the sequence of Fig. 4.25(a), 4.25(b), and 4.25(c), respectively. The output voltage is reduced to a two-level waveform when the modulation index is equal to 0.2.

4.10 FULL-BRIDGE AND OTHER CIRCUITS (FOUR-LEVEL CONFIGURATION)

The full-bridge single-phase converter using four-level NPC legs is shown in Fig. 4.26(a). This topology could be of interest in single-phase applications when the quality of the output voltage is an important aspect of the design since the load voltage ($v_o = v_{a0} - v_{b0}$) has almost sinusoidal behavior. Figure 4.26(b) shows (from top to bottom) pole voltage v_{a0}, pole voltage v_{b0}, load voltage, and load current, respectively.

A three-phase three-level full-bridge NPC configuration is depicted in Fig. 4.27(a), while Fig. 4.27(b) shows some variables associated with this converter, that is, (from top to bottom) pole voltage (v_{a0}), voltage between load NP and dc-link midpoint connection (v_{n0}), load phase voltage (v_{an}), and load current (i_a), respectively.

Figure 4.28 shows a half-bridge three-phase four-level NPC topology. In this case, a similar approach employed for the three-level topology, in terms of the PWM circuit, is also considered for this configuration.

A four-level topology can be obtained for a three-phase machine with open-end windings following the pattern of the circuit given in Fig. 4.12, and shown in Fig. 4.29. This is an open-end winding motor drive system, which requires access to six terminals of the machine. A four-level voltage appears in each phase of the machine.

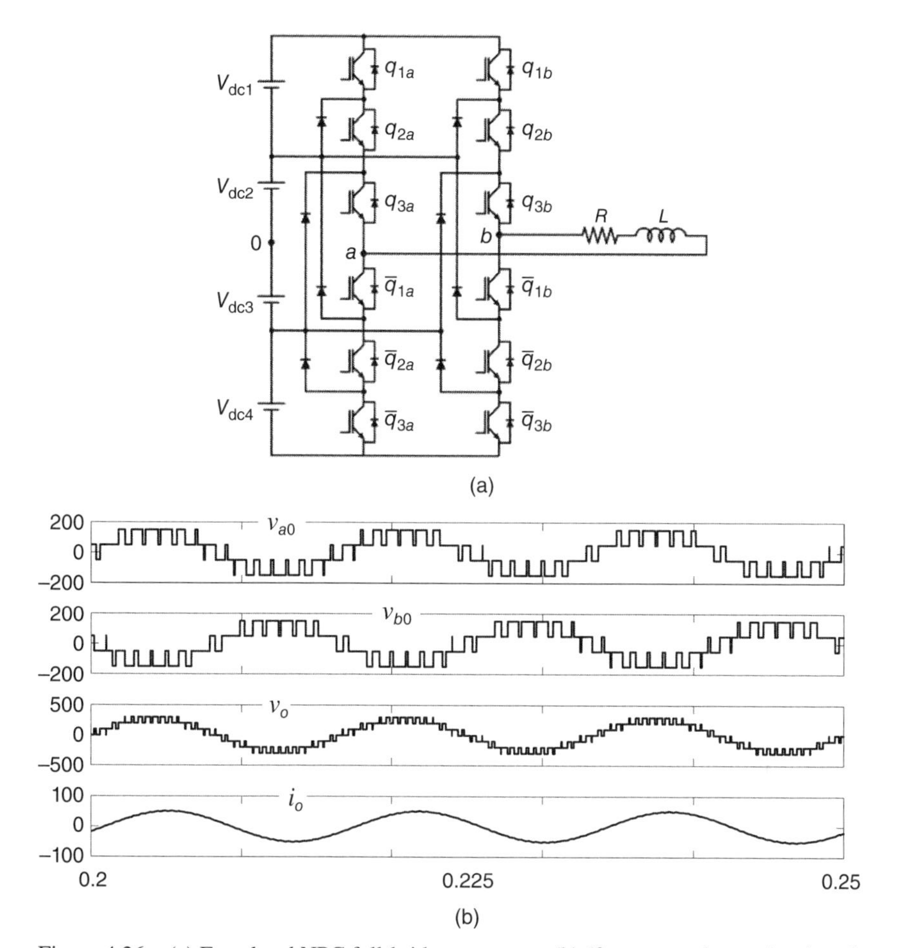

Figure 4.26 (a) Four-level NPC full-bridge converter. (b) (from top to bottom) pole voltage v_{a0}, pole voltage v_{b0}, load voltage, and load current.

4.11 FIVE-LEVEL CONFIGURATION

Following the same approach presented before for the three- and four-level configurations, the five-level converter is shown in Fig. 4.30(a). As done for the previous NPC topologies. Notice that just two types of PBs are employed. Figure 4.30(b) shows the five-level configuration with the traditional representation, using switches.

In addition to the capacitor voltage balancing problem, another challenge appears when the number of levels increases for an NPC topology. The blocking voltage for each clamping diode changes and it depends on its position in the circuit. The diode voltage can be written as

$$v_D = (N - 1 - x)(N - 1)^{-1} V_{\text{dc}} \tag{4.29}$$

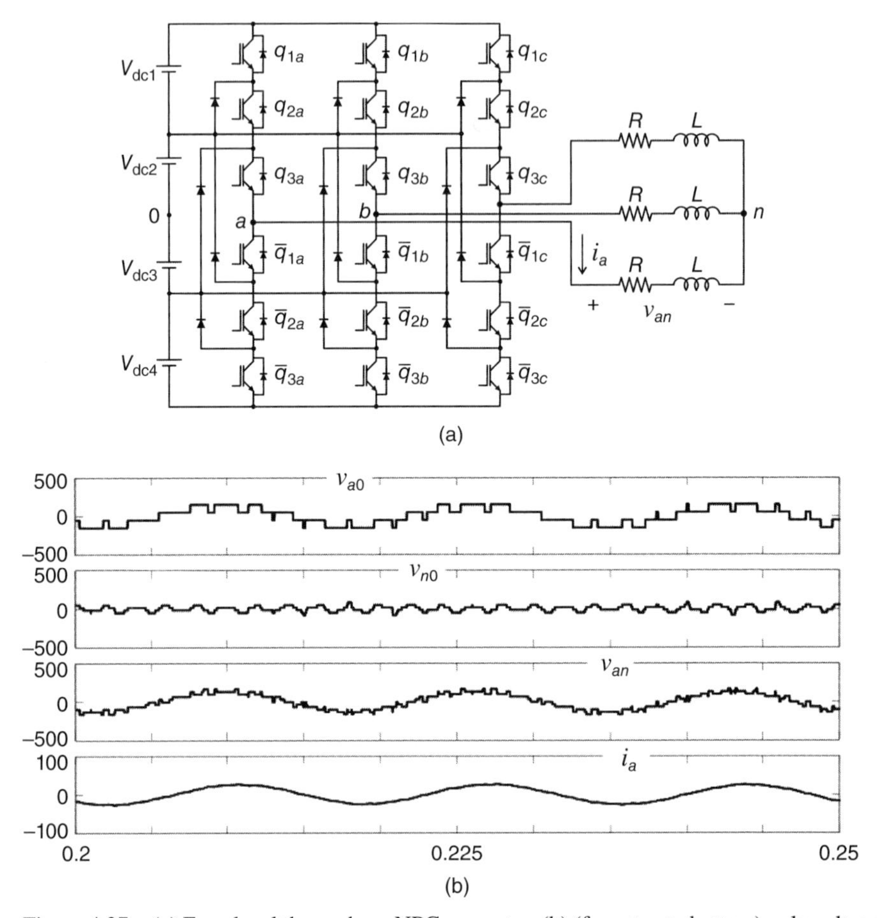

Figure 4.27 (a) Four-level three-phase NPC converter. (b) (from top to bottom) pole voltage, voltage between load NP and dc-link midpoint connection, load voltage, and load current.

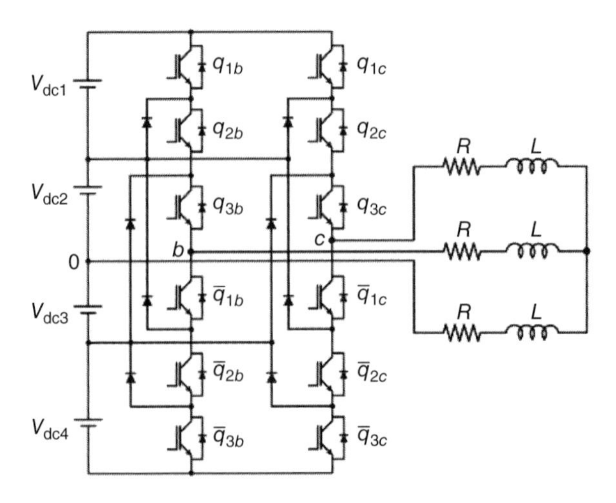

Figure 4.28 Half-bridge four-level three-phase NPC converter.

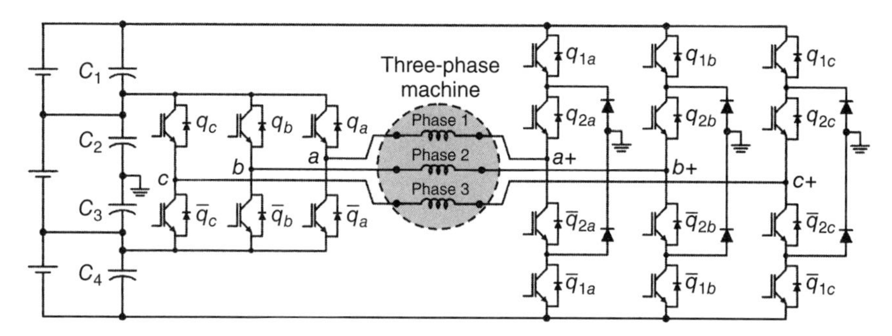

Figure 4.29 Four-level open-end motor drive system.

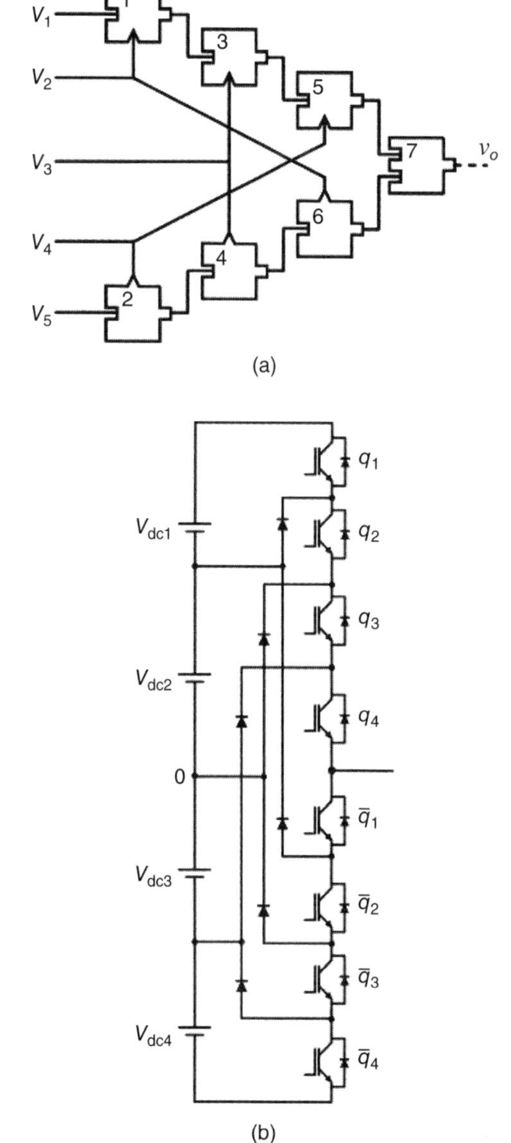

Figure 4.30 Five-level NPC converter: (a) PB representation and (b) representation with switches.

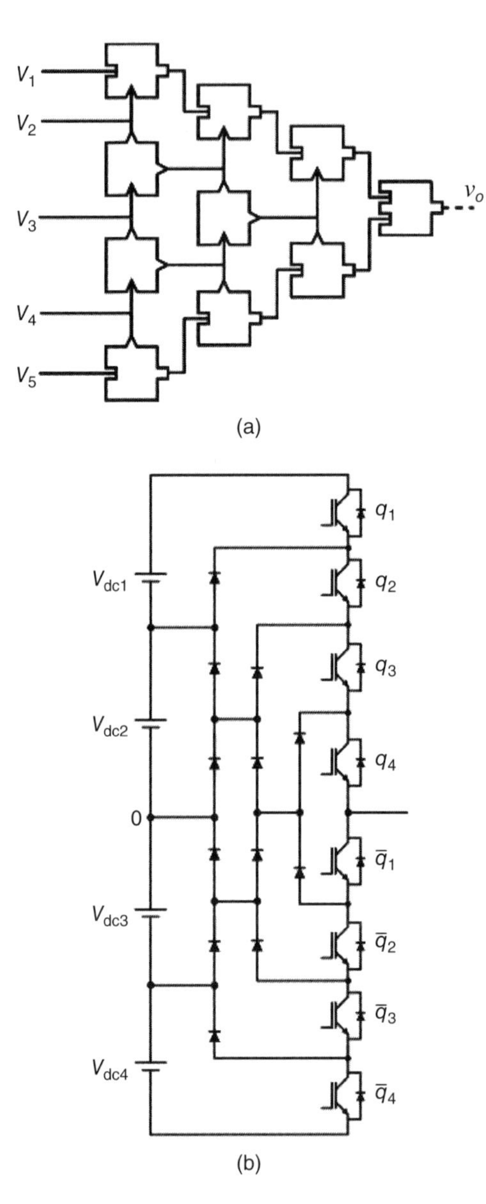

Figure 4.31 Five-level NPC converter with pyramid structure: (a) PB representation and (b) representation with switches.

where N is the number of levels, x changes from 1 to $N-2$ and V_{dc} is the dc-link voltage of the whole dc source. For high voltage applications, such different blocking voltage characteristics could bring up operational limits for the converter. To solve this problem and guarantee the same blocking voltages for all diodes, it is necessary to add more clamping diodes in series.

By adding the first PB-dc shown in Fig. 4.31(a), it is possible to find a topology with the same blocking voltages for all diodes with a pyramid structure as proposed by Yuan and Barbi [14] and depicted in Fig. 4.31(b).

Theoretically, the number of levels in an NPC topology can be increased as necessary, and to do so the PB-cc blocks must be stacked as in Fig. 4.32.

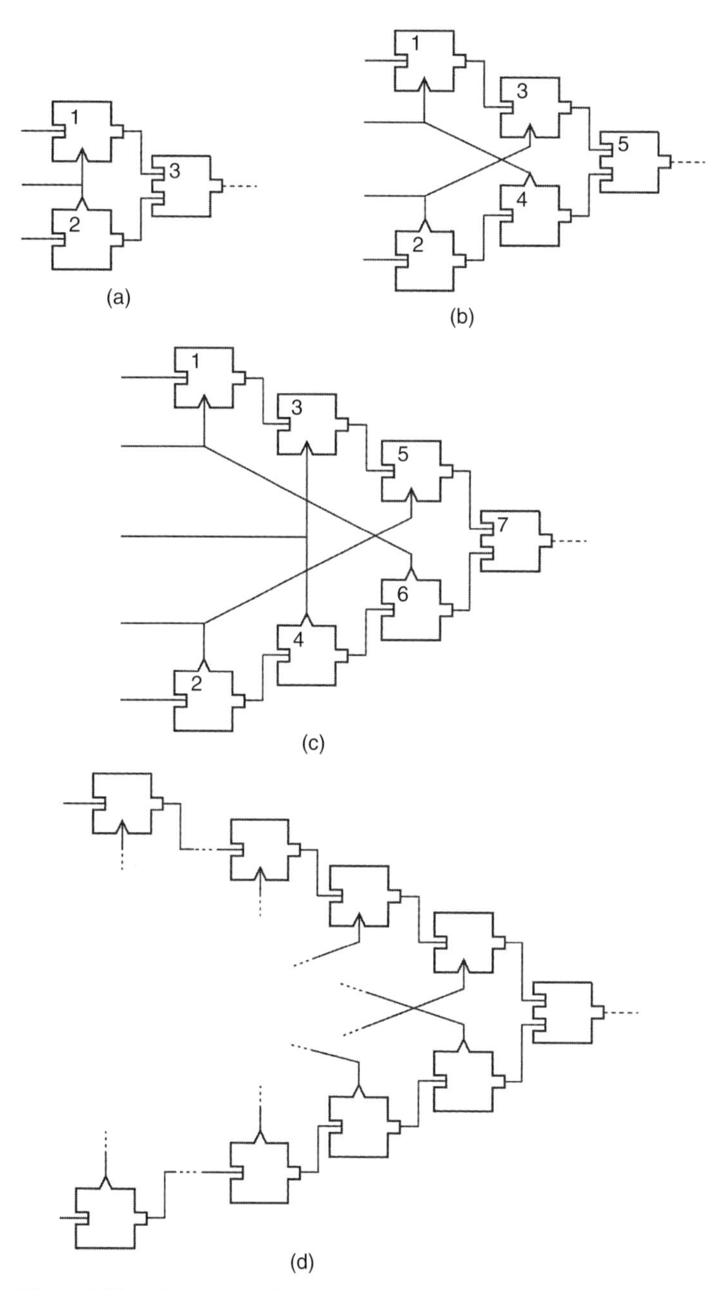

Figure 4.32 Generation of the n-level NPC topologies: (a) three-level, (b) four-level, (c) five-level, and (d) n-level.

4.12 SUMMARY

This chapter is dedicated to NPC configurations. Further details about NPC topologies can be found in References from 1 to 14. The topologies are presented in a systematic way in the following sequence: (i) PBs connections obtained from the rules presented in Chapter 3; (ii) once the topology is established, the Table of Variables is obtained; (iii) the next step is modeling the converter by writing the output voltage to describe what was obtained in the Table of Variables; and (iv) and finally the PWM approach is developed from the Table of Variables. The same systematic approach has been applied for both conventional and nonconventional topologies.

More than bringing new possibilities for power electronics converters in different applications, nonconventional converters play an important role in the learning process and ensure a deep understanding of the subject.

REFERENCES

[1] Baker RH. Bridge converter circuit. US patent 4270163. 1981 May (filed in Aug. 79).
[2] Nabae A, Takahashi I, Akagi H. A new neutral-point-clamped PWM inverter. IEEE Trans Ind Appl 1981;IA-17(5):518–523.
[3] Steinke JK. Control of a neutral-point-clamped PWM inverter for high power AC traction drives. Int'l Conf on Power Electron and Variable-Speed Drives on; 1988 Jul 13–15. p 214–217.
[4] Holtz J. Self-commutated inverters with a stepped output-voltage waveform suitable for high powers and frequencies. Siemens Forsch-U Entwick1-Ber 1977;6(3):164–171.
[5] Carrara G, Gardella S, Marchesoni M, Salutari R, Sciutto G. A new multilevel PWM method: a theoretical analysis. IEEE Trans Power Electron 1992;7:497–505.
[6] Lin B-R, Wei T-C. Space vector modulation strategy for an eight-switch three-phase NPC converter. IEEE Trans Aerosp Electron Syst 2004;40(2):553–566.
[7] Holmes DG, Lipo TA. *Pulse Width Modulation for Power Converters*. Piscataway: IEEE/Wiley Interscience; 2003.
[8] E. C. dos Santos Jr., Muniz JHG, and da Silva ERC. 2L3L inverter. Proc of Brazilian Power Electron Conf (COBEP); 2011, p 924–929
[9] Mihalache L. A hybrid 2/3 level converter with the minimum switch count. Industry Applications Conference. 41st IAS Annual Meeting. Conference Record of the 2006 IEEE; 2006; vol. 2. p 611–618.
[10] da Silva ER, Muniz JG, da Nobrega RB, dos Santos Jr. EC. An improved pulse-width-modulation for the modified hybrid 2/3-level converter. Proc of Brazilian Power Electron Conf (COBEP); 2013, p 248–253.
[11] Choi NS, Cho JG, Cho GH. A general circuit topology of multilevel inverter. Proc. Rec. IEEE PESC; 1991. p 96–103.
[12] Carpita M, Tenconi S. A novel multilevel structure for voltage source inverter. Proc. Rec. EPE Conf.; 1991. p 1.90–1.94.
[13] Hammond PW. Medium voltage PWM drive and method. US patent 5,625,545. 1997.
[14] Yuan X, Barbi I. A new diode clamping multilevel inverter. Proceedings of APEC99; 1999. p 495–501.

CHAPTER 5

CASCADE CONFIGURATION

5.1 INTRODUCTION

The PBG is employed in this chapter to describe the cascade multilevel configurations, also known as H-bridge series connected configurations. The introduction and fundamental concepts are furnished for a single H-bridge topology, followed by the connection of other H-bridge converters for generation of topologies with more levels. Such configurations deal with attractive characteristics common for multilevel circuits, such as low electromagnetic interference (EMI), reduced total harmonic distortion and high relationship between the number of levels per number of switches. Comparing this multilevel configuration with the neutral point clamped (NPC) topology described earlier, it is possible to sort some advantages such as (i) simple packaging and physical structure with no neutral point clamping circuits, (ii) modular structure (more or fewer H-bridge cells can be cascaded depending on the desired power quality), and (iii) possibility to operate even with partial failure. On the other hand, the main disadvantages are associated with (i) greater number of dc sources and (ii) control complexity.

The previous chapter was organized with the sections divided in terms of the number of levels of the converter, that is, three levels, four levels, and so on. In this chapter, on the other hand, the organization (labeling scheme) of the sections considers the number of H-bridge converters instead of number of levels. The main reason for that is, depending on the values of the dc voltage sources and modulation approach, the same topology can generate different number of levels.

Accordingly, Section 5.2 presents the description of a single H-bridge converter along with its Table of Variables. The pulse width modulation (PWM) implementation is derived from the Table of Variables, as highlighted in Section 5.3. The following section presents the three-phase version with a single H-bridge converter per phase. Sections 5.5, 5.6, and 5.7 deal with two H-bridge converters in series, with the sequence given respectively by power converter construction, PWM approach, and three-phase version. After showing how it is possible to guarantee post-fault operation with this type of converter in Section 5.7, Section 5.8 demonstrates that two H-bridge converters can be employed to generate an output voltage with either seven or nine levels, by employing the correct relationship between the dc-link voltages.

Advanced Power Electronics Converters: PWM Converters Processing AC Voltages,
Forty Fifth Edition. Euzeli Cipriano dos Santos Jr. and Edison Roberto Cabral da Silva.

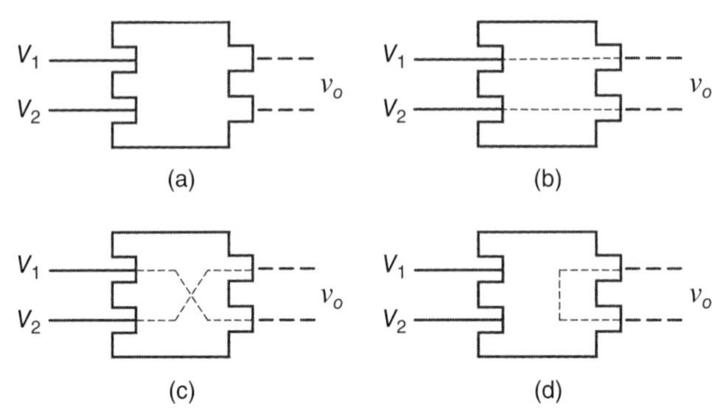

Figure 5.1 (a) One H-bridge configuration with PBs. (b) $v_o = V_1 - V_2$. (c) $v_o = V_2 - V_1$. (d) $v_o = 0$.

Section 5.9 presents different arrangements and PWM solutions for three H-bridge converters connected in series. Finally, Section 5.10 introduces the generalization and other possible converters characterized by series arrangements of H-bridge circuits. This chapter has been written with the help of References from 1 to 14.

5.2 SINGLE H-BRIDGE CONVERTER

Considering the power blocks (PBs), no connection is needed among PBs to obtain an H-bridge topology, since there is a block that represents this converter directly, as observed in Fig. 5.1(a). Figure 5.1(b), 5.1(c) and 5.1(d) illustrates the methods used to obtain three levels at the output converter side ($V_1 - V_2$, $V_2 - V_1$, 0). In general terms, to generate an output waveform with three levels as in Fig. 5.2 (bottom), it is necessary to activate the PBs in a sequence as observed in this figure (top).

Notice that instead of selecting one of the input voltages V_1, V_2, and V_3 as done for the NPC topology considered in the previous chapter, the three-level output voltage in Fig. 5.2 is a combination of two voltages V_1 and V_2 available at the input converter side. Despite the three-level voltage in Fig. 5.2, it is convenient to start the analysis of the cascade multilevel topologies by assuming just two levels ($V_1 - V_2$, $V_2 - V_1$) at the output converter side, as in Fig. 5.1(b) and 5.1(c).

Since v_o can assume just one combination of the input voltages each time, it is reasonable to write an equation to describe the output voltage v_o as a function of both input voltages and an independent binary variable. As the number of input voltages to be selected is two, the minimum number of binary control variable is one. This binary variable should be used to select one combination of the inputs to appear at the output converter side, which means that one possible solution to achieve this goal is given in Table 5.1.

From q_1 and its complementary signal $\overline{q}_1 = 1 - q_1$, the output voltage can be obtained as follows:

$$v_o = q_1(V_1 - V_2) + \overline{q}_1(V_2 - V_1) \tag{5.1}$$

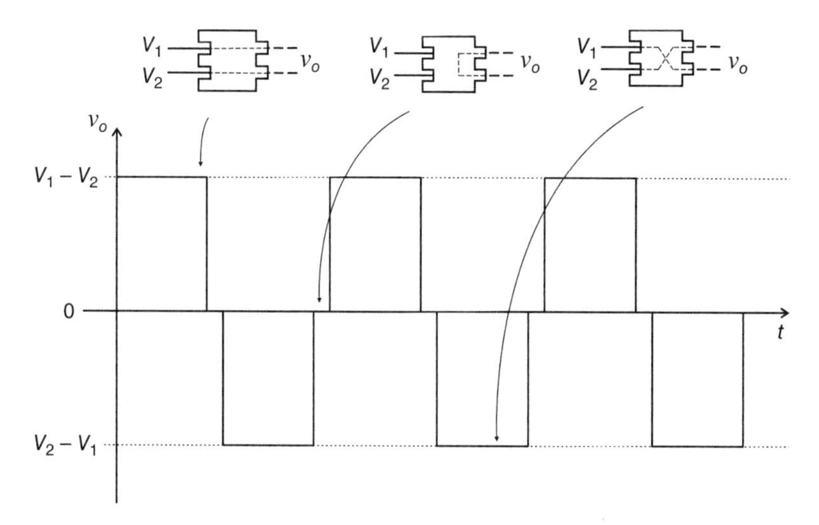

Figure 5.2 General three-level output voltage.

TABLE 5.1 Converter in Fig. 5.1(a) Operating as a Two-Level Converter

State	q_1	v_o
1	1	$V_1 - V_2$
2	0	$V_2 - V_1$

If $V_1 - V_2 = V_{dc}$, then (5.1) can be written as follows:

$$v_o = (2q_1 - 1)V_{dc} \tag{5.2}$$

As seen in the previous chapter, these control variables (q_1 and $\overline{q}_1$) are in fact the states of the switches inside the blocks. Comparing Table 5.1 with Fig. 5.2 (bottom) it is evident that just one binary control variable (q_1) and its complementary signal ($\overline{q}_1$) are not enough to select three levels at the output converter side.

If the number of levels for the output voltage is expected to be three ($\pm V_{dc}$,0), the minimum number of binary variables in this case should be two, which are used to select a desired output voltage, meaning that one possible combination to achieve this goal is given in Table 5.2.

The output voltage can be written as

$$v_o = -\overline{q}_1 q_2 V_{dc} + \overline{q}_2 q_1 V_{dc} \tag{5.3}$$

where $\overline{q}_2 = 1 - q_2$. Substituting the complementary variables, the previous equation becomes

$$v_o = -(1 - q_1)q_2 V_{dc} + (1 - q_2)q_1 V_{dc} \tag{5.4}$$

$$v_o = (q_1 - q_2)V_{dc} \tag{5.5}$$

TABLE 5.2 Converter in Fig. 5.1(a) Operating as a Three-Level Converter

State	q_1	q_2	v_o
1	0	0	0
2	0	1	$-V_{dc}$
3	1	0	V_{dc}
4	1	1	0

Figure 3.10 shows the same configuration presented in Fig. 5.1, along with binary variables describing the state of the switches for positive and negative output currents (i_o). Notice that there is no restriction in terms of state of the switches and the values of currents, which means that each voltage value can be obtained at the output converter independent of the current values ($i_o > 0$ and $i_o < 0$).

Example 5.1

Considering the previous analysis, place each switch (q_1, $\overline{q}_1$, q_2, and $\overline{q}_2$) in the correct position in the following circuit to guarantee the generation of the levels as shown in Table 5.2.

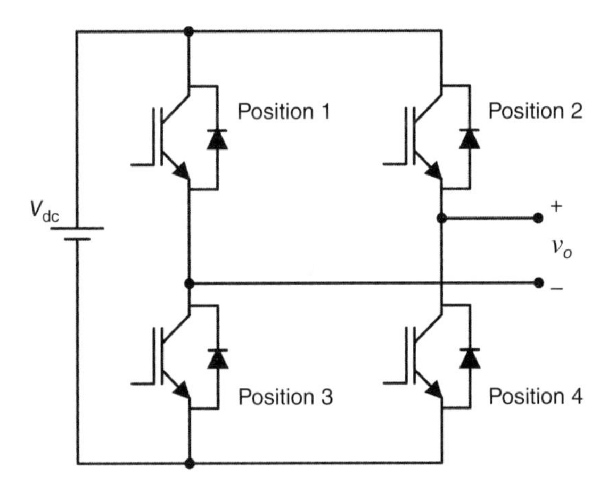

Solution

For the generation of a voltage v_o with three levels, as in Table 5.2, q_1 should be placed, for example, in position 2, while q_2 should be placed in position 1. Consequently $\overline{q}_1$ and $\overline{q}_2$ should be placed in positions 4 and 3, respectively.

5.3 PWM IMPLEMENTATION OF A SINGLE H-BRIDGE CONVERTER

If a single H-bridge converter is operating with two or three levels independently, the ultimate control goal is to guarantee a regulated voltage at its output side. The output voltage is regulated in terms of amplitude and frequency, obtained through PWM strategies. Many works in the technical literature have proposed solutions in terms of modulation approach applied to cascade converters. A simple sinusoidal carrier-based PWM strategy is considered in this chapter.

Figure 5.3 shows an analog solution for implementation of PWM, in which the states of the four switches are determined by a comparison between one high frequency triangular waveform (v_t^*) and a sinusoidal waveform (v_{sin}^*). Note that this modulation strategy generates the waveforms with two levels as seen in Table 5.1. In Fig. 5.3(a) (top) the same gating signals are sent to q_1 and $\overline{q}_2$, while the same complementary signals are sent to q_2 and $\overline{q}_1$, which implies that the pair of switches $q_1 - \overline{q}_2$ and $q_2 - \overline{q}_1$ are always turned *on* and *off* simultaneously. Figure 5.3(a) (bottom) shows the power part of the converter, and Fig. 5.3(b) shows the waveforms of the PWM and gating signals.

While Fig. 5.4(a) shows an H-bridge converter supplying a single-phase *RL* load, Fig. 5.4(b) shows (from top to bottom) PWM reference waveforms (v_{sin}^* and v_t^*), gating signal of switch q_1, gating signal of switch q_2, load voltage and load current, respectively. The parameters employed in these simulation results were $V_{\text{dc}} = 200\,\text{V}$, $R = 5\,\Omega$, and $L = 5\,\text{mH}$, and switching frequency was equal to 1000 Hz. These same parameters have been used for other simulation outcomes in this chapter.

The modulation strategy presented in Figs 5.3 and 5.4 seems to have low practical interest due to a two-level voltage generation, which means reduced THD as compared to the three-level option. However, it permits applicability in photovoltaic systems due to common mode voltage and current reduction.

As indicated by Fig. 5.2 and Table 5.2, an H-bridge converter is also able to generate a three-level voltage by just changing the PWM control signals.

Figure 5.5 shows two PWM approaches used to control the H-bridge converter. A direct comparison between the results obtained in Fig. 5.5 highlights some advantages and drawbacks for each method used to generate three levels at the output converter side. For instance, with two high frequency waveforms (v_{t1}^* and v_{t2}^*) [see Fig. 5.5(a) and 5.5(c) – Method 1] it is possible to operate the converter with reduced power converter losses, since one leg will modulate (turning the switches *on* and *off*) while the other one will be clamped. Method 1 is also known as level-shift PWM strategy. On the other hand, the operation with one high frequency waveform (v_t^*) and two modulating signals ($v_{\text{sin}\,1}^*$ and $v_{\text{sin}\,2}^*$) [see Fig. 5.5(b) and 5.5(d) – Method 2] guarantees lower harmonic distortion, which can be verified by the reduced current ripple. In this case, both legs modulate during the whole period of the sinusoidal waveforms; as a consequence, the harmonic content of the high frequency components is twice that of the switching frequency.

In Method 1 [Fig. 5.5(a) and 5.5(c)] just two states in Table 5.2 are employed per half of the sinusoidal period, that is, states 1 and 3 for the positive half of the

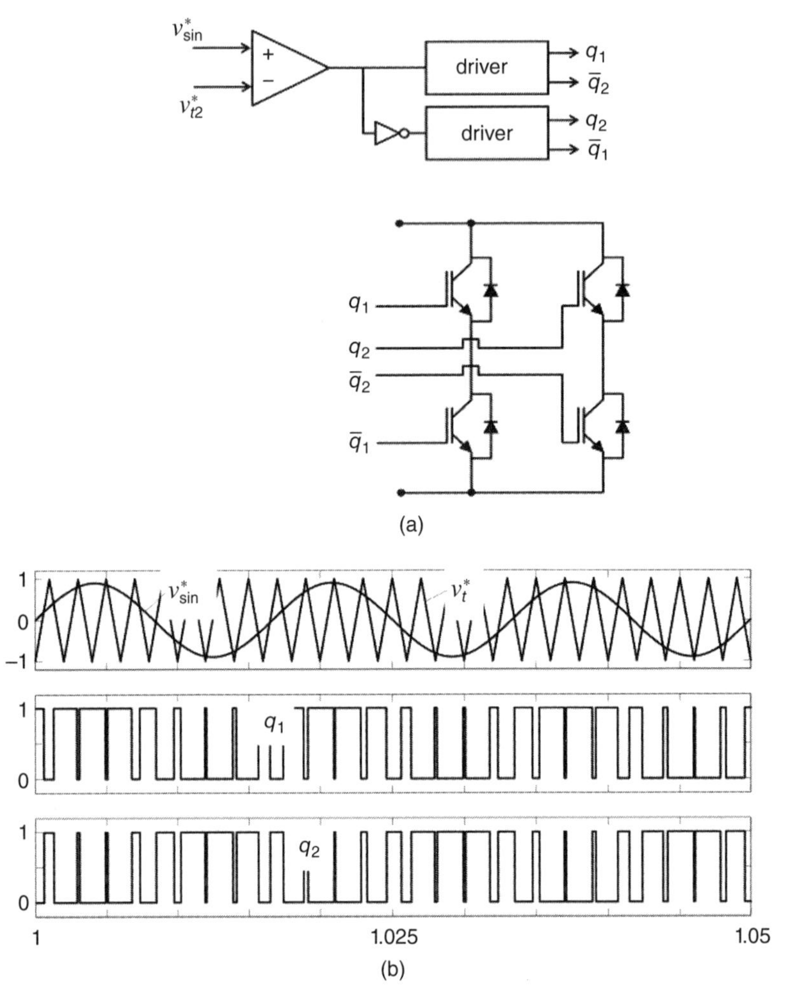

Figure 5.3 (a) PWM generation for switches q_1, q_2, $\overline{q}_1$, and $\overline{q}_2$. (b). (from top to bottom) – reference voltages, gating signal of switch q_1, and gating signal of switch q_2.

sine, and states 1 and 2 for the negative half of the sine. In this method, state 4 with both switches *on* is never employed. On the other hand, in Method 2 all states are employed.

Figures 5.4 and 5.5 highlight that the same average voltage and consequently the same amplitude (in terms of average values) of the load current is obtained at the output converter side, independent of the modulation strategies employed.

The frequency-domain representation, using the Fourier transformation of the output voltage, is an important tool to understand the differences between the PWM strategies in Fig. 5.5.

Figure 5.6(a) and 5.6(b) shows the waveforms presented in Fig. 5.5, with both time- and frequency-domain for Methods 2 and 1, respectively. Notice that the pole

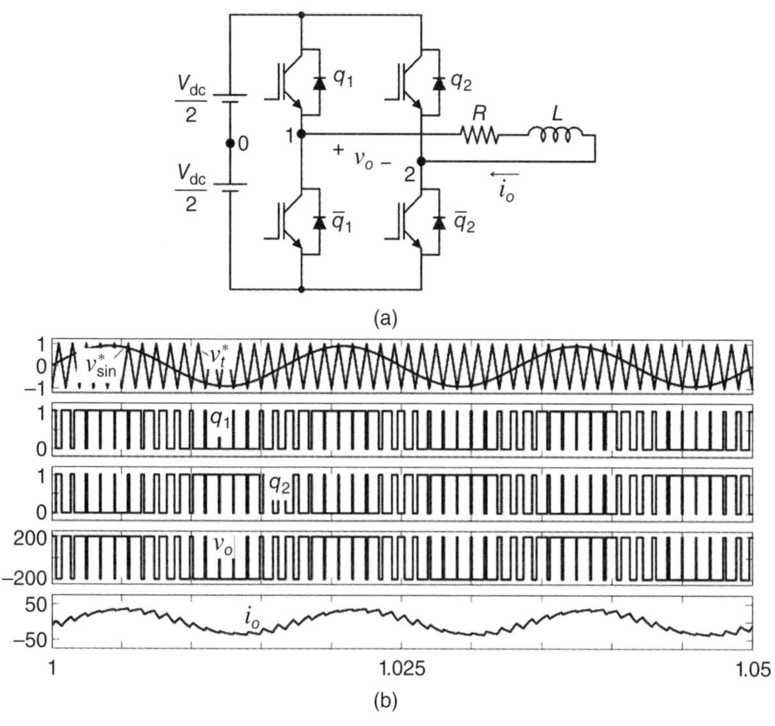

Figure 5.4 (a) Two-level H-bridge converter. (b) (from top to bottom) v^*_{sin}, v^*_t; gating signal of switch q_1; gating signal of switch q_2; load voltage and load current.

voltages for Method 2 have two main frequency components, that is, one related to the sinusoidal and another related to the triangular waveform. Since $v_o = v_{10} - v_{20}$ and there is just one triangular waveform, such a frequency component is cancelled for v_o. Although not shown in this figure, there are higher frequency components. On the other hand, there are two sinusoidal waveforms with phase shift equal to 180° for Method 2, which explains why the amplitude of the sine frequency component for v_o is twice that for v_{10} and v_{20}. Such a characteristic also explains the higher switching frequency observed for v_o in the time domain. For Method 1 [Fig. 5.6(b)], each pole voltage has dc components that are cancelled for v_o, but the triangular frequency components are not cancelled. Since this waveform was obtained by employing different carrier signals, it is clear why the switching frequency components for v_o in Method 1 are equal to the switching frequency component observed in the pole voltages.

Exercise 5.1

Determine all topological states (equivalent circuits) for the switching states in both strategies presented in Fig. 5.5. Assume both positive and negative currents.

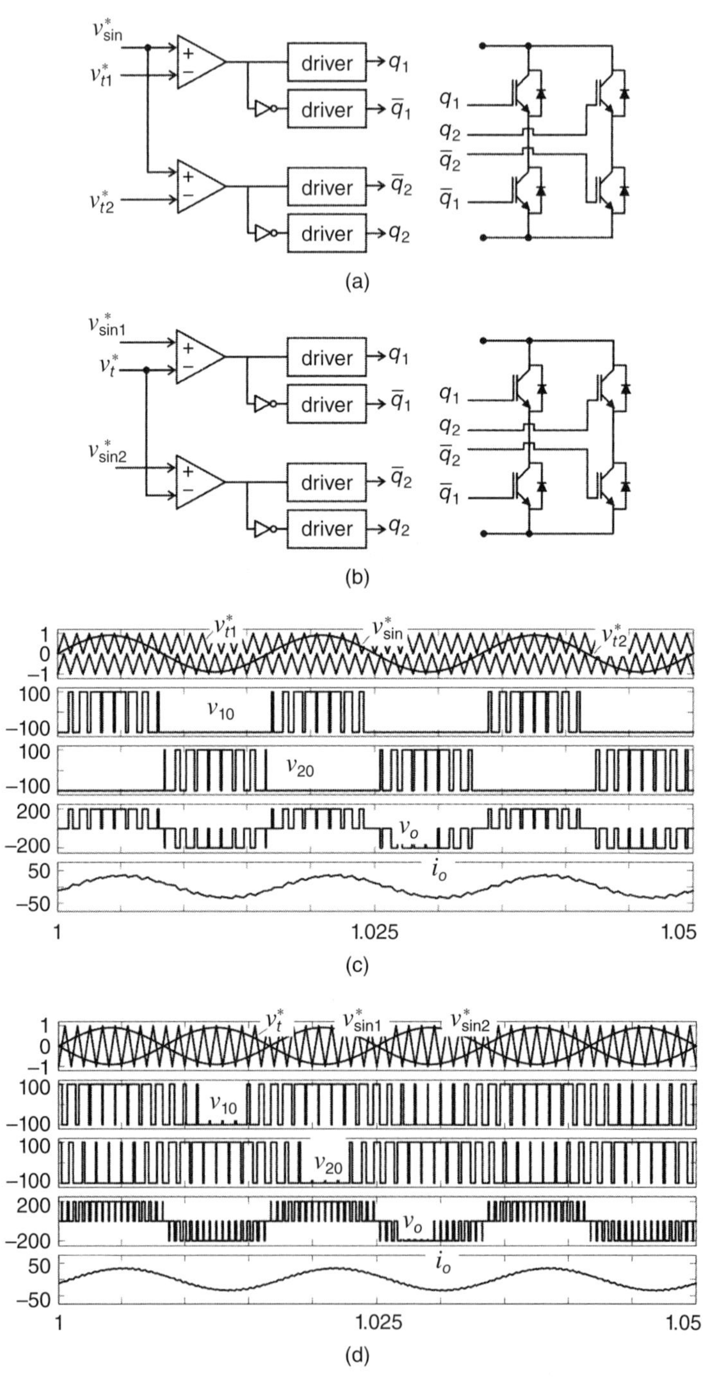

Figure 5.5 PWM strategy: (a) Method 1 and (b) Method 2. (c) Waveforms for method 1: (from top to bottom) v^*_{sin}, v^*_{t1}, and v^*_{t2}; pole voltage v_{10}; pole voltage v_{20}, load voltage v_o; and load current i_o. (d) Waveforms for method 2: (from top to bottom) $v^*_{sin\,1}$, $v^*_{sin\,2}$, and v^*_t; pole voltage v_{10}; pole voltage v_{20}, load voltage v_o; and load current i_o.

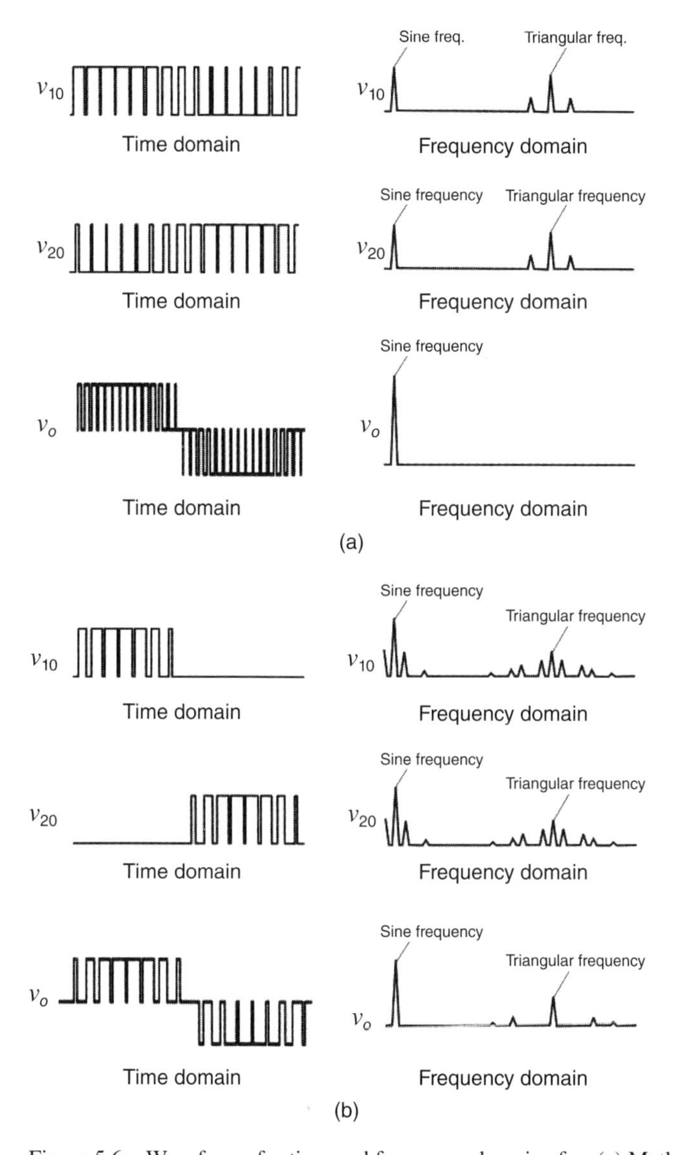

Figure 5.6 Waveforms for time and frequency domains for: (a) Method 2 and (b) Method 1.

In general terms, for a PWM strategy obtained from a comparison between sinusoidal and triangular waveforms as presented in Figs 5.3 and 5.5(b), the peak amplitude of the fundamental component for the pole voltage $V_{10(1)}$ [see equation (2.10)] is given by

$$V_{10(1)} = \frac{V_{\sin}^*}{V_t}\frac{V_{dc}}{2} = m_a\frac{V_{dc}}{2} \tag{5.6}$$

It turns out that the fundamental component can be written by $v_{10(1)}(t) = m_a(V_{dc}/2)\sin(\omega t)$ with $m_a \leq 1$. The harmonic components can be calculated using the Fourier series as follows:

$$v_{10}(t) = V_{10(0)} + \sum_{h=1}^{\infty} v_{10(h)}(t) \tag{5.7}$$

with $\sum_{h=1}^{\infty} v_{10(h)}(t) = \sum_{h=1}^{\infty} \{a_h \cos(h\omega t) + b_h \sin(h\omega t)\}$. Note that a_h are the cosine components in the Fourier series, while b_h are the sine components, see equations (2.1)–(2.3).

Indeed the harmonic for the pole voltage comes out as side-bands centered on the switching frequency and its multiples. The figure below shows the amplitude of the normalized harmonic component $V_{10(h)}/(V_{dc}/2)$ versus the harmonics, obtained from equation (5.6) for three values of modulation indexes: $m_a = 1,\ 0.6$, and 0.4, and with $m_f > 21$ for all cases.

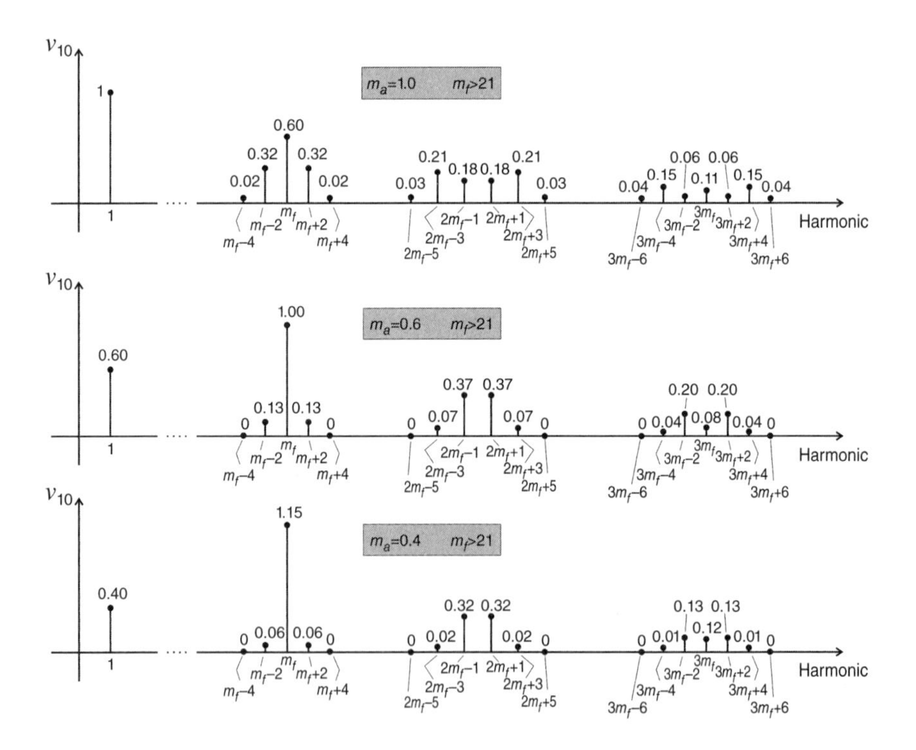

It can be seen from this figure that the pole voltage has an interesting characteristic, that is, there is only odd harmonics. The even harmonics disappear if m_f is an odd integer bigger than 9. This is due to odd symmetry with $v_{10}(-t) = -v_{10}(t)$. Also, the a_h (cosine elements in the Fourier series) are equal to zero, while the terms b_h (sine elements) can be different from zero.

The pole voltage is plotted below to demonstrate the effect of an even value for m_f as compared to an odd one. Note that when $m_f = 8$, an asymmetric waveform is obtained at the pole voltage, since the comparison of the peaks and valleys of the sinusoidal voltage along with the triangular one is different. On the other hand, when $m_f = 9$ (an odd number for m_f) a symmetric waveform is obtained since the comparison is symmetric for all periods, as highlighted in the peaks and valleys.

- $m_a = 0.9$ and $m_f = 8$

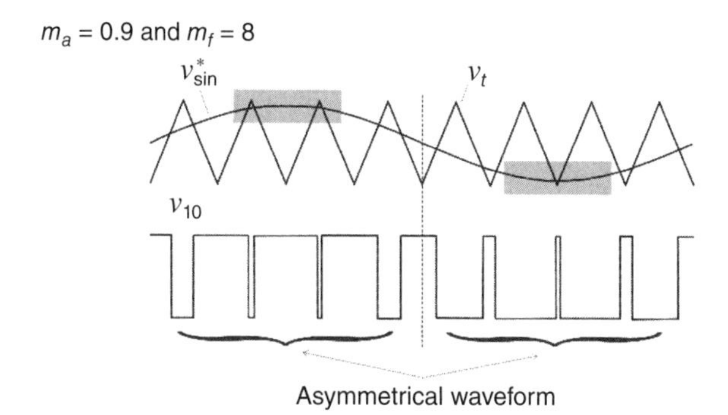

- $m_a = 0.9$ and $m_f = 9$

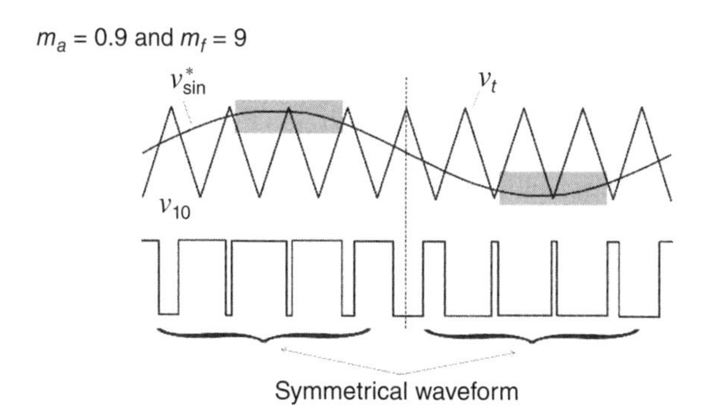

Other harmonic component elimination can be obtained when both sinusoidal (v^*_{sin}) and triangular (v_t) waveforms are synchronized for $m_f < 21$. If $m_f > 21$, no synchronization is needed due to low gain.

Application (Photovoltaic Systems)

A typical photovoltaic (PV) power system is composed of a set of PV arrays with one central inverter, which can be implemented transformerless or with an isolated transformer as shown below. The last solution is sometimes avoided due to size, weight, and price of the low frequency transformer.

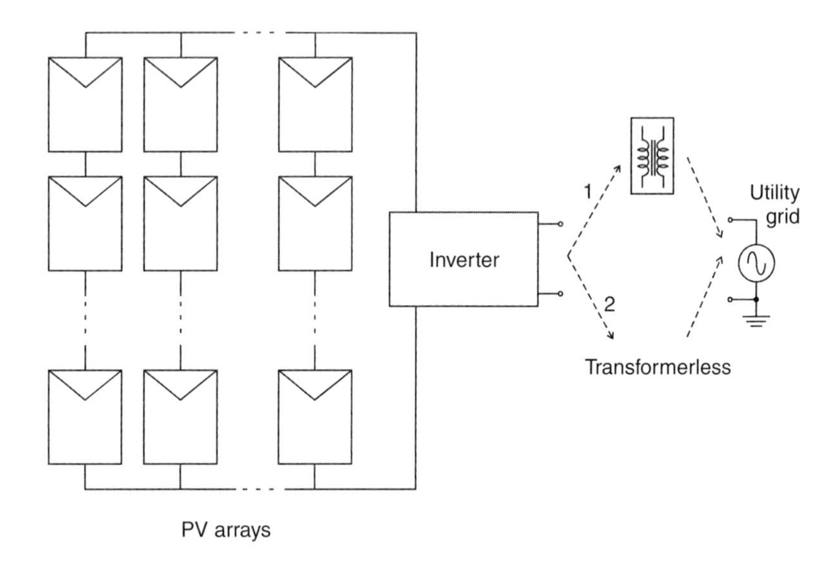

Due to high frequency on the inverter, a common mode voltage (v_{CM}) is generated with respect to the ground, which will induce a circulating current between the PV panels and the grid through the inverter. Notice that such a current does not exist when a transformer is used.

Due to safety requirements the PV module frame is connected to the ground, as represented in figure below (left side). Since the panels have a planar structure and module glass, it results in capacitance to the ground. In fact, the individual PV cells have distributed stray capacitance to the ground, as depicted in the middle of the figure below. The distributed capacitance throughout the module can be characterized by an equivalent lumped capacitance highlighted in the right figure below.

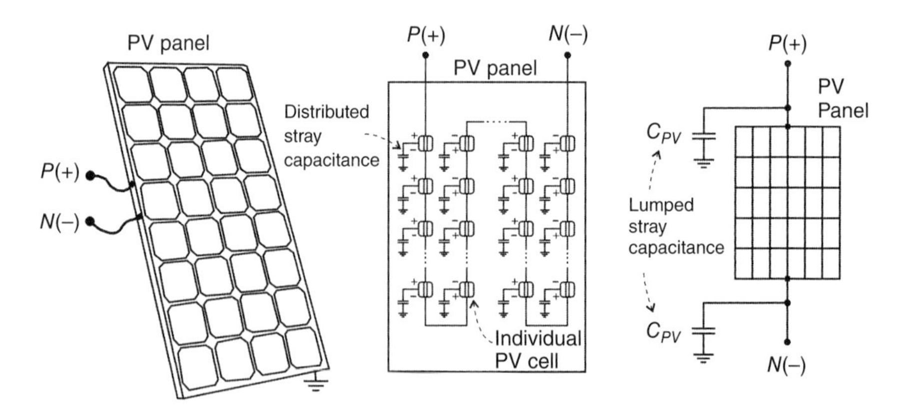

Assuming this equivalent circuit for the PV along with the stray capacitances, it is possible to obtain the circuit below (top). Without an isolated transformer there is a galvanic connection between the grid and PV arrays represented in this figure by the circulating current i_{Leakage}. The bottom-left-side figure below shows a simplification of this circuit with a block representing the panels and the inverter. In this case,

for simplification purposes, just one C_{PV} *is considered between the point* N and the ground. *If an H-bridge topology (bottom-right-side figure below) is chosen to implement the grid-tie inverter, it is possible to define two voltages as function of* $v_{1N} = q_1 V_{PN}$ *and* $v_{2N} = q_2 V_{PN}$ *as follows:*

- *Differential voltage:* $v_{DM} = v_{1N} - v_{2N}$;
- *Common mode voltage:* $v_{CM} = (v_{1N} + v_{2N})/2$.

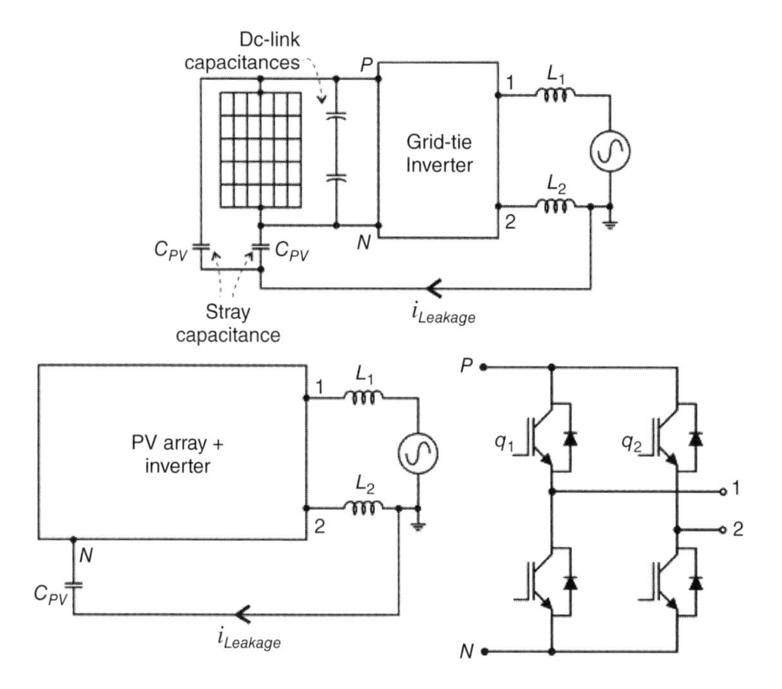

Using both voltages v_{DM} *and* v_{CM} *it is possible to model the block describing PV array + Inverter above with the circuit presented below.*

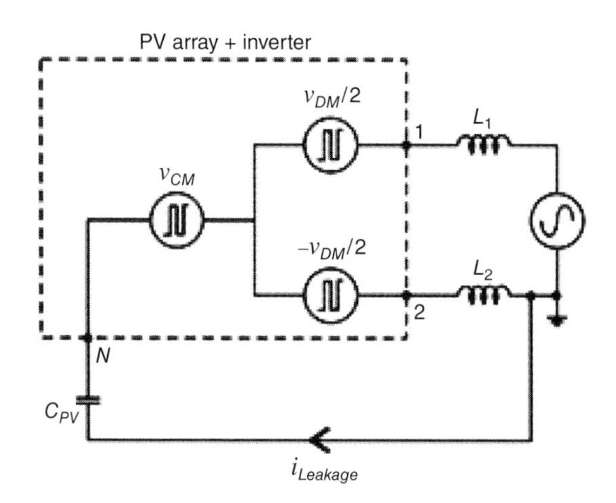

As observed in the figures above, the leakage current varies particularly with v_{CM}, or by using other words, $i_{Leakage}$ is a function of the common mode voltage given by $v_{CM} = (q_1 + q_2)V_{PN}/2$. From the Table presented in the sequence, it is quite clear that if the modulation strategy selects only states 2 and 3, v_{CM} is always equal to $V_{PN}/2$, which means constant voltage applied to the capacitor C_{PV} and consequently null circulating current (i.e., $i_{Leakage} = 0$). *The direct consequence of cancelling i_{CM} is that the number of levels at the output converter side will be two, instead of three. See Table* 5.1 *as well as Fig.* 5.3.

State	q_1	q_2	v_{CM}
1	0	0	0
2	0	1	$V_{PN}/2$
3	1	0	$V_{PN}/2$
4	1	1	V_{PN}

Example 5.2

For the H-bridge topology shown in Fig. 5.4(a) with $V_{dc} = 600\,\text{V}$, $m_a = 0.6$, switching frequency: 1950 Hz and $v^*_{sin} = 180\sin(2\pi 50t)$, calculate the root mean square (RMS) values of the fundamental and most important harmonics (with at least 10% of the amplitude of the biggest harmonic) for the pole voltage v_{10} assuming Method 2 in Fig. 5.5.

Solution

The amplitude of the fundamental and harmonics can be obtained from the figure below that shows the relationship between $V_{10(h)}/(V_{dc}/2)$ and the harmonics.

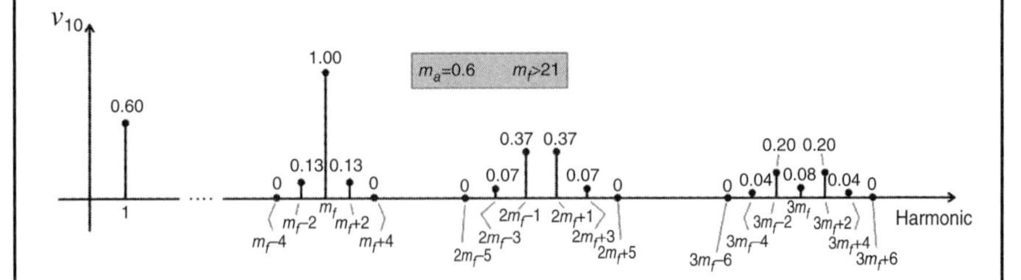

So, the amplitude of the harmonic $hV_{10(h)}$ is given by

$$V_{10(h)} = \left(\frac{V_{10(h)}}{V_{dc}/2}\right)\frac{V_{dc}}{2}$$

The RMS value can be obtained directly as follows:

$$V_{10(h)\text{RMS}} = \frac{1}{\sqrt{2}}\left(\frac{V_{10(h)}}{V_{dc}/2}\right)\frac{V_{dc}}{2}$$

Since $m_a = 0.6$ and $m_f = f_t/f_{\sin} = 1950/50 = 39$, it is possible to write:

$$h = 1 \rightarrow V_{10(1)\text{RMS}} = \frac{1}{\sqrt{2}} \cdot 0.6 \cdot \frac{600}{2} = 127\,\text{V}$$

$$h = 39 \rightarrow V_{10(39)\text{RMS}} = \frac{1}{\sqrt{2}} \cdot 1 \cdot \frac{600}{2} = 212\,\text{V}$$

$$h = 37, 41 \rightarrow V_{10(37,41)\text{RMS}} = \frac{1}{\sqrt{2}} \cdot 0.13 \cdot \frac{600}{2} = 27\,\text{V}$$

$$h = 77, 79 \rightarrow V_{10(77,79)\text{RMS}} = \frac{1}{\sqrt{2}} \cdot 0.37 \cdot \frac{600}{2} = 78\,\text{V}$$

$$h = 115, 119 \rightarrow V_{10(115,119)\text{RMS}} = \frac{1}{\sqrt{2}} \cdot 0.2 \cdot \frac{600}{2} = 42\,\text{V}$$

All the other harmonics will not be considered since their amplitude is less than 10%.

Example 5.3

Determine the RMS voltage value for the fundamental and for the harmonic number 39 of the output voltage assuming the conditions of the last example; assume both PWM strategies shown in Figs 5.4 and 5.5(b).

Solution

Since the output voltage of the PWM approach shown in Fig. 5.4 has the same shape of the pole voltage [see Fig. 5.4(b)] it is possible to keep using the template furnished in the figure $\frac{V_{10(h)}}{V_{\text{dc}}/2}$ versus harmonics. However, the RMS values are given by

$$V_{o(h)} = \left(\frac{V_{10(h)}}{V_{\text{dc}}/2}\right) V_{\text{dc}}$$

$$h = 1 \rightarrow V_{o(1)\text{RMS}} = \frac{1}{\sqrt{2}} \cdot 0.6 \cdot 600 = 254\,\text{V}$$

$$h = 39 \rightarrow V_{o(39)\text{RMS}} = \frac{1}{\sqrt{2}} \cdot 1 \cdot 600 = 424\,\text{V}$$

On the other hand, for the PWM technique given in Fig. 5.5(b) the output voltage presents a different shape from the pole voltage, which means that it is possible to use the template in figure $V_{10(h)}/(V_{\text{dc}}/2)$ versus harmonics. Although v_o has three levels instead of two, the fundamental component ($h = 1$) should be the same as for the PWM in Fig. 5.3 since in both cases the converter is operating under the same conditions in terms of V_{dc}, m_a, and m_f. Then: $h = 1 \rightarrow V_{o(1)\text{RMS}} = 254\,\text{V}$.

For the harmonic 39, it turns out that $V_{o(39)\text{RMS}} = 0$, [see Fig. 5.6(a)].

Exercise 5.2

Repeat Examples 5.1 and 5. with $m_a = 1.0$ instead of 0.6. The other operation conditions are the same as in the previous examples.

5.4 THREE-PHASE CONVERTER—ONE H-BRIDGE CONVERTER PER PHASE

A three-phase configuration by using one H-bridge converter per phase is depicted in Fig. 5.7(a). It is worth mentioning that both converter and three-phase load are connected in a Y arrangement. The converter pole voltages can be written as a function of the states of switches, as follows:

$$v_{a0} = (q_{1a} - q_{2a})V_{\text{dc}} \tag{5.8}$$

$$v_{b0} = (q_{1b} - q_{2b})V_{\text{dc}} \tag{5.9}$$

$$v_{c0} = (q_{1c} - q_{2c})V_{\text{dc}} \tag{5.10}$$

Considering a balanced three-phase system, it turns out that the voltage between the neutral of the load and the point "0" is given by

$$v_{n0} = \frac{V_{\text{dc}}}{3}(q_{1a} - q_{2a} + q_{1b} - q_{2b} + q_{1c} - q_{2c}) \tag{5.11}$$

Compiling equations (5.8)–(5.11), it is possible to write the load voltages v_{an}, v_{bn}, and v_{cn} in a matrix representation, as shown below:

$$\begin{bmatrix} v_{an} \\ v_{bn} \\ v_{cn} \end{bmatrix} = V_{\text{dc}} \begin{bmatrix} \frac{2}{3} & \frac{-2}{3} & \frac{-1}{3} & \frac{1}{3} & \frac{-1}{3} & \frac{1}{3} \\ \frac{-1}{3} & \frac{1}{3} & \frac{2}{3} & \frac{-2}{3} & \frac{-1}{3} & \frac{1}{3} \\ \frac{-1}{3} & \frac{1}{3} & \frac{-1}{3} & \frac{1}{3} & \frac{2}{3} & \frac{-2}{3} \end{bmatrix} \begin{bmatrix} q_{1a} \\ q_{2a} \\ q_{1b} \\ q_{2b} \\ q_{1c} \\ q_{2c} \end{bmatrix} \tag{5.12}$$

The power converter constituted by the arrangement of H-bridges topologies, as in Fig. 5.7(a) allows different connections for the H-bridge converters. Such a characteristic is not observed for the converters shown in the previous chapter. In fact, for a given three-level NPC topology, different arrangements can be addressed for the load side with either Y or Δ connections of the three-phase load, but the connections of the converter itself cannot be changed, since all legs are connected to the same dc-link source. On the other hand, due to its modular feature, H-bridge converters can be changed in order to deal with specific requirements. In this sense, Fig. 5.7(b)

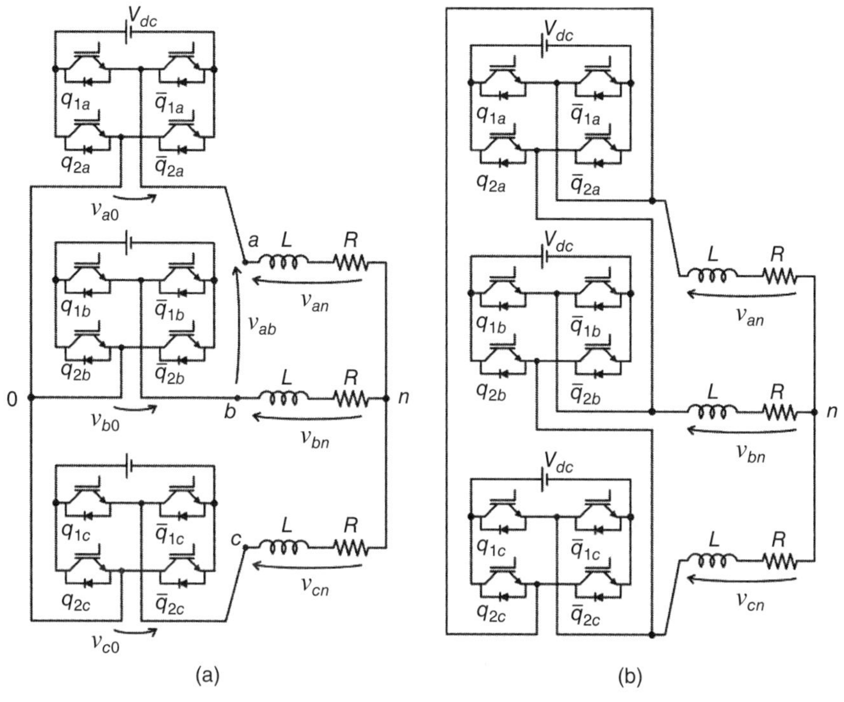

Figure 5.7 (a) Three-phase H-bridge configuration connected in a Y arrangement. (b) Three-phase H-bridge configuration connected in a Δ arrangement.

shows a solution in which the converters are arranged in a Δ connection. There is also a possibility to connect a three-phase load in Δ for the converters depicted in Fig. 5.7(a) and 5.7(b).

Exercise 5.3

Fill a table to show the number of levels of the load phase voltage v_{an} for both topologies given in Fig. 5.7 as a function of the switching states.

To highlight the influence of the PWM strategies presented in Fig. 5.5 (Methods 1 and 2) for the three-phase configuration as in Figs 5.7(a), 5.8(a), 5.8(b) show the waveforms related to Methods 1 and 2, respectively. In the three-phase case, the waveforms for the line-to-line voltage obtained through Method 1 have an interesting characteristic with the same dv/dt, while Method 2 presents a line-to-line voltage with the levels not well defined as in Fig. 5.8(a). As a consequence v_{an} in Fig. 5.8(a) is closer to a sine waveform than in Fig. 5.8(b). Note that Fig. 5.8(c) highlights the advantages of Method 1 with a voltage presenting the same dv/dt.

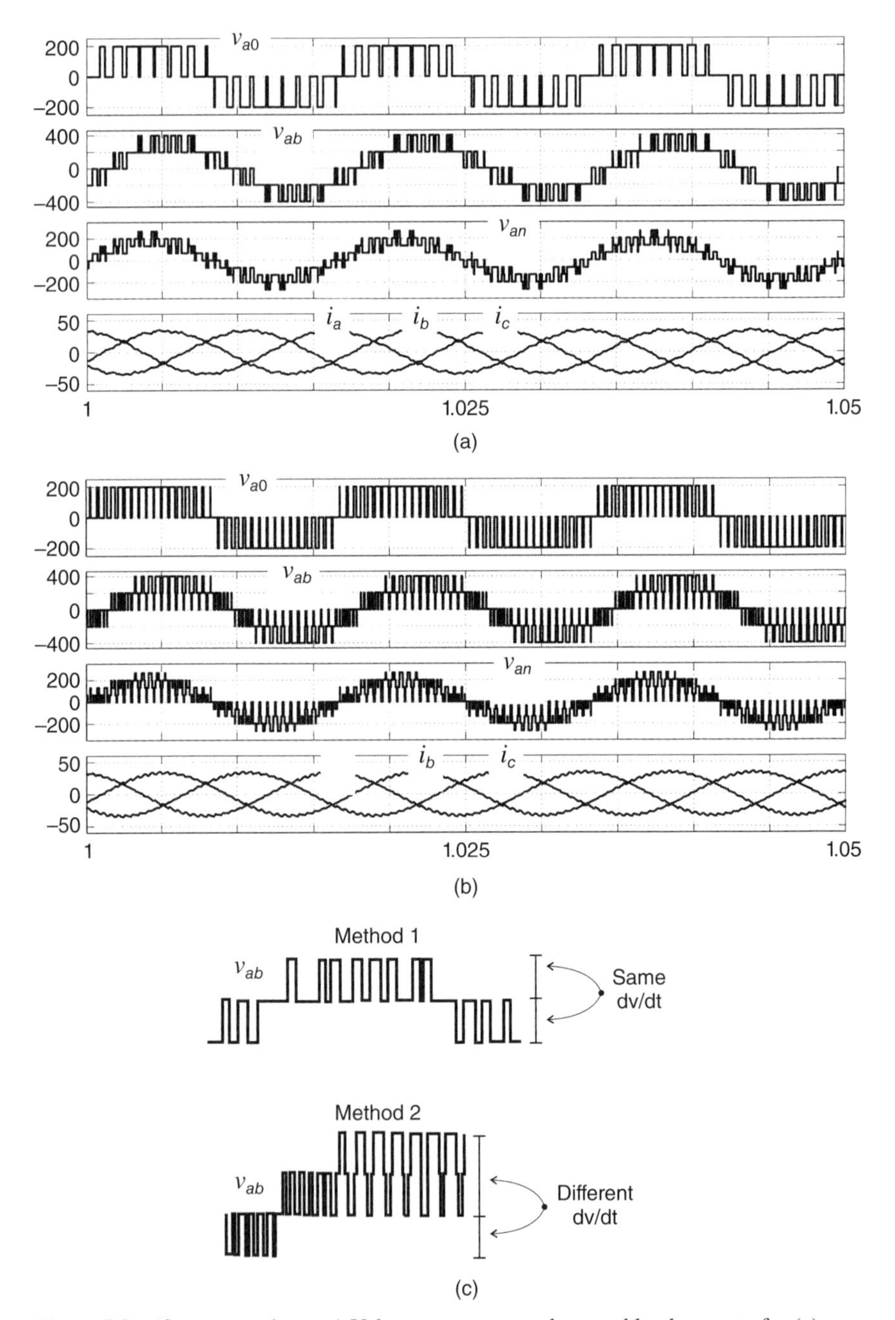

Figure 5.8 (from top to bottom) Voltages: v_{a0}, v_{ab}, and v_{an} and load currents for (a) Method 1, (b) Method 2, and (c) levels of line-to-line voltage.

As described in the previous chapter, there is a possibility to supply a three-phase load by using a converter able to supply only two phases, since the third phase will be indirectly controlled, due to the redundancy inherent in the balanced three-phase systems ($i_a + i_b + i_c = 0$ and $v_a + v_b + v_c = 0$).

Exercise 5.4

Assuming all possibilities in terms of converter and three-phase load connections as given in the table below, specify the switches of the converters in terms of voltage and current requirement, assuming the same *RL* load and dc-link voltage for all connections below. Also, disregard any safety margin while specifying the components.

Converter Connection	Load Connection	Figure
Y	Y	5.7(a)
Δ	Y	5.7(b)
Y	Δ	–
Δ	Δ	–

Figure 5.9(a) shows the configuration with only two H-bridges connected to phases *a* and *b*, while Fig. 5.9(b) shows the waveforms related to this configuration. This kind of topology seems to have low practical interest, but in a fault-tolerant system it could be interesting due to its ability to maintain the three-phase voltage applied to the load on the post-fault operation. In this scenario, the "healthy" configuration is changed to the post-fault one after fault identification in the third phase of the converter.

The PWM control of the configuration presented in Fig. 5.9(a) needs to be adapted. The amplitude of voltage applied to the load is reduced by $\sqrt{3}$ and its distortion harmonic increased. However, the converter is still able to supply a three-phase load with balanced currents as in Fig. 5.9(b) with i_a, i_b, and i_c having the same amplitude and a phase shift equal to 120°.

Exercise 5.5

Does the connection shown in Fig. 5.9 affect the number of levels of the load phase voltage? To answer this question, write a table to show the load voltages as a function of the switching states and compare with the results obtained in Exercise 5.3.

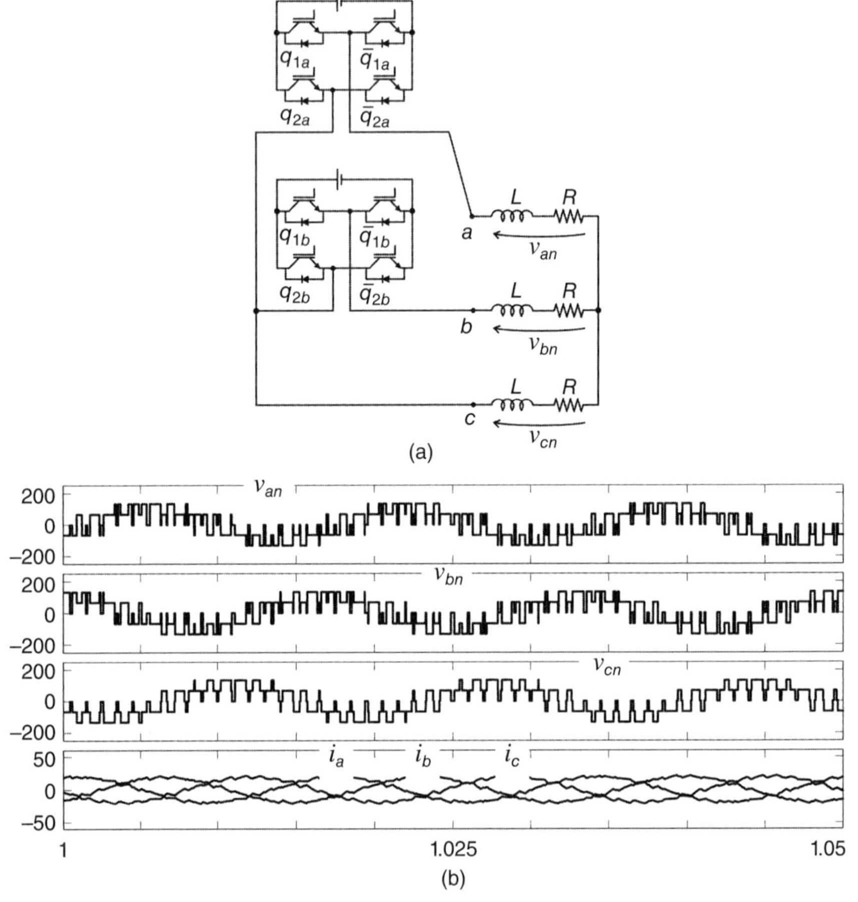

Figure 5.9 (a) Configuration supplying only two phases of the three-phase load. (b) (from top to bottom) v_{an}, v_{bn}, v_{cn}, and load currents.

5.5 TWO H-BRIDGE CONVERTERS

Connecting H-bridge converters in series, as in Fig. 5.10(a), allows the generation of voltage with a higher number of levels. Such a cascade converter is in fact a very popular topology in high voltage motor drive applications, especially due to its modular characteristic.

As discussed previously, each H-bridge converter can generate a maximum of three levels at its output, which means a maximum of nine levels at the output of two connected H-bridges series. All levels are considered in Table 5.3 for each switching state. The levels are: $\{0, \pm V_a, \pm V_b, -V_a - V_b, -V_a + V_b, V_a - V_b, V_a + V_b\}$. Figure 5.10(b)–5.10(j) show the combination used to obtain each specific level at the output converter side.

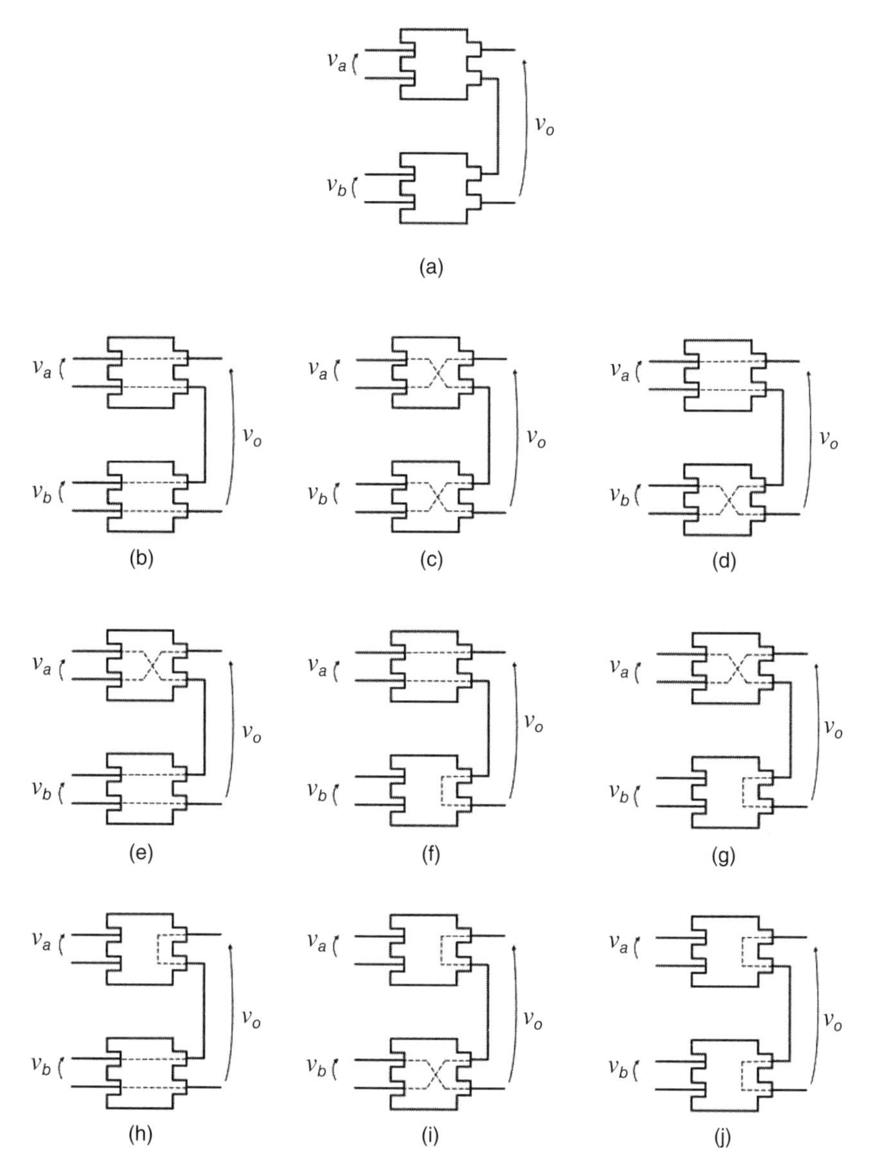

Figure 5.10 (a) Two H-bridges configuration with PBs. (b) $v_o = V_a + V_b$. (c) $v_o = -V_a - V_b$. (d) $v_o = V_a - V_b$. (e) $v_o = -V_a + V_b$. (f) $v_o = V_a$. (g) $v_o = -V_a$. (h) $v_o = V_b$. (i) $v_o = -V_b$. (j) $v_o = 0$.

From the expression for the output voltage of one H-bridge converter given in (5.5), the output voltage v_o of the converter presented in Fig. 5.10(a) can be written as a function of the binary variables presented in Table 5.3, as follows:

$$v_o = (q_{1a} - q_{2a})V_a + (q_{1b} - q_{2b})V_b \tag{5.13}$$

TABLE 5.3 Table of Variables for the Converter in Fig. 5.10

State	q_{1a}	q_{2a}	q_{1b}	q_{2b}	v_o
1	0	0	0	0	0
2	0	0	0	1	$-V_b$
3	0	0	1	0	V_b
4	0	0	1	1	0
5	0	1	0	0	$-V_a$
6	0	1	0	1	$-V_a - V_b$
7	0	1	1	0	$-V_a + V_b$
8	0	1	1	1	$-V_a$
9	1	0	0	0	V_a
10	1	0	0	1	$V_a - V_b$
11	1	0	1	0	$V_a + V_b$
12	1	0	1	1	V_a
13	1	1	0	0	0
14	1	1	0	1	$-V_b$
15	1	1	1	0	V_b
16	1	1	1	1	0

Although it is possible to generate nine levels at the output of the converter in Fig. 5.10, this section will start with the analysis to obtain five levels followed by seven and nine.

Example 5.4

What should be the relationship between V_a and V_b to reach the maximum nine levels (the maximum number of levels) at the output converter side?

Solution

To generate the maximum number of levels at the output converter side it is necessary to guarantee either $V_1 = 3V_2$ or $V_2 = 3V_1$. Other numbers of levels can be obtained with other relationships of the input voltages. For instance, seven levels is obtained with $V_1 = 2V_2$ or $V_2 = 2V_1$, while five levels is obtained with $V_1 = V_2$.

5.6 PWM IMPLEMENTATION OF TWO CASCADE H-BRIDGES

PWM control strategies are also employed for a converter with two cascade H-bridge topology to guarantee controlled voltage and frequency at its output. Since there are more switches to generate the desired output voltage as compared to the single H-bridge topology, it is also expected to have ways to implement the PWM strategies. This section explores some of these PWM approaches.

Figure 5.11 shows the analog solution with phase shift multicarrier modulation technique, in which the states of the eight switches are determined by a comparison between two high frequency triangular waveforms (v^*_{t1} and v^*_{t2}, carrier signals) and two sinusoidal waveforms (v^*_{sin} and $-v^*_{\text{sin}}$, control signals) for the case where $V_a = V_b = V_{\text{dc}}$. The voltages v^*_{t1} and v^*_{t2} are shifted by 90°. In this case the converter is able to generate a five-level waveform at its output. As mentioned previously, the

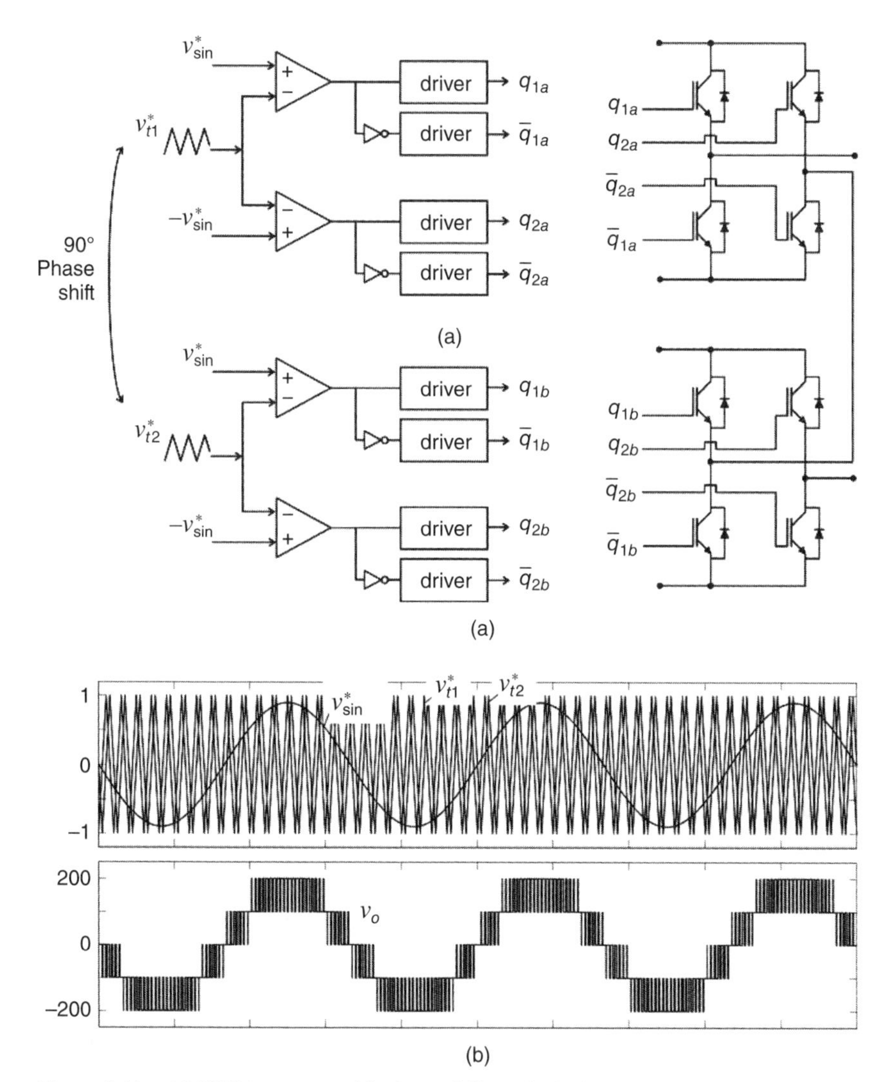

Figure 5.11 (a) PWM strategy with phase-shift method. (b) (top) PWM waveforms and (bottom) output converter voltage.

block named *driver* must guarantee the levels of voltage and current needed to operate (turning on and off) the power switches adequately, as well as isolate the control (signal) and the power circuits.

It is evident that the phase shift between v^*_{t1} and v^*_{t2} plays an important role in the PWM approach as in Fig. 5.11. In fact, the influence of this parameter is depicted in Fig. 5.12, which shows triangular (v^*_{t1} and v^*_{t2}) and sinusoidal (v^*_{sin}) voltages (top); load voltage (middle); and load current (bottom). The parameters employed in these simulation results were: input voltage – $V_a = 100\,\text{V}$ and $V_b = 100\,\text{V}$; load – $R = 5\,\Omega$ and $L = 5\,\text{mH}$; and switching frequency – 1 kHz. Figure 5.12(a) shows those variables with phase shift between v^*_{t1} and v^*_{t2} equal to 0°. Other phase shifts are considered, that is, 45°, 90°, and 180°, as in Fig. 5.12(b), 5.12(c) and 5.12(d), respectively. Considering the waveforms of the load voltage and load current, it is evident that 90° reduces the harmonic distortion for both voltage and current. Although changing the high frequency components and consequently the THD of the variables, the phase shift does not influence the value of voltage applied to the load.

In fact, the phase angle displacement of 90° between v^*_{t1} and v^*_{t2} guarantees an optimum output voltage because it generates a sequence of switching states with the

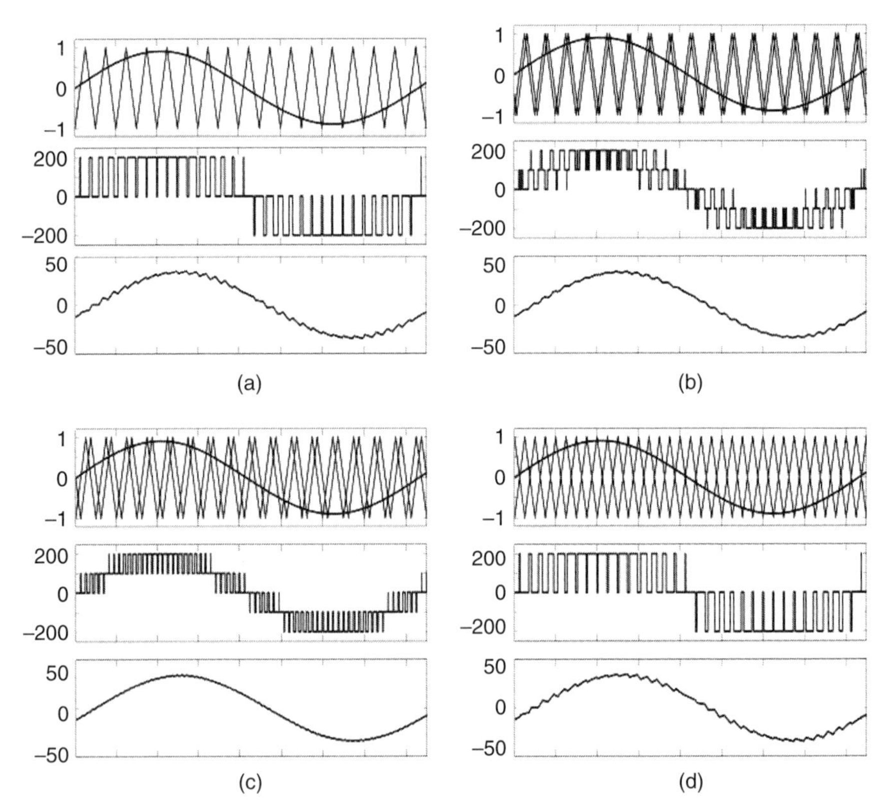

Figure 5.12 Triangular (v^*_{t1}, v^*_{t2}) and sinusoidal (v^*_{sin}) waveforms (top); load voltage (middle); and load current (bottom): with phase shift of (a) 0°, (b) 45°, (c) 90°, and (d) 180°.

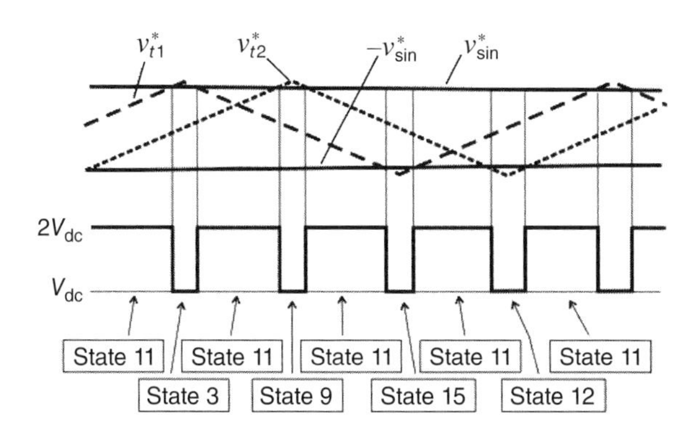

Figure 5.13 Switching states when v_{sin}^* is in its peak with 90° of the phase displacement between v_{t1}^* and v_{t2}^*.

levels well defined as in Fig. 5.8. For instance, during the peaks it is expected that the output voltages change from V_{dc} to $2V_{dc}$, while during the valleys the voltages should change from $-2V_{dc}$ to $-V_{dc}$. Figure 5.13 shows the sequence of the switching states when v_{sin}^* is around its peak. Since $V_a = V_b = V_{dc}$, States 3, 9, 15, and 12 in Table 5.3 generate the same output voltage equal to V_{dc}. On the other hand, there is only one way to generate $2V_{dc}$, which is through State 11. Note that any other phase shift will generate a sequence different from the one in Fig. 5.13.

Another way to implement the PWM strategy for the converter employing two H-bridge cells in series is presented in Fig. 5.14. Such a technique is called level-shift (or level dispositions), where all carrier signals are in phase but with different offsets.

For a symmetrical (same dc-link voltages) cascade multilevel converter, the quantity of the carrier signals is defined by the number of levels desired at the output converter side, that is, for N levels $(N-1)$ triangular carriers are required.

Figure 5.15 shows a comparison between two-level shift PWM approaches. Note that other outcomes are obtained when all triangular carriers have the same peak-to-peak voltage, but there is a phase shift of 180° between any two adjacent carrier waves.

5.7 THREE-PHASE CONVERTER – TWO CASCADE H-BRIDGES PER PHASE

A three-phase configuration employing two H-bridge converters per phase is depicted in Fig. 5.16(a), with both converter and load connected in a Y arrangement. The waveforms associated with this configuration are depicted in Fig. 5.16(b). These load voltages present high quality with reduced THD due to the high number of levels. The PWM strategy employed in the results of Fig. 5.16 were obtained with the level shift approach, as in Fig. 5.14.

The converter voltages can be written as a function of the states of switches, as follows:

$$v_{a0} = (q_{1a} - q_{2a} + q_{3a} - q_{4a})V_{dc} \tag{5.14}$$

$$v_{b0} = (q_{1b} - q_{2b} + q_{3b} - q_{4b})V_{dc} \tag{5.15}$$

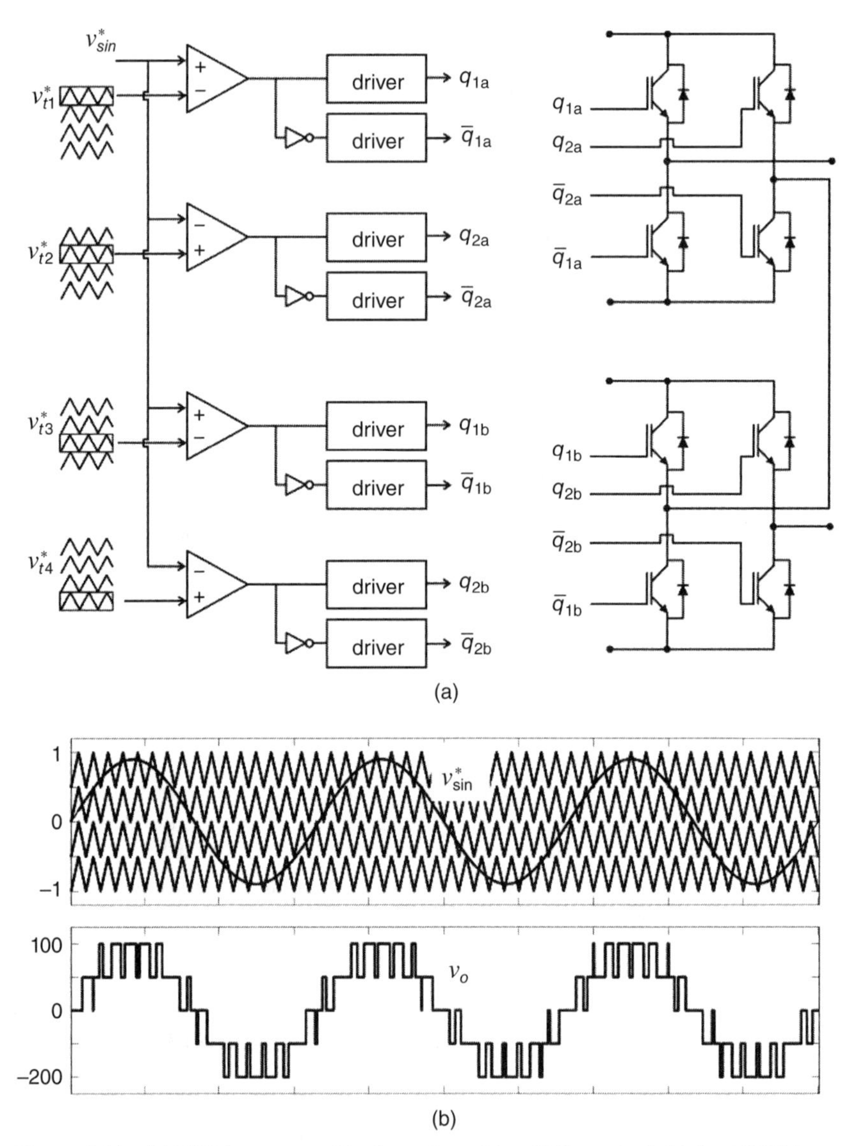

Figure 5.14 (a) PWM strategy with phase disposition (PD) method. (b) (top) PWM waveforms and (bottom) output converter voltage.

$$v_{c0} = (q_{1c} - q_{2c} + q_{3c} - q_{4c})V_{dc} \tag{5.16}$$

Considering a balanced three-phase system, the voltage between the neutral of the load "n" and the point "0" is given by

$$v_{n0} = \frac{V_{dc}}{3}\sum_{x=a}^{c}(q_{1x} - q_{2x} + q_{3x} - q_{4x}) \tag{5.17}$$

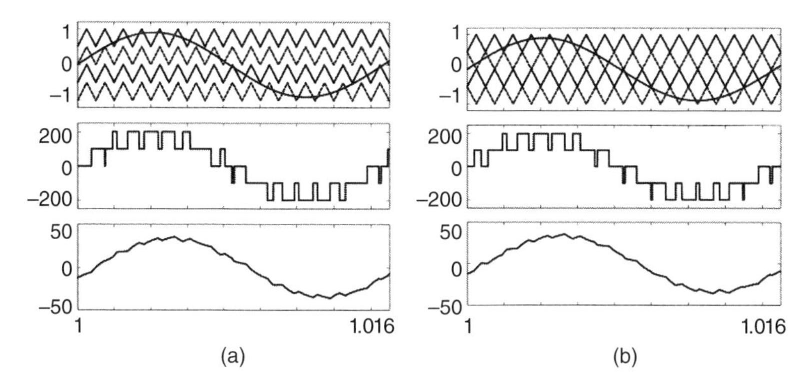

Figure 5.15 Comparison between two-level shift PWM techniques.

Compiling equations (5.14)–(5.17), and assuming a balanced scenario, it is possible to write the load voltages v_{an}, v_{bn}, and v_{cn} as a function of both switching states and dc-link voltages.

As mentioned previously, one of the main advantages of the cascade configurations is their modularity, which brings benefits in terms of fault tolerance capability. The fault compensation is achieved by reconfiguring the power converter topology to keep the system in operation, even with a different number of H-bridges per phase. Figure 5.17(a) shows one phase of the three-phase converter as in Fig. 5.16(a) under "healthy" operation and generating a five-level voltage at the output. For the sake of illustration, let us assume that a fault occurs in the switch q_{1a}, as highlighted in Fig. 5.17(b). Such a fault can be characterized as either a permanent short-circuit [see Fig. 5.17(c)] or a permanent open-circuit [see Fig. 5.17(d)]. The fault detection system must be able to identify where the fault has occurred (determining which switch is at fault), and also the type of fault (short- or open-circuit). Sometimes it is necessary to use fuses to deal with the reconfiguration of the power converter, which will help to isolate part of converter under fault. For simplification purposes, the fuse devices are not presented in Fig. 5.17.

The first action to be taken for a short-circuit failure [Fig. 5.17(c)] in the switch q_{1a} is to keep the complementary switch $\overline{q}_{1a}$ open. As a consequence, the voltage generated at the output of this H-bridge converter $v_{1a2a} = v_{1a} - v_{2a}$ can assume only two values 0 or V_{dc}, that is, $v_{1a2a} = \{0, V_{dc}\}$. Another scenario is with an open-circuit failure as shown in Fig. 5.17(d). In this case, the complementary switch $\overline{q}_{1a}$ must be on to avoid current discontinuity. Consequently, the voltage generated at the output of this H-bridge converter can also assume only two values, that is, $v_{1a2a} = \{0, -V_{dc}\}$. Both cases of failure lead to a voltage generation at the output of the H-bridge under fault with just one polarity: positive polarity $\{0, V_{dc}\}$ for short-circuit failure and negative polarity $\{0, -V_{dc}\}$ for open-circuit failure. This represents a restriction for the amount of voltage generated by the phase under fault. However, a sinusoidal voltage with smaller amplitude and only three levels $\{0, \pm V_{dc}\}$ can still be generated by these two H-bridge converters in series.

Although the faults presented in Fig. 5.17, there are some failures that damage the whole H-bridge converter. For example, it is quite common to have

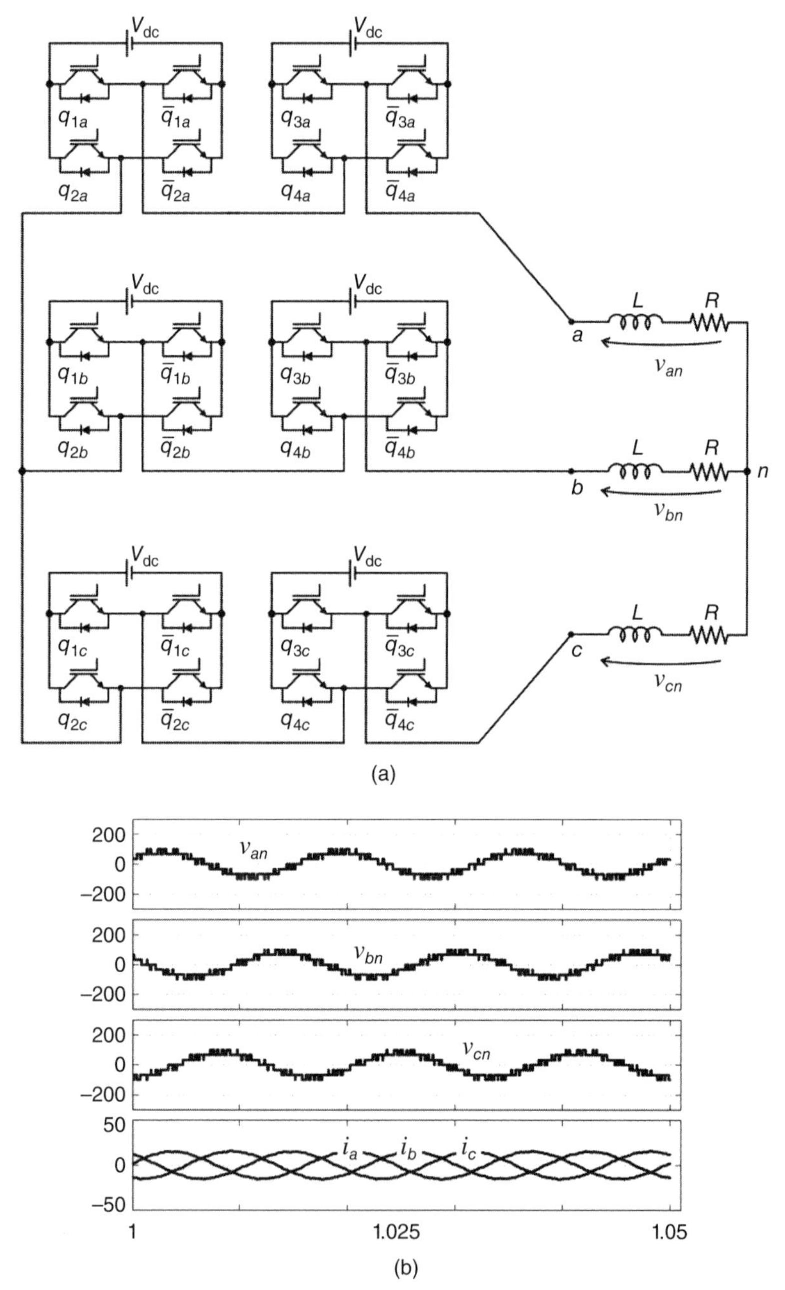

Figure 5.16 (a) Three-phase H-bridge configuration connected in a Y arrangement (b) Voltages and currents.

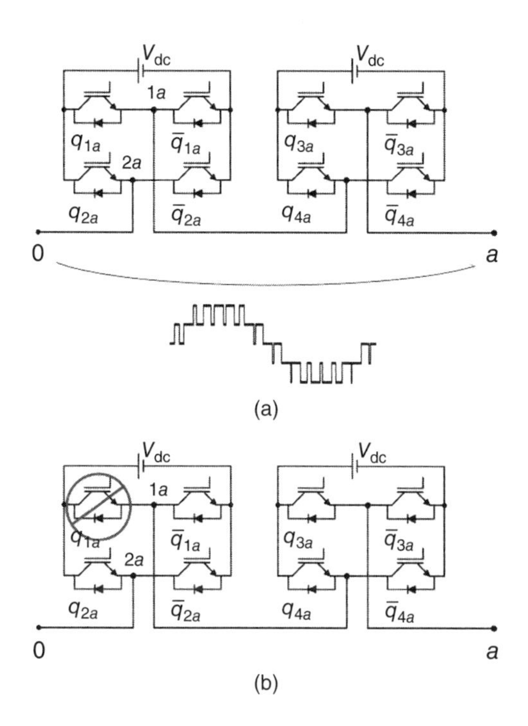

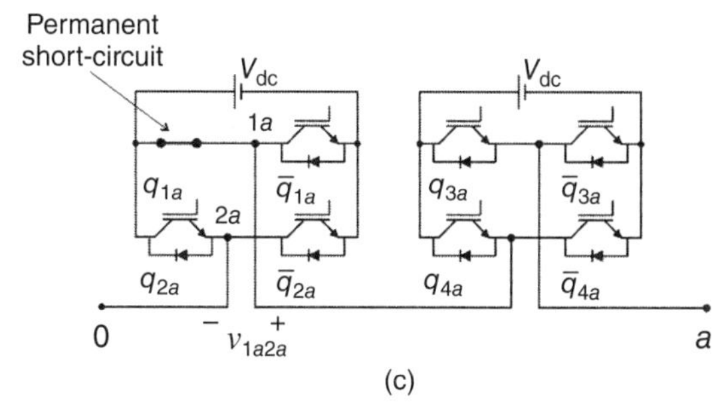

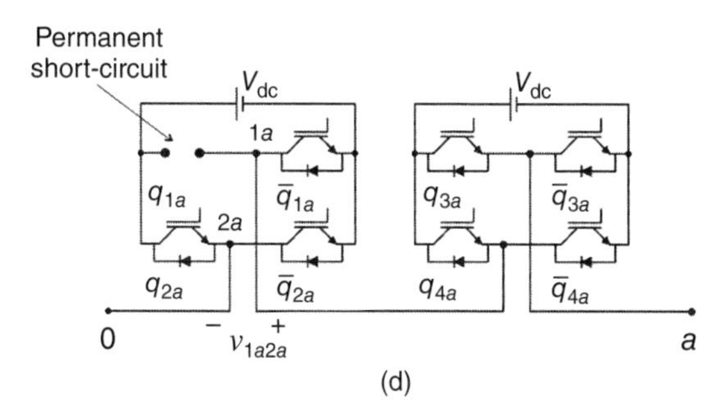

Figure 5.17 (a) "Healthy" operation for the leg of the cascade topology. (b) Fault occurring in the switch q_{1a}. (c) Short-circuit type of fault. (d) Open-circuit type of fault.

two switches inserted in the same package constituting a leg, and likely one switch will be affected if a malfunction has happened to another switch in the same leg. Another example of malfunction that deteriorates the whole H-bridge cell (in terms of capacity to generate any voltage) is when a fault occurs in the dc-link. There is no way to generate any desired voltage at the output of any H-bridge without dc-link voltage. The fault compensation when the entire H-bridge is suffering from a fault is achieved by reconfiguring the power converter topology along with additional devices, such as triacs, as shown in Fig. 5.18(a). The triacs t_1 and t_2 are open under "healthy" operation and must be turned on if a fault is detected, to isolate the whole H-bridge under fault. Figure 5.18(b) shows the post-fault topology with the H-bridge converter isolated by using the triac t_1.

The post-fault operation requires a new PWM approach to keep a set of balanced three-phase voltages applied to the load. There exist different ways to generate the PWM signals for the post-fault topology as in Fig. 5.18(b), since there are two converters per phase (in phases b and c) and just one converter in the phase a to synthesize the desire three-phase load voltage. One possibility to maintain the three-phase balanced voltage is to turn on the triacs t_3 and t_5 to make this topology equal to that presented in Fig. 5.7(a). This means that the two H-bridge per phase configuration will be changed to a single H-bridge per phase. Also the rated voltage is reduced to half of the pre-fault operation. Another possibility is to keep the post-fault topology as presented in Fig. 5.18(b), that is, phase a with one H-bridge and phases b and c with two H-bridges. To guarantee a balanced three-phase voltage among the phases, the modulation index for phases b and c should be reduced to 50% of the modulation index employed for phase a.

Exercise 5.6

Determine the number of levels of the post-fault topology in Fig. 5.18(b) considering the case in which the modulation index for phases b and c is reduced by 50%.

As done before, the connection of the converters can be changed in order to deal with specific requirements of the load. For instance, Fig. 5.19 shows a solution in which the converters are arranged in a Δ connection. There is also a possibility to connect a three-phase load in the Δ connection.

If the three-phase load is an electrical machine, it is also possible to connect this machine with an open-end winding arrangement as seen in Fig. 5.20. Open-end winding motor structure is obtained by opening the neutral point of the conventional induction or synchronous machine and does not require any design change for the machine itself. In fact, a three-phase machine with six terminals available is needed. In this case the model of the converter is given by

$$v_{a1n1} = (q_{1a} - q_{2a})V_{dc} \tag{5.18}$$

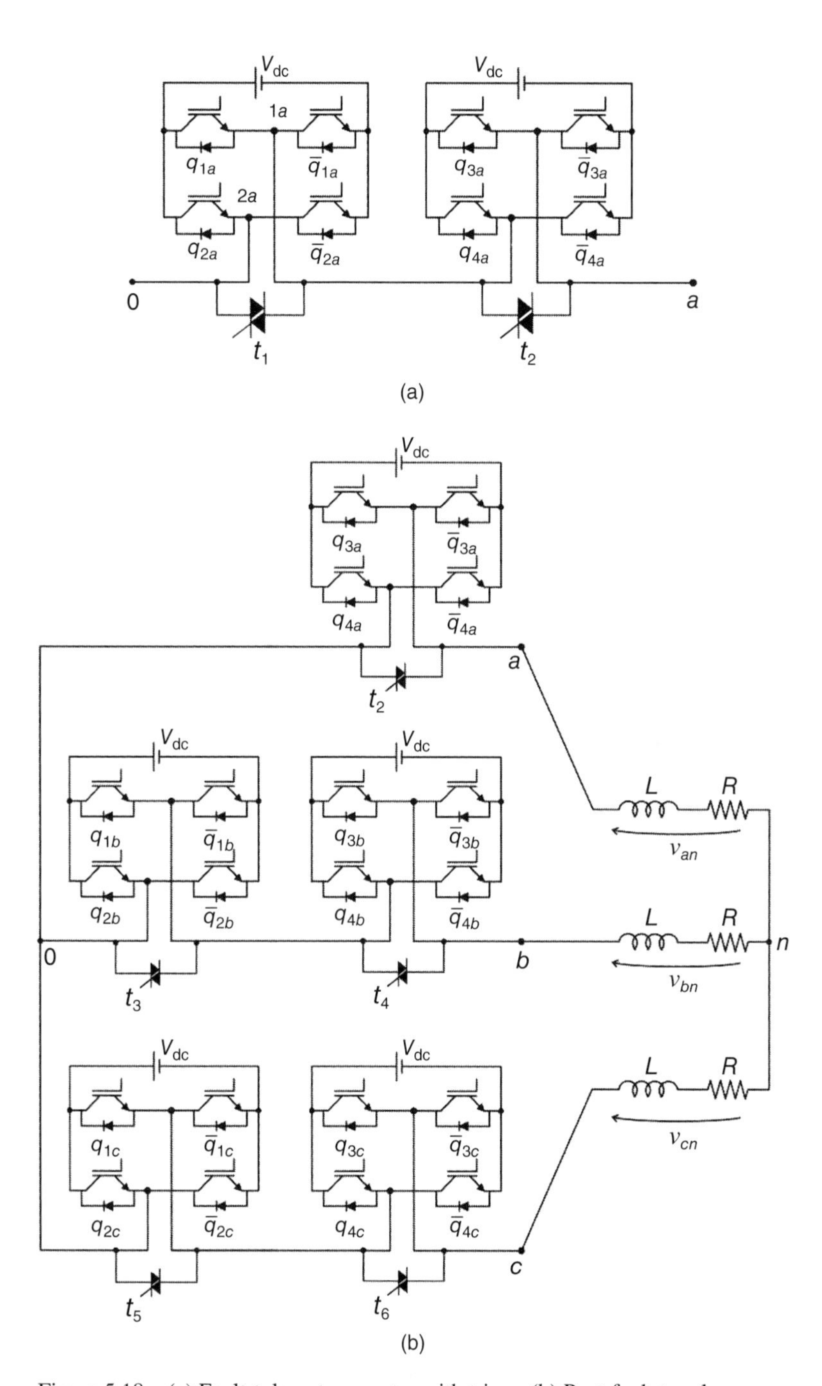

Figure 5.18 (a) Fault-tolerant converter with triacs. (b) Post-fault topology.

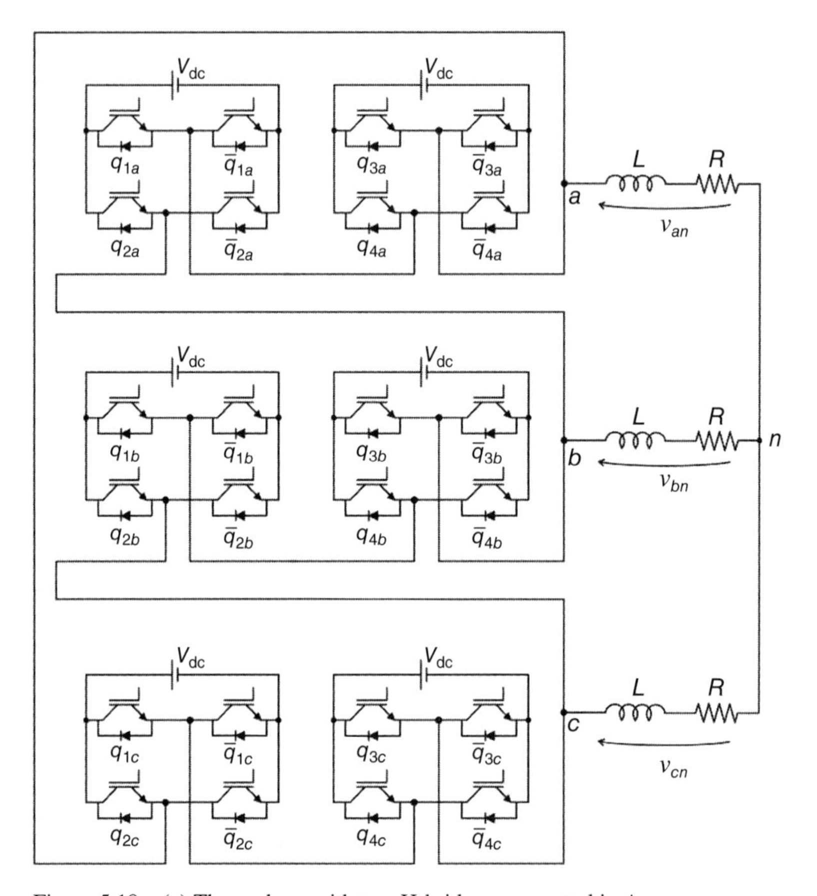

Figure 5.19 (a) Three-phase with two H-bridges connected in Δ.

$$v_{b1n1} = (q_{1b} - q_{2b})V_{\text{dc}} \tag{5.19}$$

$$v_{c1n1} = (q_{1c} - q_{2c})V_{\text{dc}} \tag{5.20}$$

and

$$v_{a2n2} = (q_{3a} - q_{4a})V_{\text{dc}} \tag{5.21}$$

$$v_{b2n2} = (q_{3b} - q_{4b})V_{\text{dc}} \tag{5.22}$$

$$v_{c2n2} = (q_{3c} - q_{4c})V_{\text{dc}} \tag{5.23}$$

Then the voltage applied to the machine's phases are given by

$$v_{a1a2} = v_{a1n1} - v_{a2n2} + v_{n1n2} \tag{5.24}$$

$$v_{b1b2} = v_{b1n1} - v_{b2n2} + v_{n1n2} \tag{5.25}$$

$$v_{c1c2} = v_{c1n1} - v_{c2n2} + v_{n1n2} \tag{5.26}$$

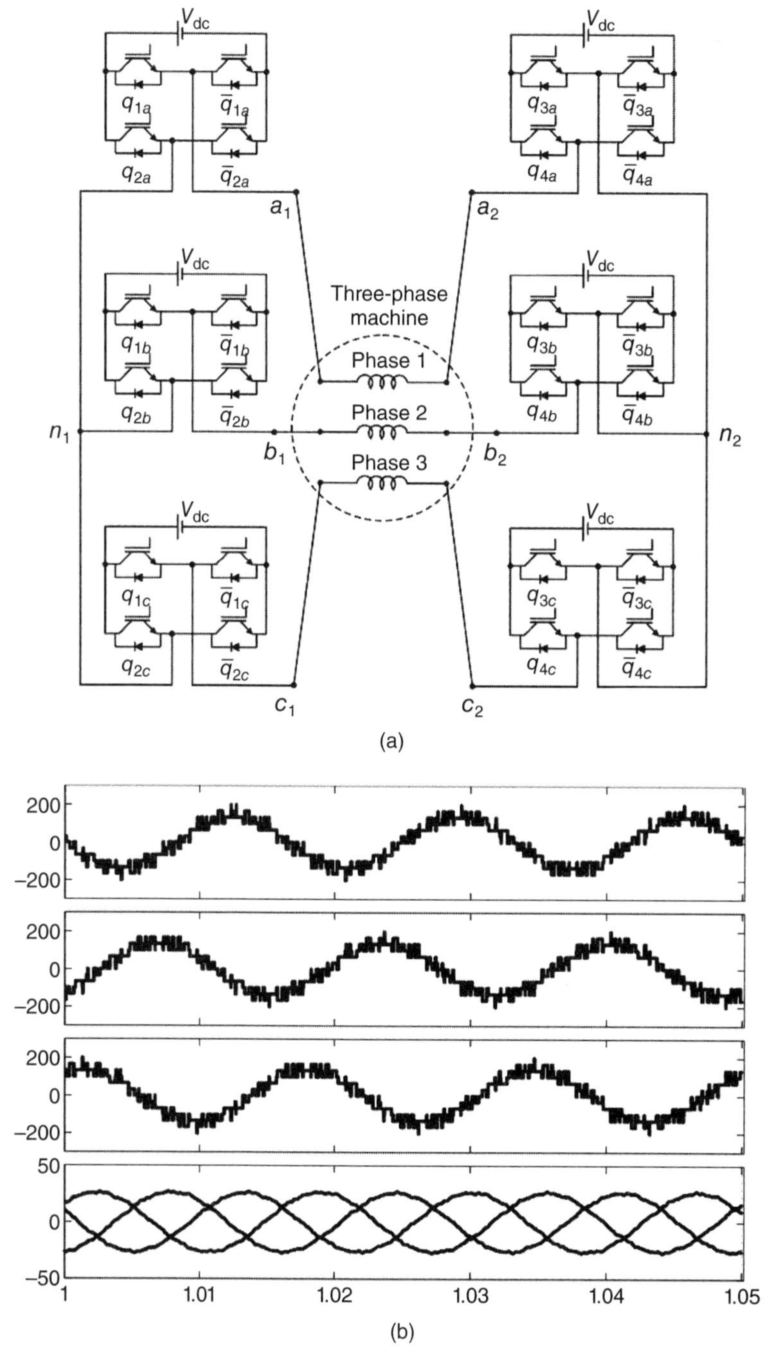

Figure 5.20 (a) Motor drive system with open-end winding topology. (b) Voltages and currents.

with v_{n1n2} given by

$$v_{n1n2} = v_{a1n1} + v_{b1n1} + v_{c1n1} - v_{a2n2} - v_{b2n2} - v_{c2n2} \tag{5.27}$$

The open-end winding arrangement of the three-phase machine allows another series connection of the converter, as presented in Fig. 5.21(a). This is a three-phase motor drive system employing two three-leg inverters. Since the three-phase machine is in series with both converters, this is also considered a series connection arrangement among converters and machine. One of the advantages of the system presented in Fig. 5.21(a) is its fault tolerance capability. Two cases can be considered:

Case I. The fault compensation is achieved by reconfiguring the power converter topology without any additional devices such as triacs. The post-fault configuration guarantees a Y connection of the motor.

Case II. The fault compensation is achieved by reconfiguring the power converter topology with devices such as triacs. The post-fault configuration guarantees a delta connection of the motor.

Both fault-tolerant strategies (Cases I and II) are based on the open-end winding machine and can be compensated for (i) dc-link capacitor failure; (ii) short-circuit failure; and (iii) open-circuit failure of the power switches. As previously presented, fault compensation is achieved by reconfiguring the power converter topology when a failure is detected.

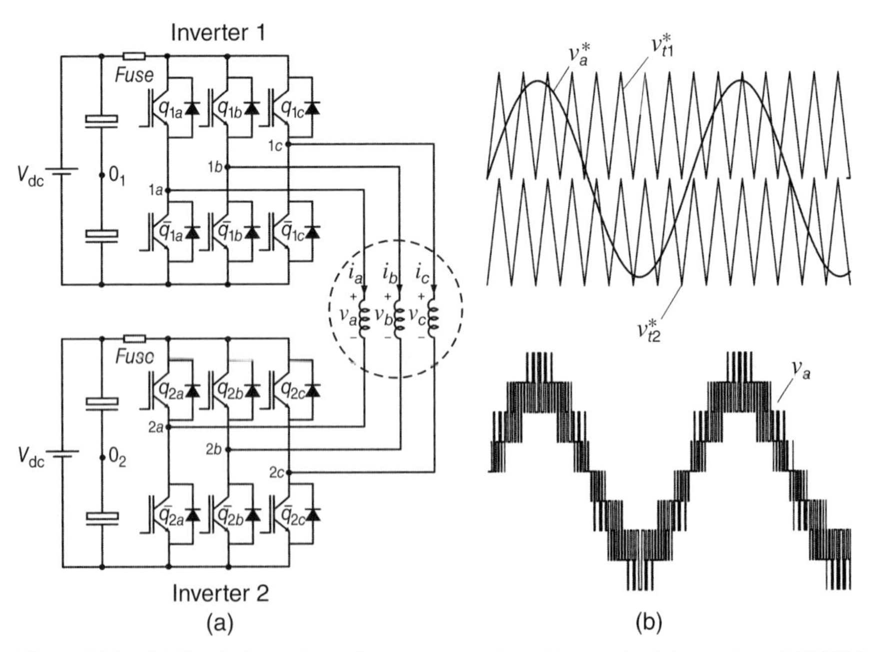

Figure 5.21 (a) Pre-fault configuration—open-end winding motor drive system. (b) PWM waveforms (top) and motor phase voltage (bottom).

A modulator with two carriers can be used for the PWM generation of the "healthy" open-end winding motor drive system [Fig. 5.21(a)], as observed in Fig. 5.21(b). The comparison between v_a^* and v_{t1}^* defines the state of the switch q_{1a}, while the comparison between v_a^* and v_{t2}^* defines the state of the switch $\overline{q}_{2a}$. Each inverter also has a fuse for isolation purposes. As seen in Fig. 5.21(b) the motor phase voltage (v_a) has also multilevel characteristics.

For Case I, the reconfiguration approach is obtained by using the other power switches not affected by the fault to create a neutral for the machine. Figure 5.22(a), 5.22(b), and 5.22(c) show the procedure to create a post-fault configuration in the case of open-circuit failure, short- circuit failure, and dc-link failure, respectively. Note that, in the case of dc-link failure, the creation of the neutral of the machine can be accomplished by either upper or lower switches of the converter under fault. In this strategy (Case I) the motor voltage value after the fault is reduced by 50%.

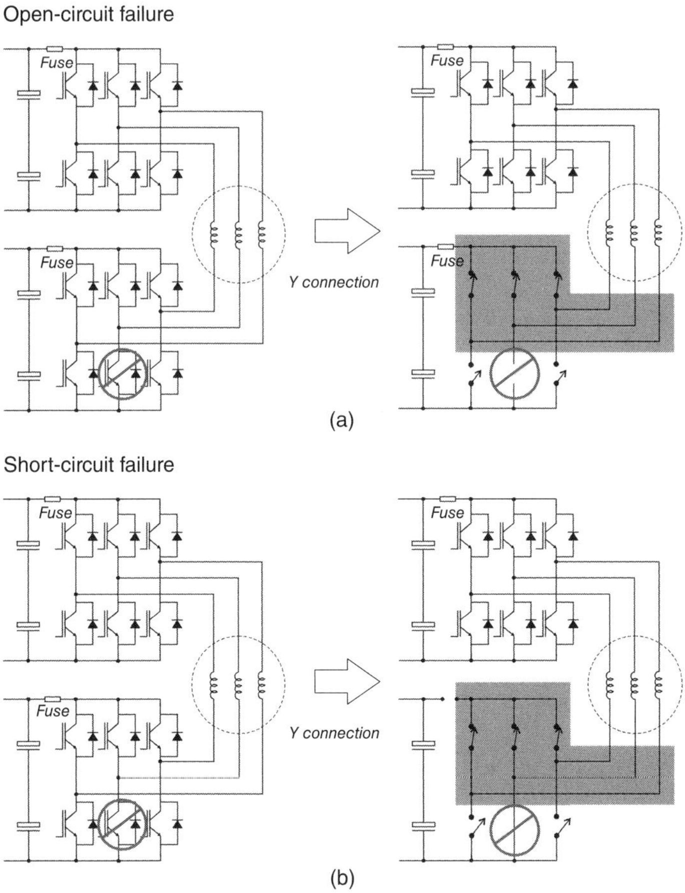

Figure 5.22 Post-fault configuration for: (a) open-circuit failure, (b) short-circuit failure, and (c) dc-link failure.

DC-link failure

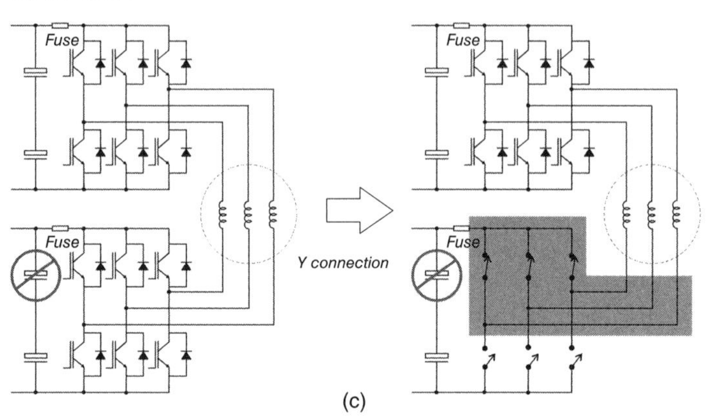

Figure 5.22 (*Continued*).

On the other hand, for Case II, the fault compensation is achieved by reconfiguring the power converter topology along with reconfiguration devices (triacs). The pre-fault configuration is depicted in Fig. 5.23(a). In order to increase the voltage applied to the machine, the reconfiguration approach aims to guarantee a delta connection of the motor after the fault. Two actions must be managed by the control system: isolate the source of the converter under fault and turn on all triacs, as depicted in Fig. 5.23(b). Despite using more devices (three triacs – t_a, t_b, and t_c) than the solution presented in Case I, the values for the motor phase voltage after the fault are reduced by just 13.4%, due to the Δ connection.

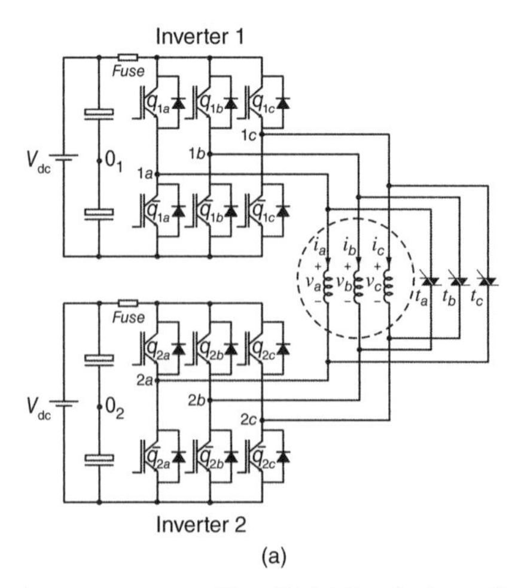

Figure 5.23 Fault-tolerant strategy—Case II. (a) Pre-fault configuration. (b) Fault reconfiguration procedure to make a delta connection of the machine considering a short-circuit failure.

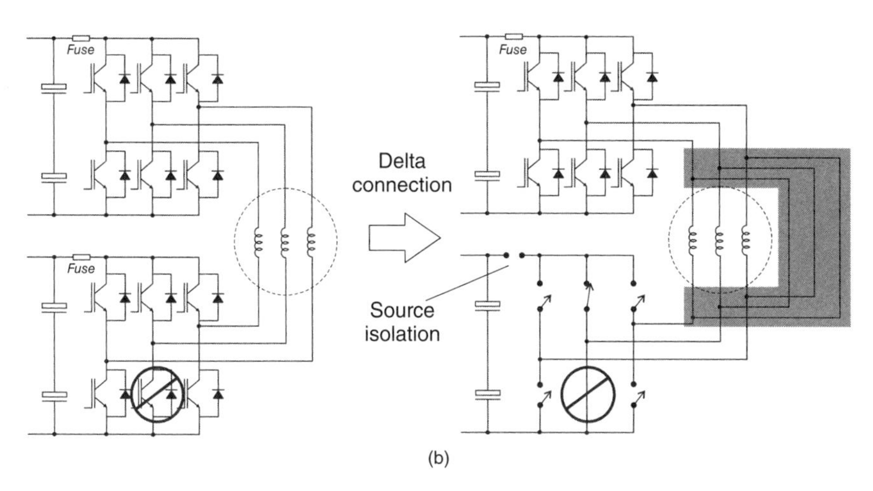

Figure 5.23 (*Continued*).

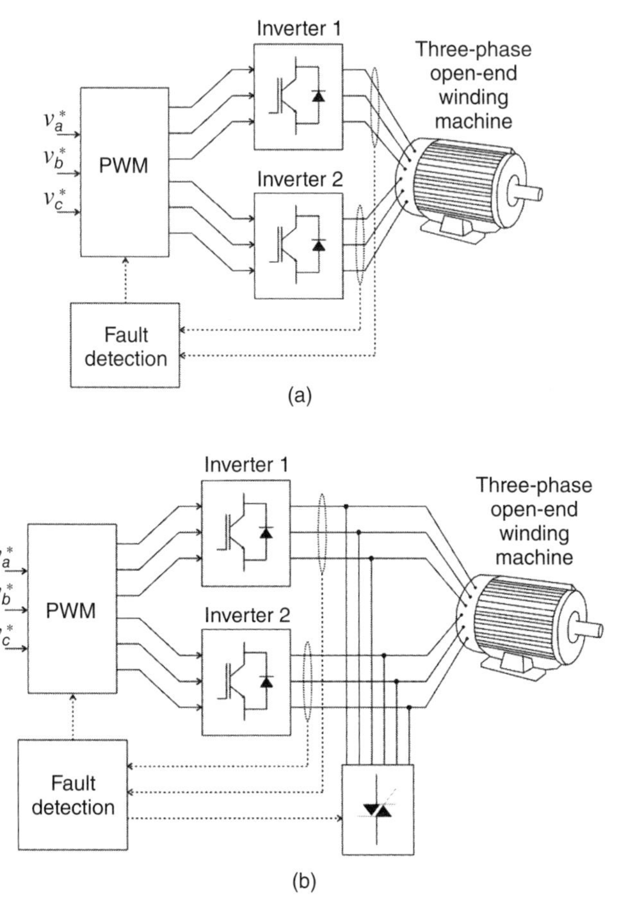

Figure 5.24 Control block diagram. (a) Case I. (b) Case II.

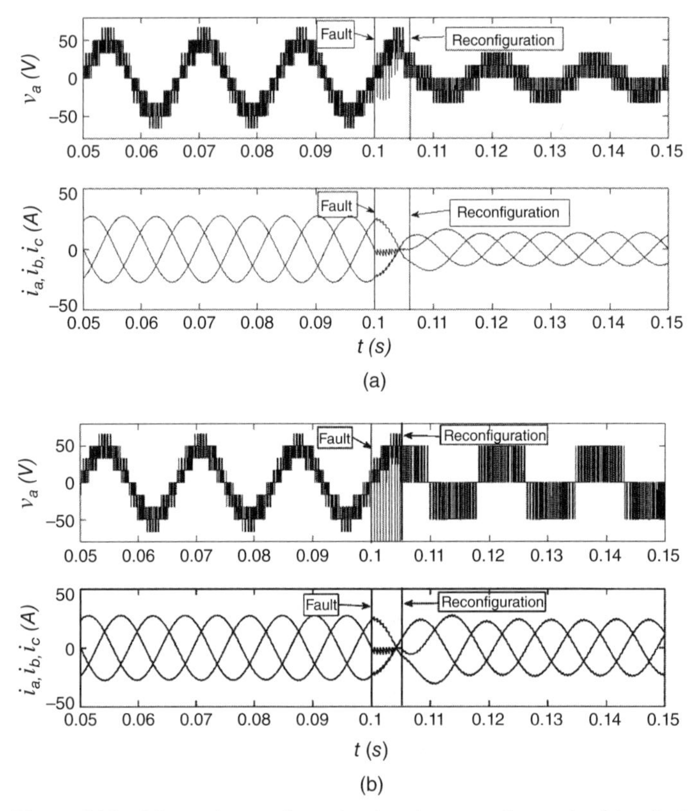

Figure 5.25 Motor phase voltage (top) and current (bottom) with a fault occurring at $t = 0.1$ s. (a) Case I and (b) Case II.

Figure 5.24 shows a simplified schematic of the control strategy for Cases I and II. The fault can be identified by comparing the reference voltage with the measured ones. After the fault occurrence, the reconfiguration procedure changes from the open-end winding motor ("healthy" configuration) to a post-fault converter. Figure 5.25 shows simulated results with a fault (open-circuit failure) occurring in switch $\overline{q}_{2b}$ at $t = 0.1$ s. The variables in this figure are motor phase voltage (v_a) and motor currents (i_a, i_b, and i_c). Figure 5.25(a) shows the results for Case I, while Fig. 5.25(b) shows the results for Case II. In this case, the switching frequency was 2 kHz and the dc-link voltage of each dc source was 50 V.

5.8 TWO H-BRIDGE CONVERTERS (SEVEN- AND NINE-LEVEL TOPOLOGIES)

The PWM approach previously presented for two H-bridge converters connected in series generates the gating signals to guarantee a five-level voltage at the output of the converter. Such a converter structure is also known as symmetric cascade multilevel

TABLE 5.4 Table of Variables for the Cascade Converter Operating with Seven Levels

State	q_{1a}	q_{2a}	q_{1b}	q_{2b}	v_o
1	0	0	0	0	0
2	0	0	0	1	$-V_b$
3	0	0	1	0	V_b
4	0	0	1	1	0
5	0	1	0	0	$-2V_b$
6	0	1	0	1	$-3V_b$
7	0	1	1	0	$-V_b$
8	0	1	1	1	$-2V_b$
9	1	0	0	0	$2V_b$
10	1	0	0	1	V_b
11	1	0	1	0	$3V_b$
12	1	0	1	1	$2V_b$
13	1	1	0	0	0
14	1	1	0	1	$-V_b$
15	1	1	1	0	V_b
16	1	1	1	1	0

inverter, since it uses the same dc-link voltages. An asymmetric cascade multilevel inverter using unequal dc sources in each phase can be employed to generate either a seven-level or a nine-level waveform.

From Table 5.3 and considering $V_a = 2V_b$, it turns out that the number of levels for v_o will be seven: $\{0,\pm V_b,\pm 2V_b,\pm 3V_b\}$, as presented in Table 5.4. Note that equation (5.28) describes the voltage at the output converter side as a function of the switching states.

$$v_o = (2q_{1a} - 2q_{2a} + q_{1b} - q_{2b})V_b \tag{5.28}$$

Although increasing the number of levels, asymmetric cascade multilevel inverters operate with switches dealing with different blocking voltages and also require different dc-link voltages, which could restrict their applications.

A higher number of levels can be obtained when $V_a = 3V_b$. Table 5.5 shows nine levels for the output voltage as a function of the state of the switches. In this case the nine levels are $\{0,\pm V_b,\pm 2V_b,\pm 3V_b,\pm 4V_b\}$. Another possible restriction of the unsymmetrical cascade multilevel topology is related to the lack of redundancy for generation of each level. Such a redundancy brings a degree of freedom to optimize the converter's operation.

A comparison between Tables 5.4 and 5.5 shows the generation of the level $2V_b$ is obtained with two switching states (States 9 and 12 in Table 5.4) for the seven-level configuration, while the generation of the same level for a nine-level configuration is done with just one switching state (State 10 in Table 5.5).

TABLE 5.5 Table of Variables for the Cascade Converter Operating with Nine Levels

State	q_{1a}	q_{2a}	q_{1b}	q_{2b}	v_o
1	0	0	0	0	0
2	0	0	0	1	$-V_b$
3	0	0	1	0	V_b
4	0	0	1	1	0
5	0	1	0	0	$-3V_b$
6	0	1	0	1	$-4V_b$
7	0	1	1	0	$-2V_b$
8	0	1	1	1	$-3V_b$
9	1	0	0	0	$3V_b$
10	1	0	0	1	$2V_b$
11	1	0	1	0	$4V_b$
12	1	0	1	1	$3V_b$
13	1	1	0	0	0
14	1	1	0	1	$-V_b$
15	1	1	1	0	V_b
16	1	1	1	1	0

TABLE 5.6 Different Switching States to Generate the Same Output Level $2V_{dc}$

State	q_{1a}	q_{2a}	q_{1b}	q_{2b}	q_{1c}	q_{2c}	v_o
1	1	0	1	0	0	0	$2V_{dc}$
2	1	0	1	0	1	1	$2V_{dc}$
3	1	0	1	1	1	0	$2V_{dc}$
4	1	0	0	0	1	0	$2V_{dc}$
5	0	0	1	0	1	0	$2V_{dc}$
6	1	1	1	0	1	0	$2V_{dc}$

5.9 THREE H-BRIDGE CONVERTERS

Multilevel configurations with three H-bridges can be obtained easily just following the axioms and postulates of the power blocks geometry (PBG). Figure 5.26(a) shows a single-phase topology with three H-bridges series connected. Figure 5.26(b), 5.26(c), and 5.26(d) present three-phase solutions with H-bridges Y-connected, Δ-connected, and with open-winding machine arrangement, respectively. Figure 5.26(e) shows another possibility of connections with Δ arrangements for the outer converters.

With the increase of the number of H-bridge cells, the number of switching states and levels increases. For example, assuming that V_a, V_b, and V_c have the same value V_{dc}, there are six ways to generate the level $2V_{dc}$, as presented in Table 5.6.

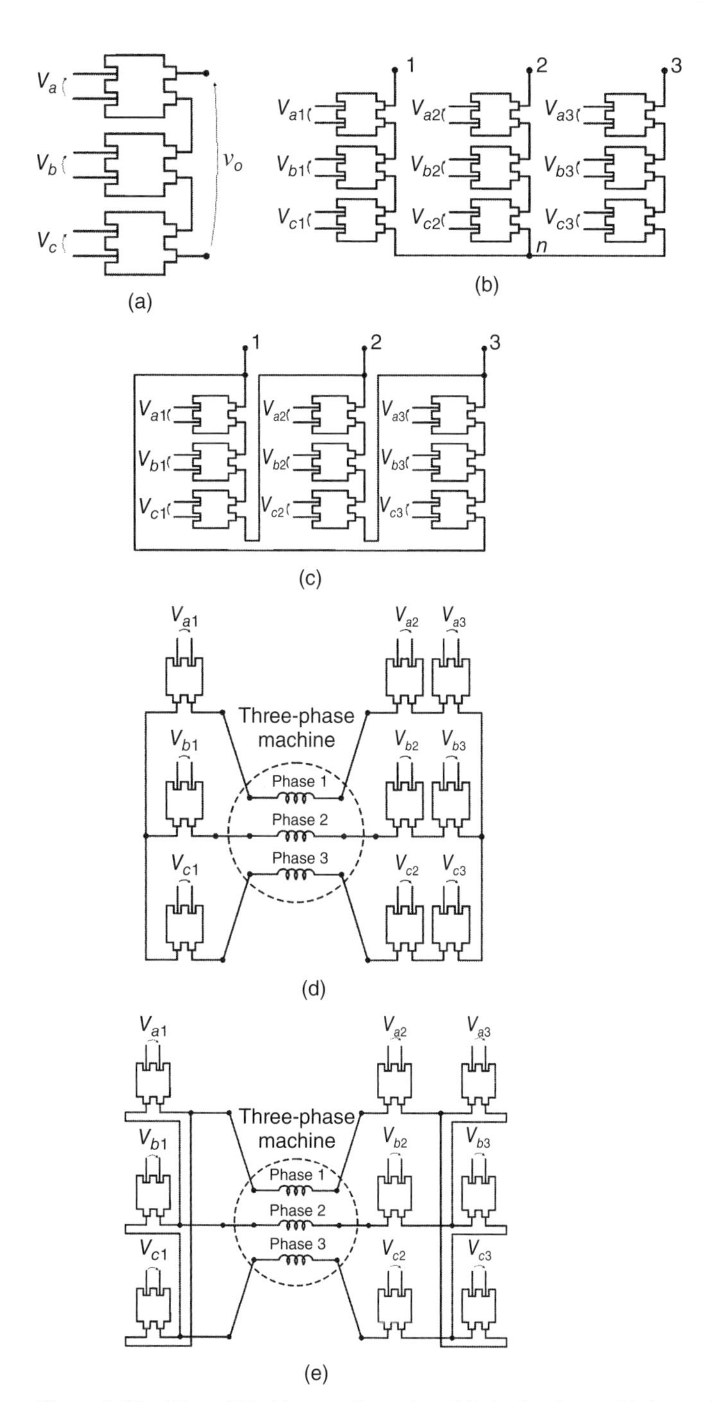

Figure 5.26 Three H-bridge configuration: (a) single-phase, (b) three-phase Y-connected, (c) three-phase Δ-connected, and (d) open-winding machine. (e) Δ connections for the outer converters.

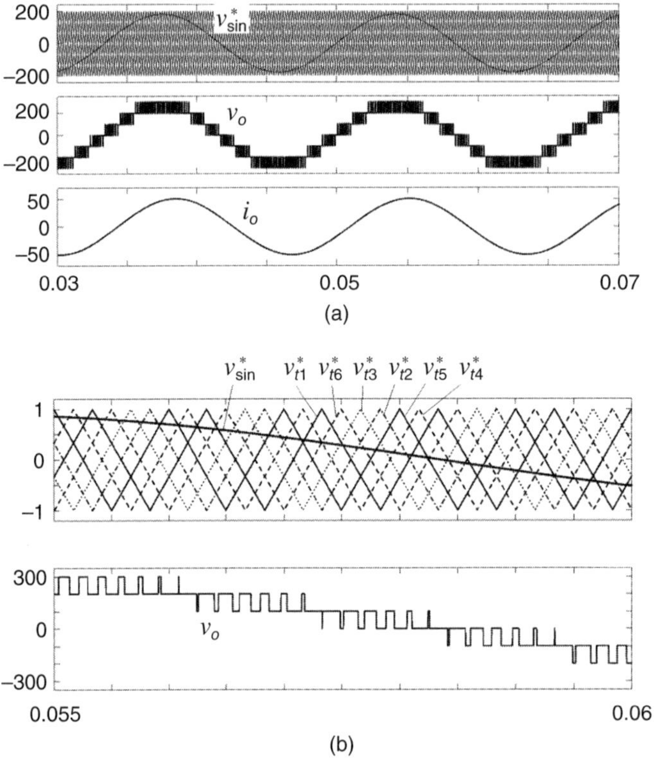

Figure 5.27 (a) Phase-shift modulation with (top) PWM signals, (middle) output voltage and (bottom) output current. (b) Zoom of the PWM and output voltages.

In a general way, the output voltage of the three symmetrical H-bridge configuration can be written as

$$v_o = (q_{1a} - q_{2a})V_a + (q_{1b} - q_{2b})V_b + (q_{1c} - q_{2c})V_c \tag{5.29}$$

Figure 5.27 shows the waveforms for the configuration in Fig. 5.26(a) supplying a single-phase *RL* load. Figure 5.27(a) top presents the PWM signals with phase-shift multicarrier modulation. Six triangular carrier signals (v^*_{t1}, v^*_{t2}, v^*_{t3}, v^*_{t4}, v^*_{t5}, and v^*_{t6}) are required since 7 levels are expected at the output. There is a phase shift between any two adjacent carrier waveforms given by $360°/(N-1)$, where N is the number of levels.

Figure 5.27(b) shows details (zoom) for the voltages presented in Fig. 5.27(a). The control circuitries as well as the power part of the converter (with the power switches) are presented in Fig. 5.28.

Note that the control waveforms ($v^*_{\sin}$ and v^*_{t1}, v^*_{t6}, v^*_{t3}, v^*_{t2}, v^*_{t5}, v^*_{t4}) and the analog circuit as presented in Fig. 5.28 generate a desired sequence of pulses at the output

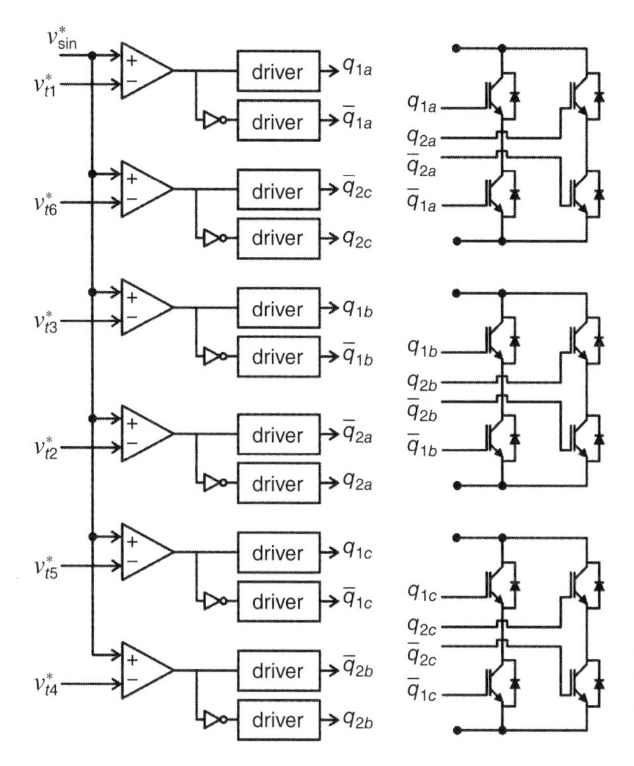

Figure 5.28 Analog control circuitries (left side) and the power part of the converter (right side).

voltage with seven levels, all of them with the same dv/dt. If another PWM signal is employed (different from that presented in Fig. 5.27), changes are also expected in the analog circuit solution. For example, Fig. 5.29 shows the level-shift multicarrier PWM modulation for the three H-bridge converter with six triangular carrier signals vertically disposed. In this case, all carrier signals are in phase with each other. Such a technique is also known as in-phase disposition (IPD). In addition to the approach presented in Fig. 5.29 there are other possibilities: alternate phase opposite disposition (APOD) and phase opposite disposition (POD). In the APOD approach all carriers are alternately in opposite disposition, while for the POD all carriers below zero are in phase, but in phase opposition with those above the zero axis.

Figure 5.30 shows the control and power part of the converter to generate a seven-level output voltage as presented in Fig. 5.29. A direct comparison between phase shift and level shift strategies (presented in Figs 5.27 and 5.29, respectively) reveals that the phase shift approach has advantages in terms of reduced harmonic distortions.

It is also possible to combine the solutions presented in Figs 5.27 and 5.29, as presented in Fig. 5.31. Such a combination results in a phase shift and level shift multicarrier approach.

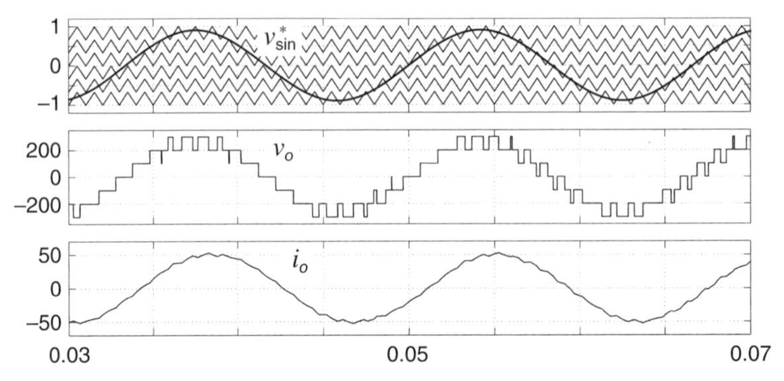

Figure 5.29 Level shift multicarrier modulation: (top) PWM signals (from top to bottom): v_{t1}, v_{t3}, v_{t5}, v_{t6}, v_{t4}, and v_{t2}. (middle) output voltage v_o, and (bottom) output current i_o.

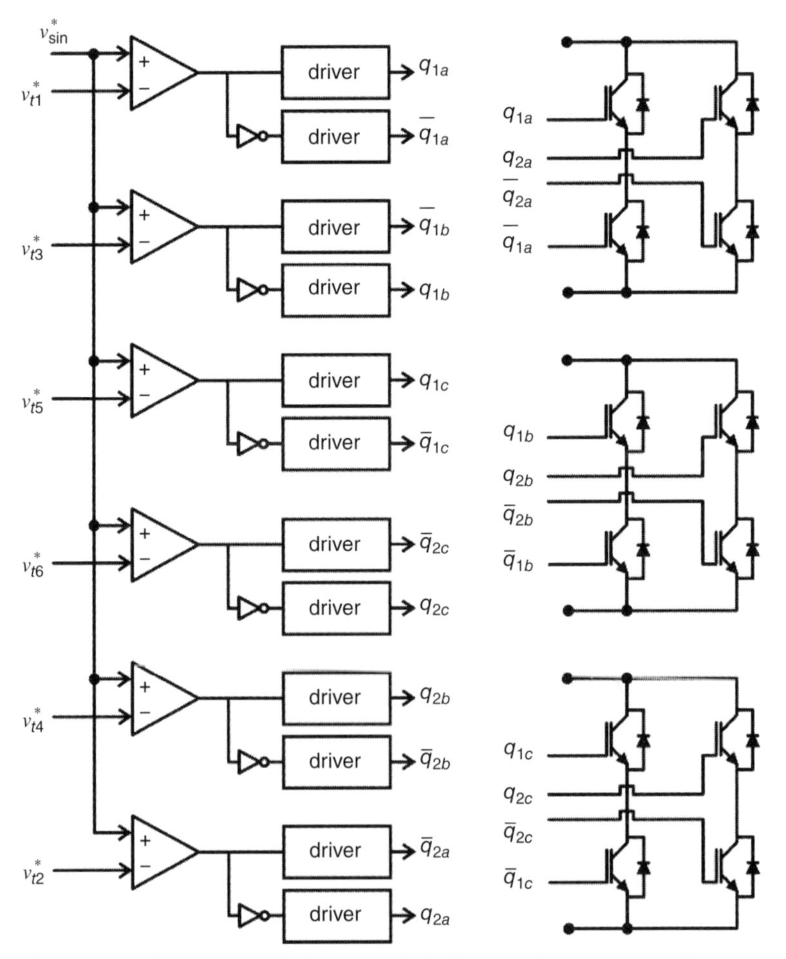

Figure 5.30 Analog control circuitries (left side) and the power part of the converter (right side) for the level-shift modulation.

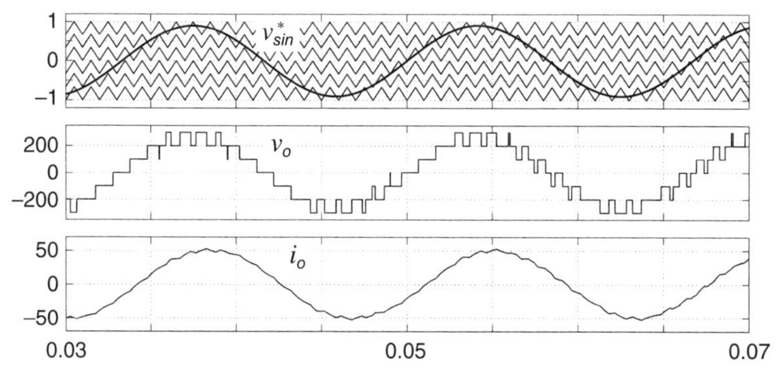

Figure 5.31 Phase shift/level shift multicarrier modulation: (top) PWM signals: v_{t1}, v_{t3}, v_{t5}, v_{t6}, v_{t4}, and v_{t2}. (b) Output voltage v_o, and (c) output current i_o.

5.10 FOUR H-BRIDGE CONVERTERS AND GENERALIZATION

A converter with four H-bridges and also the generalization of the cascade converters are presented in Fig. 5.32(a). Figure 5.32(b) shows another type of series connection converter with a combination of the conventional three-phase converter and H-bridges topologies.

5.11 SUMMARY

This chapter dealt with configurations and PWM techniques for connected H-bridge converters series. After discussing the single-phase converter, the three-phase version was presented using a single H-bridge converter per phase. Also, it was shown how to increase the number of levels with two H-bridge converters connected in series, bringing up the sequence given respectively by power converter construction, model, PWM approach, and three-phase version. Different multicarrier based schemes were presented and various associated aspects discussed. For instance, the direct comparison between phase shift and level shift strategies revealed that the phase shift approach has advantages in terms of reduced harmonic distortions. One special issue treated in this chapter was the possibility of guaranteeing post-fault operation with cascaded H-bridge converters. In addition, it was shown that two H-bridge converters can be employed to generate an output voltage with seven and nine levels, employing the correct relationship between the dc-link voltages. Finally, different arrangements and PWM solutions for three connected H-bridge converter series were introduced based on the PBG approach. The chapter is enriched with Table of Variables and analog control circuitries for most cases.

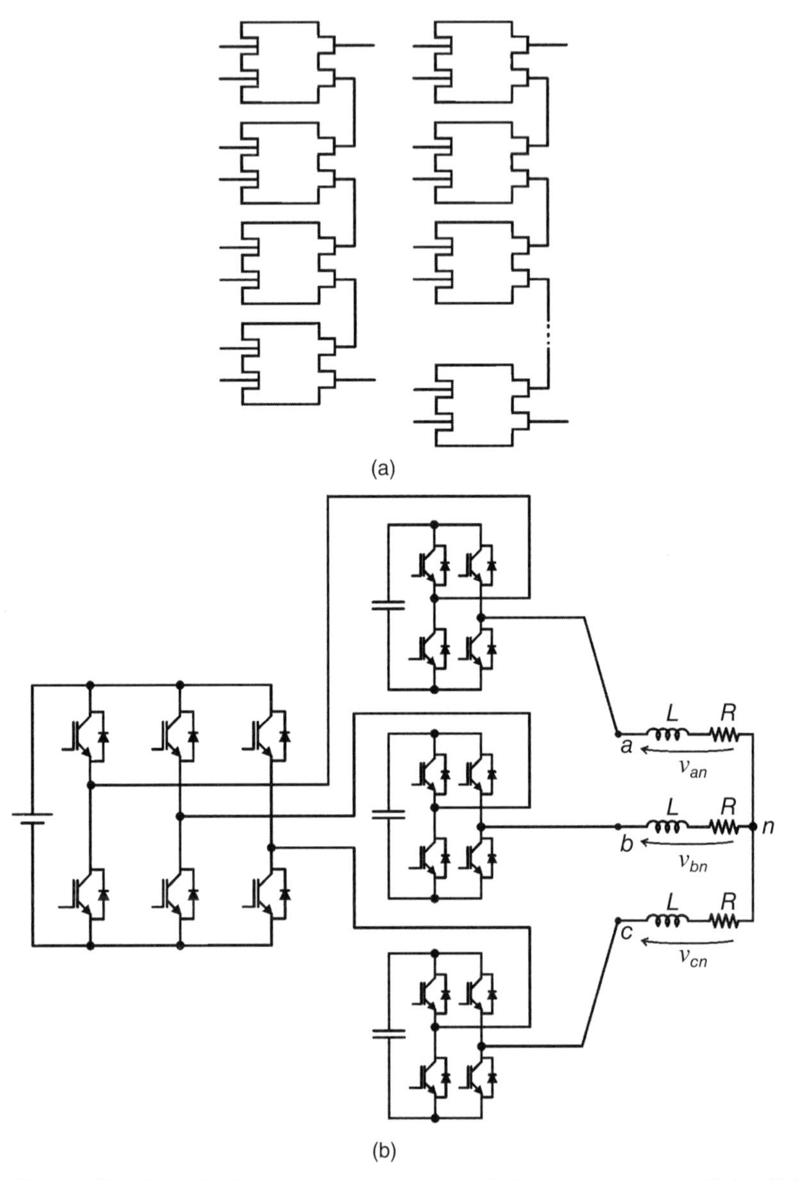

Figure 5.32 Four H-bridges and generalization. (b) Hybrid topology combining H-bridge and conventional three-phase inverter.

REFERENCES

[1] L. M. Tolbert, F. Z. Peng, and T. G. Habetler, Multilevel converters for large electric drives, IEEE Trans Ind Appl, vol. 35, no. 1, pp. 36–44, 1999.

[2] R. Rabinovici, D. Baimel, J. Tomasik, A. Zuckerberger, Series space vector modulation for multi-level cascaded H-bridge inverters, IEE Proc IET, vol. 3, no. 10, pp. 843–857, 2010.

[3] Rodriguez J, Hammond PW, Pontt JO, Musalem R, Escobar Mj. Operation of a medium-voltage drive under faulty conditions. IEEE Trans Ind Electron 2005;52(4):1080–1085.

[4] Wu B. High Power Converters and AC Drives. New Jersey: John Wiley and Sons; 2006.

[5] J. Rodriguez, J. Lai and F. Z. Peng, Multilevel inverters: a survey of topologies, controls, and applications, IEEE Trans Ind Electron, 49, No. 4, 2002, 724–738.

[6] J Rodriguez, S. Bernet, Bin Wu, J. O. Pontt, S. Kouro, Multilevel voltage-source-converter topologies for industrial medium-voltage drives, IEEE Trans Ind Electron, vol. 54, no. 6, pp. 2930–2945, 2007.

[7] Rashid MH. Power Electronics Handbook. New York: Academic; 2001.

[8] Peng FZ, Lai J-S. Multilevel cascade voltage source inverter with separate DC sources. US patent 5,642,275. Jun 1997.

[9] Hammond PW. Medium voltage PWM drive and method. US patent 5,625,545. Apr 1997.

[10] Corzine Keith, Familiant Yakov. A new cascaded multilevel H-bridge drive. IEEE Trans Power Electron, 2002, Vol. 17, No. 1, pp.125–131.

[11] Manjrekar MD, Lipo TA. Hybrid multilevel power conversion system: a competitive solution for high-power applications. IEEE Trans Ind Appl 2000;36(3):834–841.

[12] Lund R, Manjrekar MD, Steimer P, Lipo TA. Control strategies for a hybrid seven-level Inverter, Conference Record of EPE, pp. 1–10, 1999.

[13] Manjrekar MD, Lipo TA. A hybrid multilevel inverter topology for drive applications. IEEE APEC; 1998, 523–529.

[14] Carrara G, Gardella S, Marchenosi M, Salutari R, Sciutto G. A new multilevel PWM method: a theoretical analysis. IEEE Trans Power Electron, July 1992;7:497–505.

CHAPTER 6

FLYING-CAPACITOR CONFIGURATION

6.1 INTRODUCTION

A type of multilevel configuration known as flying capacitor (FC) converter is presented in this chapter under powerblock geometry (PBG) point of view. The introduction and fundamental concepts of configurations with three and more levels employing a FC approach, as well as a detailed description of these circuits, are addressed in this chapter. FC converters present higher degrees of freedom (redundant states) to synthesize a specific output voltage and employ a lesser number of semiconductor power devices than the neutral point clamped (NPC) topology. Such redundant states are used to regulate the FC voltage. Unlike the cascade configuration presented in Chapter 5, there is no need for isolated dc sources.

The main challenge of the FC configuration is to keep the FC voltage under a desired level, avoiding distortion at the ac part of the converter. Part of this chapter will be dedicated to discussing strategies to balance the FC voltage. For the sake of illustration, the analysis starts with the FC being substituted by a dc voltage source. A simple pulse width modulation (PWM) approach is employed to generate a three-level output voltage. It is demonstrated that the PWM approach affects the voltage ripple of the FC voltage. In fact, there are different strategies to regulate the capacitor voltage, which can be sorted into two main types: natural-balancing and active schemes. The natural-balancing approach employs solutions in terms of PWM strategies without any voltage and current sensors, while the active method employs feedback regulation and control algorithms to maintain the voltage constant. It is worth mentioning that different aspects affect the capacitor voltage balancing, for example, the type of load (lincar or nonlinear). This chapter is organized considering the number of levels generated by each circuit, as done in Chapter 4 for NPC circuits.

Subsequent to this introduction, Section 6.2 presents the power blocks (PBs) and the circuit with switches for the FC three-level topology, as also the model and Table of Variables. Section 6.3 deals with a simple PWM strategy applied for the three-level circuit. The FC voltage regulation is presented in Section 6.4. Sections 6.5 and 6.6 show the FC topologies for single-phase (full-bridge) and three-phase applications, respectively. Other circuits of three-phase converters appear in Section 6.7.

Advanced Power Electronics Converters: PWM Converters Processing AC Voltages,
Forty Fifth Edition. Euzeli Cipriano dos Santos Jr. and Edison Roberto Cabral da Silva.

Four-level topologies are considered in Section 6.8. Finally, Section 6.9 presents the generalization and Section 6.10 summarizes this chapter.

6.2 THREE-LEVEL CONFIGURATION

The PBs, as described in the Chapter 3, can be connected in a systematic approach to guarantee the creation of FC multilevel configurations. An arrangement using two PB-ac blocks is depicted in Fig. 6.1(a), which follows Postulate 2. Figure 6.1(b)–6.1(e) shows the ways used to obtain each specific level at the output converter side v_o referred to as the point "0", that is, $V_1, (V_1 - V_2), (V_2 - V_3), -V_3$, respectively. Figure 6.1(c) and 6.1(d) illustrates that it is necessary to activate two PBs to obtain $v_o = (V_1 - V_2)$ and $v_o = (V_2 - V_3)$, while $v_o = V_1$ and $v_o = -V_3$ are obtained activating only one PB as in Fig. 6.1(b) and 6.1(e), respectively.

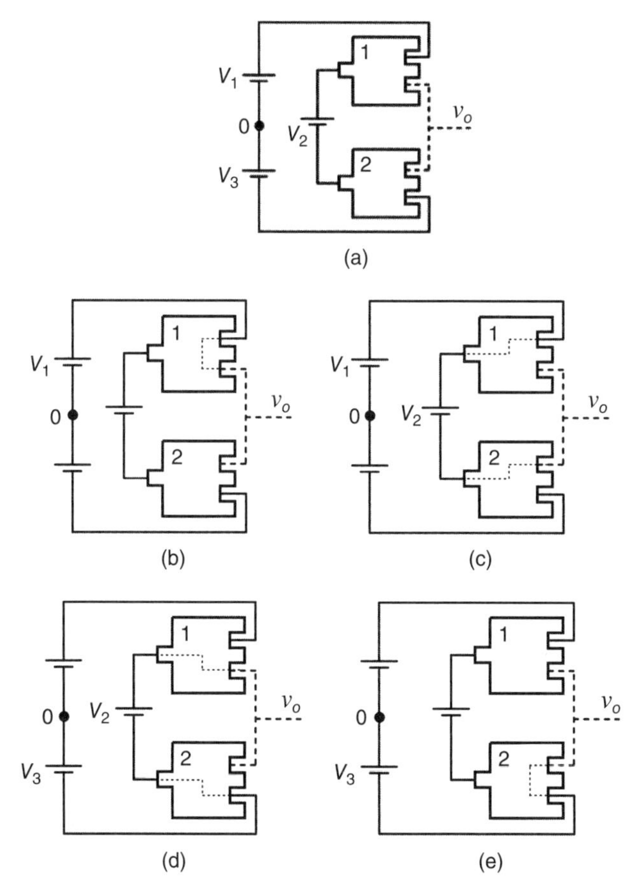

Figure 6.1 (a) Three levels FC configuration with PBs. Arrangement to obtain: (b) $v_o = V_1$. (c) $v_o = V_1 - V_2$. (d) $v_o = V_2 - V_3$. (e) $v_o = -V_3$.

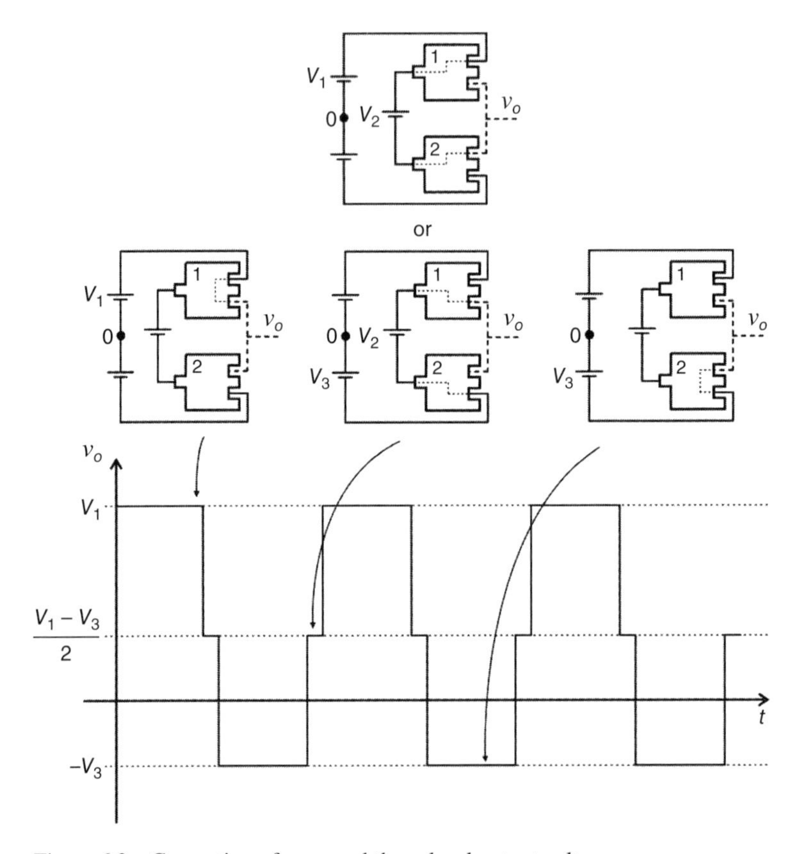

Figure 6.2 Generation of a general three-level output voltage.

In general terms, to generate an output waveform with three levels it is necessary to activate the PBs in a sequence as observed from the waveform of v_o in Fig. 6.2 (bottom). Since this is a three-level converter, $V_1 - V_2 = V_2 - V_3$, which means $V_2 = (V_1 + V_3)/2$. It is worth emphasizing that with FC topology it is possible to generate the V_2 level with two different circuit arrangements (i.e., topological states), as observed in Fig. 6.2 (top). This is an important characteristic of the FC topologies, which will be explored later in this chapter.

Since v_o can assume just one voltage value V_1, $(V_1 - V_3)/2$ or $-V_3$ at each instant of time, it is possible to establish an equation to describe the output voltage v_o as a function of the input voltages V_1', V_2', V_3' [where $V_1' = V_1$, $V_2' = (V_1 - V_3)/2$, and $V_3' = -V_3$] as well as a function of two independent binary variables (q_1 and q_2). Following the same approach employed previously in this book, these binary variables define the states of the switches inside the blocks. In this sense, since there are three inputs to be selected (V_1', V_2', and V_3'), the minimum number of binary variables employed to select those inputs is two. Considering that there are three input voltages to be selected by using these binary variables, the output voltage can be obtained as presented in Table 6.1.

TABLE 6.1 Table of Variables for the Converter in Fig. 6.1(a)

State	q_1	q_2	v_o	v_o
1	0	0	$-V_3$	V'_3
2	0	1	$\frac{V_1 - V_3}{2}$	V'_2
3	1	0	$\frac{V_1 - V_3}{2}$	V'_2
4	1	1	V_1	V'_1

An equation used to describe the behavior presented in this table is given below:

$$v_o = \overline{q}_1\overline{q}_2V'_3 + \overline{q}_1q_2V'_2 + q_1\overline{q}_2V'_2 + q_1q_2V'_1 \quad (6.1)$$

where $\overline{q}_1 = 1 - q_1$ and $\overline{q}_2 = 1 - q_2$. In this case, (6.1) becomes

$$v_o = (1 - q_1)(1 - q_2)V'_3 + (1 - q_1)q_2V'_2 + q_1(1 - q_2)V'_2 + q_1q_2V'_1 \quad (6.2)$$

Normally, there are some requirements associated with total harmonic distortion (THD) minimization for the output voltage, which ends up with a symmetrical waveform at the converter output, and consequently $V'_1 = V_{dc}$, $V'_2 = 0$, and $V'_3 = -V_{dc}$. In this case it is possible to write equation (6.2) as follows:

$$v_o = (q_1 + q_2 - 1)V_{dc} \quad (6.3)$$

The binary variables considered in Table 6.1 are in fact the switching states inside the blocks, which means that the topology presented in Fig. 6.1 can be treated at its lower level presentation (switches instead of blocks), as in Fig. 6.3.

Note that the FC configuration is bidirectional, which means that each voltage value V'_1, V'_2, V'_3 can be obtained at the output converter side independent of the current, that is, any desired output value is obtained for both cases $i_o > 0$ and $i_o < 0$, as observed in Fig. 6.3.

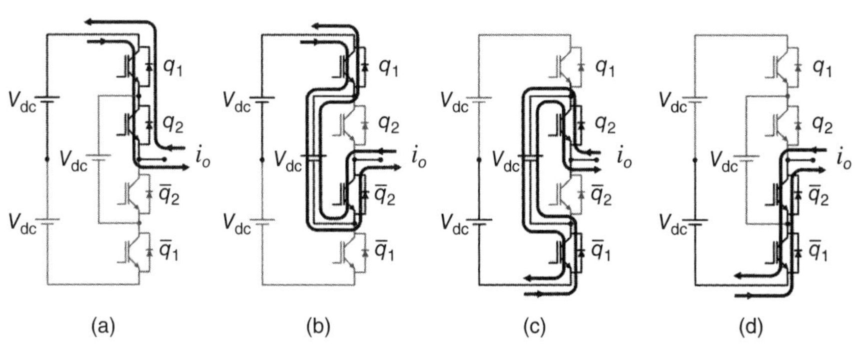

Figure 6.3 Generation of levels: (a) V'_1, (b) and (c) V'_2, and (d) V'_3 independent of the output current.

Exercise 6.1

(a) Assuming the FC configuration with four independent switches (q_1, q_2, q_3, and q_4) as presented below, fill the table below eliminating the prohibited states caused by either undefined states (e.g., all switches off) or short-circuit of the dc-link source (or FC). (b) Eliminate all undefined states in this table and find a logic relationship among q_1, q_2, q_3, and q_4.

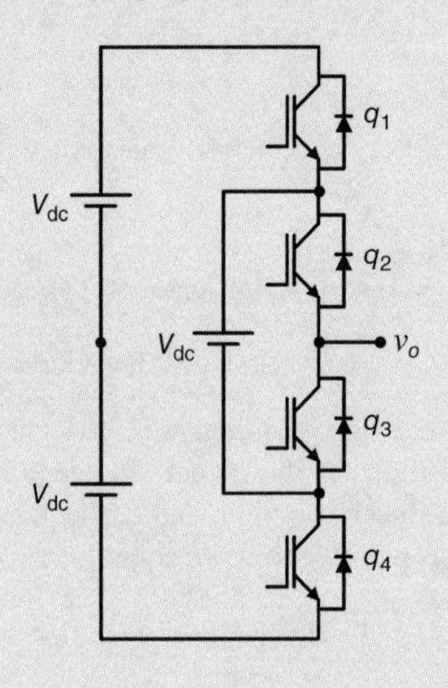

State	q_1	q_2	q_3	q_4	v_o
1	0	0	0	0	
2	0	0	0	1	
3	0	0	1	0	
4	0	0	1	1	
5	0	1	0	0	
6	0	1	0	1	
7	0	1	1	0	
8	0	1	1	1	
9	1	0	0	0	
10	1	0	0	1	
11	1	0	1	0	
12	1	0	1	1	
13	1	1	0	0	
14	1	1	0	1	
15	1	1	1	0	
16	1	1	1	1	

6.3 PWM IMPLEMENTATION (HALF-BRIDGE TOPOLOGY)

A desired output voltage with regulation of both voltage and frequency can be obtained for the FC converter with PWM strategies. As considered for the other topologies, such a strategy is able to synthesize a square waveform with an average value (at the switching frequency) equal to a sinusoidal waveform.

Figure 6.4 shows one possible analog solution for the PWM implementation known as level-shift, in which the states of the four power switches are determined by a comparison among two high frequency triangular waveforms (v^*_{t1} and v^*_{t2}) and a sinusoidal waveform ($v^*_{\sin}$). The block marked "driver" must guarantee the

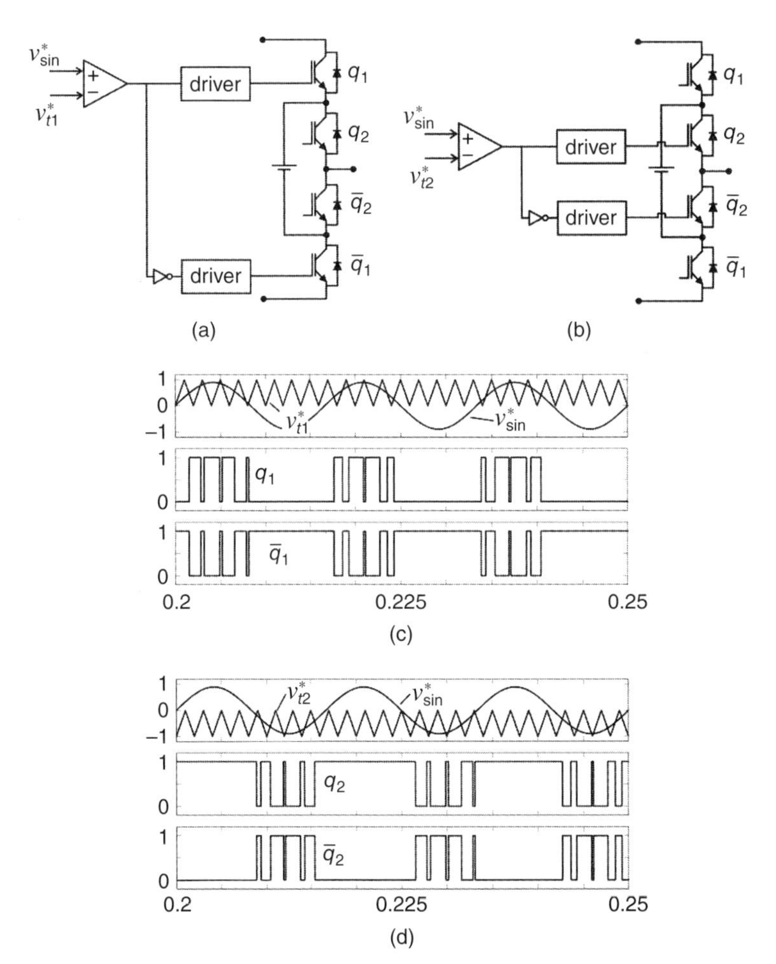

Figure 6.4 (a) PWM generation for switches q_1 and $\bar{q}_1$. (b) PWM generation for switches q_2 and $\bar{q}_2$. (c) Variables associated with switches q_1 and $\bar{q}_1$ (from top to bottom) – reference voltages, gating signals of switches q_1 and $\bar{q}_1$. (d) Variables associated with switches q_2 and $\bar{q}_2$ (from top to bottom) – reference voltages, gating signals of switches q_2 and $\bar{q}_2$.

levels of voltage and current at the gate circuit needed to operate the power switches adequately.

Note that when the switches q_1 and $\overline{q}_1$ are in operation (on and off during the positive half cycle of $v^*_{\sin}$) the switch q_2 is on and consequently the switch $\overline{q}_2$ is off for the whole half period. This is equivalent to states 2 and 4 in Table 6.1, which generates the levels V'_2 and V'_1, respectively. On the other hand, on the negative half cycle of $v^*_{\sin}$, when the switches q_2 and $\overline{q}_2$ are in operation (alternating on and off) the switch q_1 is off and consequently the switch $\overline{q}_1$ is on. This is equivalent to states 1 and 2 in Table 6.1, which generates the levels V'_3 and V'_2, respectively.

The PWM presented in Fig. 6.4(a) is able to generate the output voltage with the desired number of levels (V'_1, V'_2, and V'_3) using such a simple implementation. However, State 3 in Table 6.1 has not been used in this simple strategy, which means that there is still a degree of freedom to be explored, as seen in the following sections.

Figure 6.5(a) shows the half-bridge three-level configuration, in which the load is connected between the dc-link mid-point and the center point of the leg. Figure 6.5(b) depicts (from top to bottom) PWM reference waveforms, gating signals of switches q_1 and q_2, load voltage, and load current, respectively. The parameters employed in these simulation tests were: $V_{dc} = 200$ V, $R = 5\ \Omega$, and $L = 5$ mH, and switching frequency equal to 500 Hz.

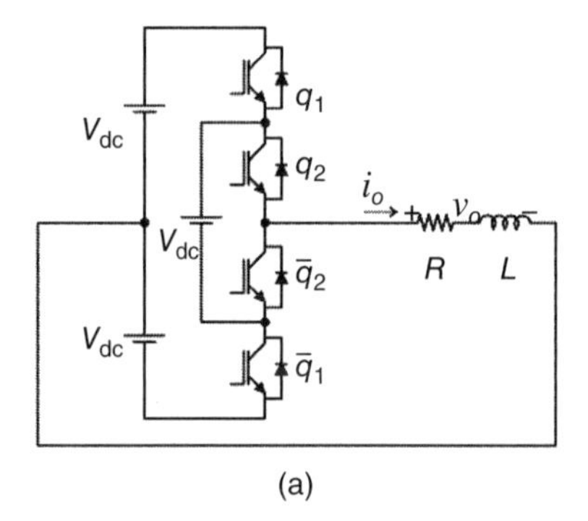

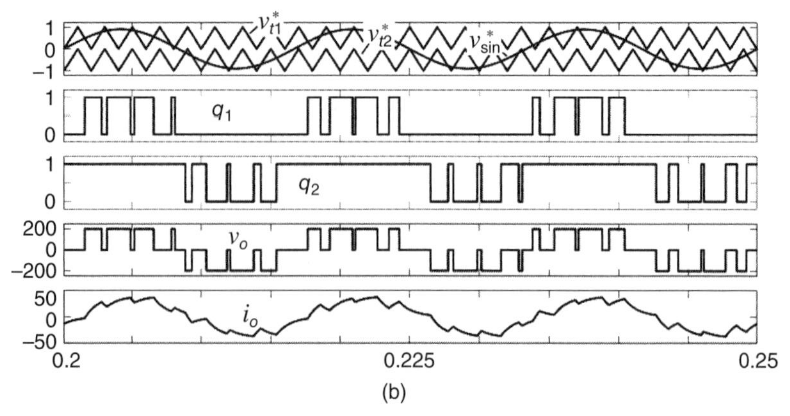

Figure 6.5 (a) Three-level FC half-bridge converter. (b) (from top to bottom) $v^*_{\sin}$, v^*_{t1}, v^*_{t2}; gating signal of switch q_1; gating signal of switch q_2; load voltage and load current.

6.4 FLYING CAPACITOR VOLTAGE CONTROL

In order to simplify the analysis of an FC converter, the FC was substituted by a dc voltage source, as in Fig. 6.5(a). However, in practical circuits the center dc voltage source in Fig. 6.5(a) is replaced by a capacitor, as in Fig. 6.6(a). In this case, it is necessary to guarantee the control voltage of this capacitor.

For example, if the level-shift PWM approach presented in Fig. 6.4 is applied to the half-bridge converter presented in Fig. 6.6(a), the results for the capacitor variables (v_C and i_C) are as presented in Fig. 6.6(b). Even with the average value of v_C equal to 200 V (as desired), there are variations for the voltage v_C which will bring up low frequency distortion at the output converter side.

From the node between the switches q_1 and q_2 in Fig. 6.6(a) it is possible to determine the capacitor current as follows: $i_C = i_{q1} - i_{q2}$, and assuming the level-shift modulation approach, it can be seen that:

When $v^*_{\text{sin}} > 0$:

$$q_2 = 1$$
$$i_{q2} = i_o$$
$$i_{q1} = q_1 i_o$$
$$\therefore i_C = (q_1 - 1) i_o$$

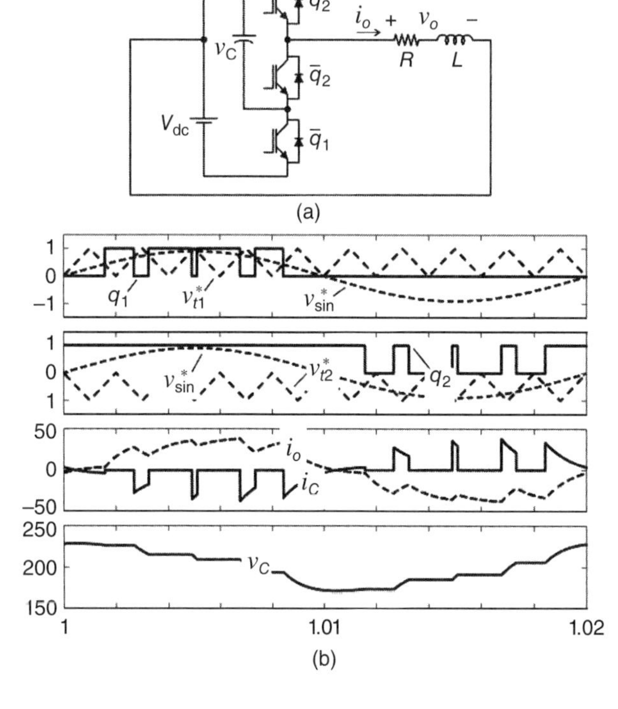

Figure 6.6 (a) Half-bridge converter with flying capacitor. (b) Control and power variables for the half-bridge circuit with level-shift PWM.

When $v^*_{\sin} < 0$:

$$
\begin{aligned}
q_1 &= 0 \\
i_{q1} &= 0 \\
i_{q2} &= q_2 i_o \\
\therefore i_C &= -q_2 i_o
\end{aligned}
$$

From the current i_C obtained previously, it is possible to determine directly the capacitor voltage, since: $v_C = \frac{1}{C}\int i_C dt$.

Exercise 6.2

In order to deeply understand the mechanism behind the FC voltage control, draw the topological states considering the switching signals and capacitor waveforms in steady-state operation for the circuit in Fig. 6.6(a), as highlighted below. Consider each of the intervals of time below (from 1 to 17).

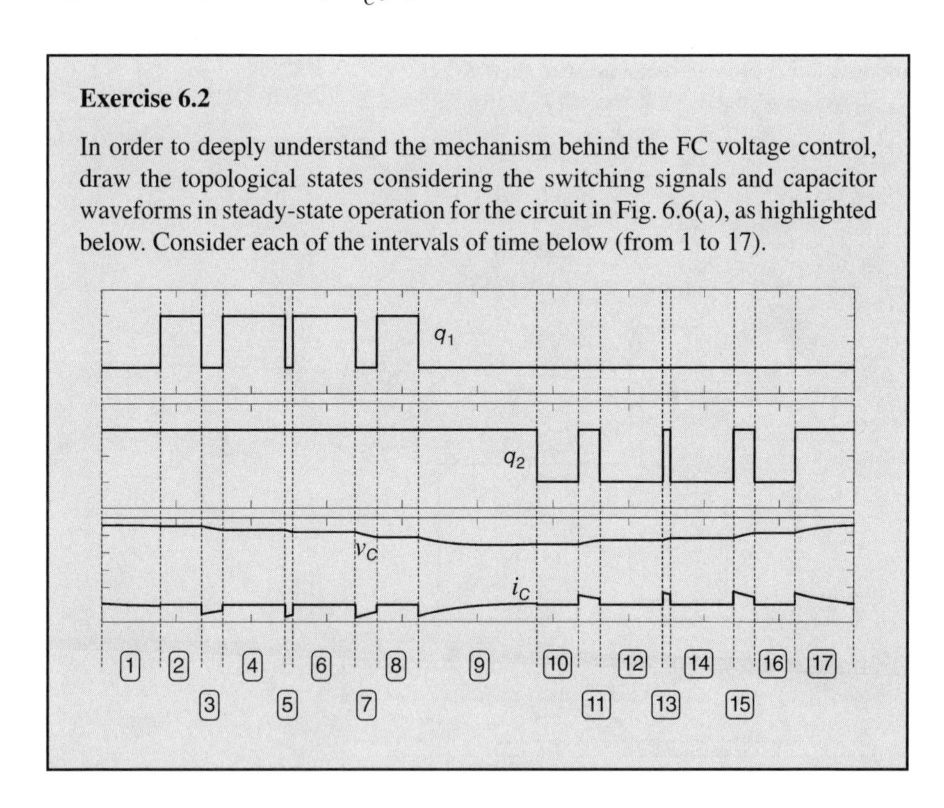

To guarantee the capacitor voltage control with reduced variation, another PWM strategy can be considered, as seen in Fig. 6.7. A phase-shift has been considered in this case, which leads to a capacitor charging and discharging at the switching frequency. This plays an important role in keeping v_C with reduced ripple. The polarity of the capacitor current in Fig. 6.6(b) has changed with the sinusoidal frequency, which increases the ripple for v_C. Note from Fig. 6.6(b) that the FC is discharged during the entire positive half of the sinusoidal voltage ($v^*_{\sin}$) since i_C is either zero or negative. On the other hand, the FC during the negative half of the sinusoidal voltage is charged because i_C is either zero or positive.

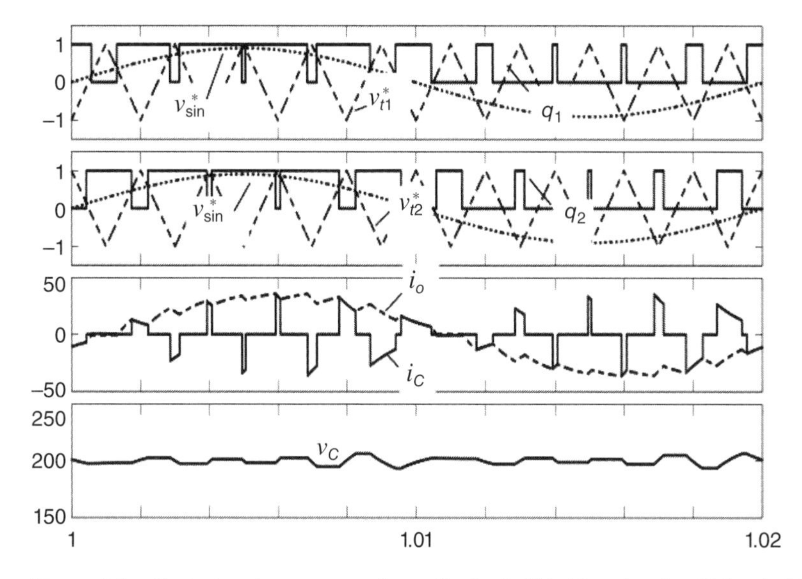

Figure 6.7 Control and power waveforms for the half-bridge circuit with level-shift PWM.

On the other hand, in the PWM technique shown in Fig. 6.7, the charging/discharging occurs on the switching frequency; this keeps the capacitor voltage ripple smaller.

The PWM techniques able to reduce the ripple on the FC are fundamental for reducing the distortion of the output voltages. Figure 6.8(a) and 6.8(b) shows the load voltage for the level-shift and phase-shift approaches, respectively. Figure 6.9 in turn shows the start-up transient for the FC voltage.

6.5 FULL-BRIDGE TOPOLOGY

Full-bridge three-level FC converters for both single-phase and three-phase loads are obtained by connecting the blocks in Fig. 6.1 as observed in Figs 6.10(a) and 6.10(b), respectively. Figure 6.10(c) shows the option in which the three-phase four-wire system is required.

The full-bridge single-phase three-level FC converter is presented in Fig. 6.11(a) by using switches instead of blocks (lower level representation). Figure 6.11(b) shows (from top to bottom) pole voltage of the first leg (with switches q_{1a} and q_{2a}), pole voltage of the second (with switches q_{1b} and q_{2b}), load voltage, and load current, respectively.

Comparing the load voltage in Figs 6.5(b) and 6.11(b), it is quite clear that the increased number of levels observed in the load voltage of the full-bridge topology guarantees the ripple reduction of the current and consequently a better quality of the voltage delivered by the converter.

In this case the output voltage is given by

$$v_o = (q_{1a} + q_{2a} - q_{1b} - q_{2b})V_{\text{dc}} \tag{6.4}$$

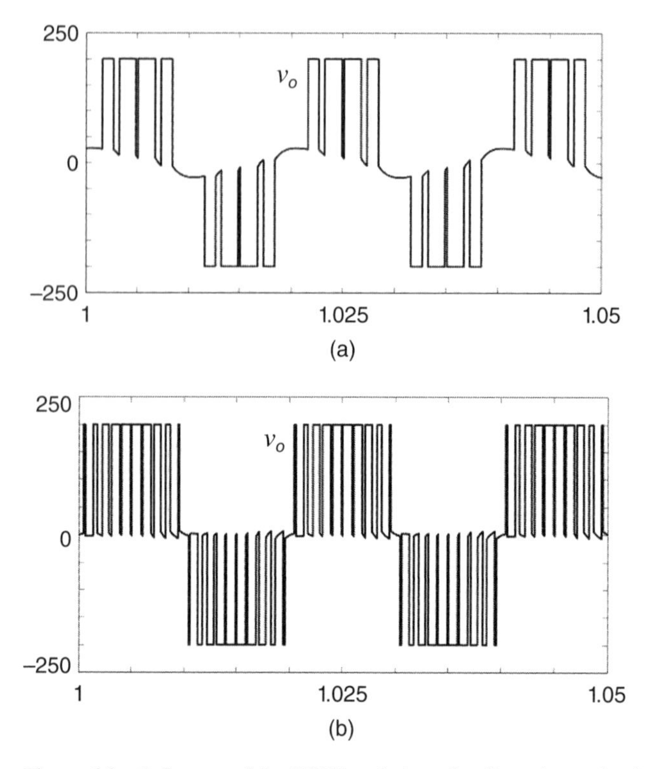

Figure 6.8 Influence of the PWM technique for distortion reduction at the output voltage.

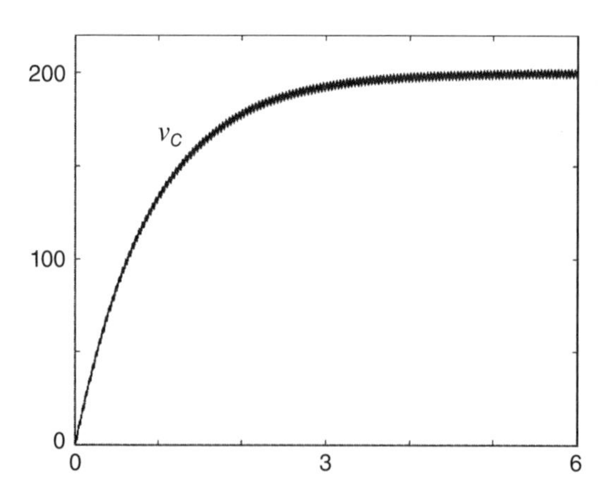

Figure 6.9 Start-up transient for the flying capacitor voltage (v_C).

As mentioned previously, the number of levels considered in this chapter is defined by the voltage between the midpoint of the leg and the dc-link midpoint, that is, through the pole voltage. For instance, both converters in Figs 6.6 and 6.11 are considered as single-phase three-level FC configurations, even though the load voltage in the full-bridge topology presents five levels.

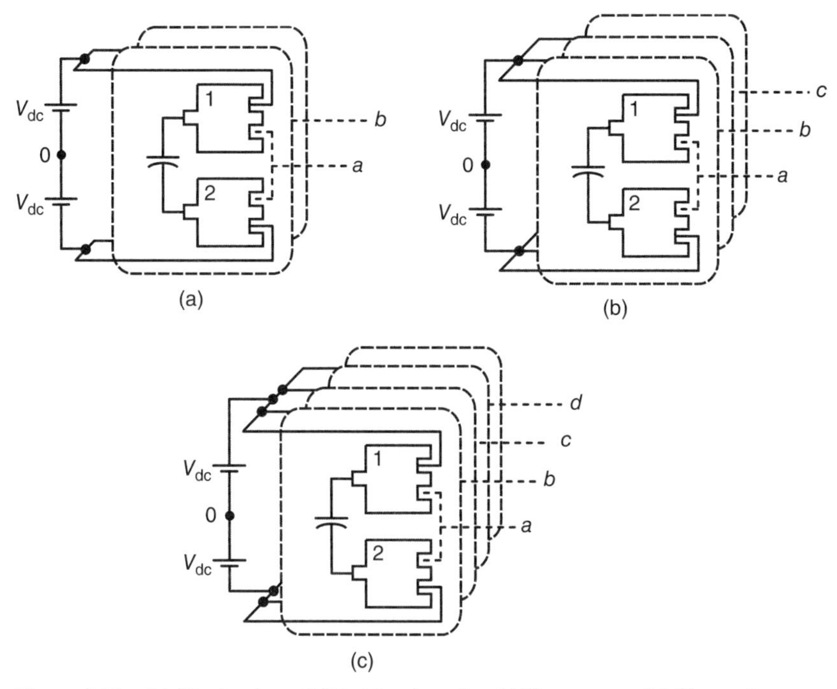

Figure 6.10 (a) Single-phase full-bridge three-level FC converter. (b) Three-phase full-bridge three-level FC converter. (c) Three-phase four-wire three-level FC converter.

6.6 THREE-PHASE FC CONVERTER

A three-phase three-level FC configuration is depicted in Fig. 6.12(a), while Fig. 6.12(b) shows some variables associated with this converter, that is, (from top to bottom) pole voltage, voltage between the load neutral point and dc-link mid-point connection, load phase voltage, and load current, respectively.

The pole voltages for the three-phase converter can be written as follows:

$$v_{a0} = (q_{1a} + q_{2a} - 1)V_{\mathrm{dc}} \tag{6.5}$$

$$v_{b0} = (q_{1b} + q_{2b} - 1)V_{\mathrm{dc}} \tag{6.6}$$

$$v_{c0} = (q_{1c} + q_{2c} - 1)V_{\mathrm{dc}} \tag{6.7}$$

Considering a balanced three-phase load, the voltage v_{n0} can be defined analytically by

$$v_{n0} = \frac{V_{\mathrm{dc}}}{3}(q_{1a} + q_{2a} + q_{1b} + q_{2b} + q_{1c} + q_{2c}) - V_{\mathrm{dc}} \tag{6.8}$$

From (6.8), and considering all 64 switching states, the voltage v_{n0} depicted in Fig. 6.12(b) is responsible for increasing the number of levels in v_{an}. The phase

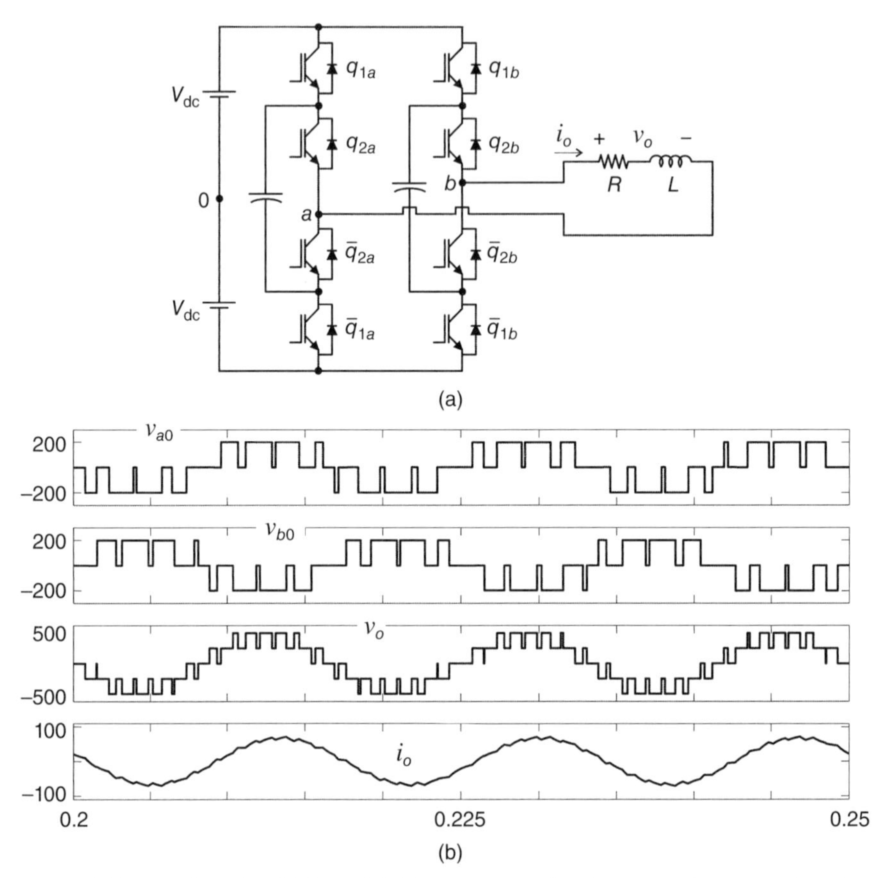

Figure 6.11 (a) Three-level NPC full-bridge converter. (b) (from top to bottom) pole voltage of the first leg (q_{1a} and q_{2a}), pole voltage of the second (q_{1b} and q_{2b}), load voltage, and load current.

voltage for each phase of a balanced three-phase load is given by

$$v_a = \left(\frac{2}{3}q_{1a} + \frac{2}{3}q_{2a} - \frac{1}{3}q_{1b} - \frac{1}{3}q_{2b} - \frac{1}{3}q_{1c} - \frac{1}{3}q_{2c}\right) V_{\mathrm{dc}} \tag{6.9}$$

$$v_b = \left(-\frac{1}{3}q_{1a} - \frac{1}{3}q_{2a} + \frac{2}{3}q_{1b} + \frac{2}{3}q_{2b} - \frac{1}{3}q_{1c} - \frac{1}{3}q_{2c}\right) V_{\mathrm{dc}} \tag{6.10}$$

$$v_c = \left(-\frac{1}{3}q_{1a} - \frac{1}{3}q_{2a} - \frac{1}{3}q_{1b} - \frac{1}{3}q_{2b} + \frac{2}{3}q_{1c} + \frac{2}{3}q_{2c}\right) V_{\mathrm{dc}} \tag{6.11}$$

In this case (balanced three-phase load) the load voltages depend only on state of the switches and dc-link voltage.

Notice that the control circuit is obtained as seen in Fig. 6.4, replicating the PWM control for the other two phases with the following sinusoidal references: $v^*_{\sin 1} = V_p \sin(wt)$, $v^*_{\sin 2} = V_p \sin(wt + 120°)$, and $v^*_{\sin 3} = V_p \sin(wt + 240°)$, where V_p is the peak voltage desired for the load.

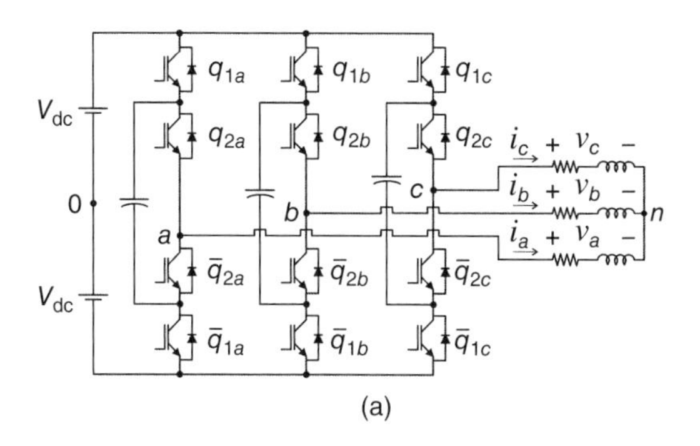

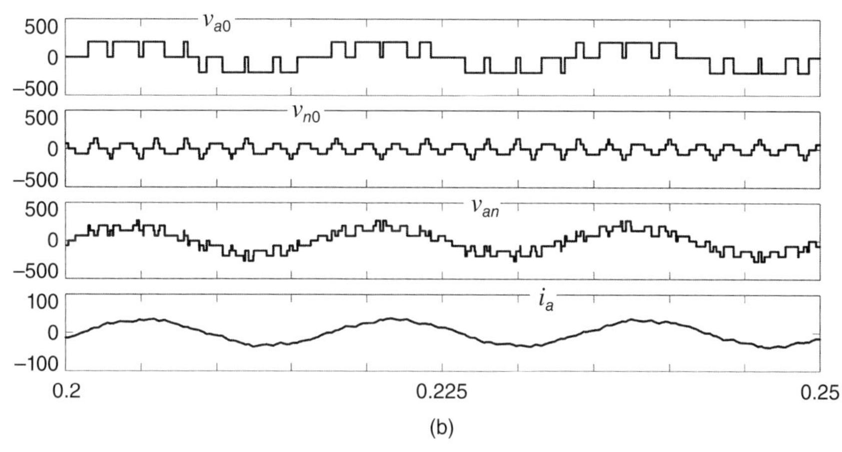

Figure 6.12 (a) Three-level NPC converter. (b) (from top to bottom) pole voltage, voltage v_{n0}, load voltage and load current.

Figure 6.13(a) shows the same power converter topology supplying a three-phase load connected in Δ arrangement. The way the load is connected to the FC topology plays an important role for the converter's design and specification. Figure 6.13(b) shows the impact of the load's connection (i.e., Δ or Y) on the FC voltage ripple of the phase a (v_{Ca}) with Fig. 6.13(b) – top showing the result for Y connection and Fig. 6.13(b) – bottom showing the same result under the same conditions for Δ connection. A FC of 100 μF was considered for both cases in Fig. 6.13. Since the Δ connection implies higher load currents, it also leads to higher FC voltage fluctuation, as highlighted in Fig. 6.13(b) – bottom. Another significant difference expected with the type of load connection (Y or Δ) is in terms of the number of levels obtained on the load phase as follows:

$$v_a = (q_{1a} + q_{2a} - q_{1b} - q_{2b})V_{\text{dc}} \tag{6.12}$$

$$v_b = (q_{1b} + q_{2b} - q_{1c} - q_{2c})V_{\text{dc}} \tag{6.13}$$

$$v_c = (q_{1c} + q_{2c} - q_{1a} - q_{2a})V_{\text{dc}} \tag{6.14}$$

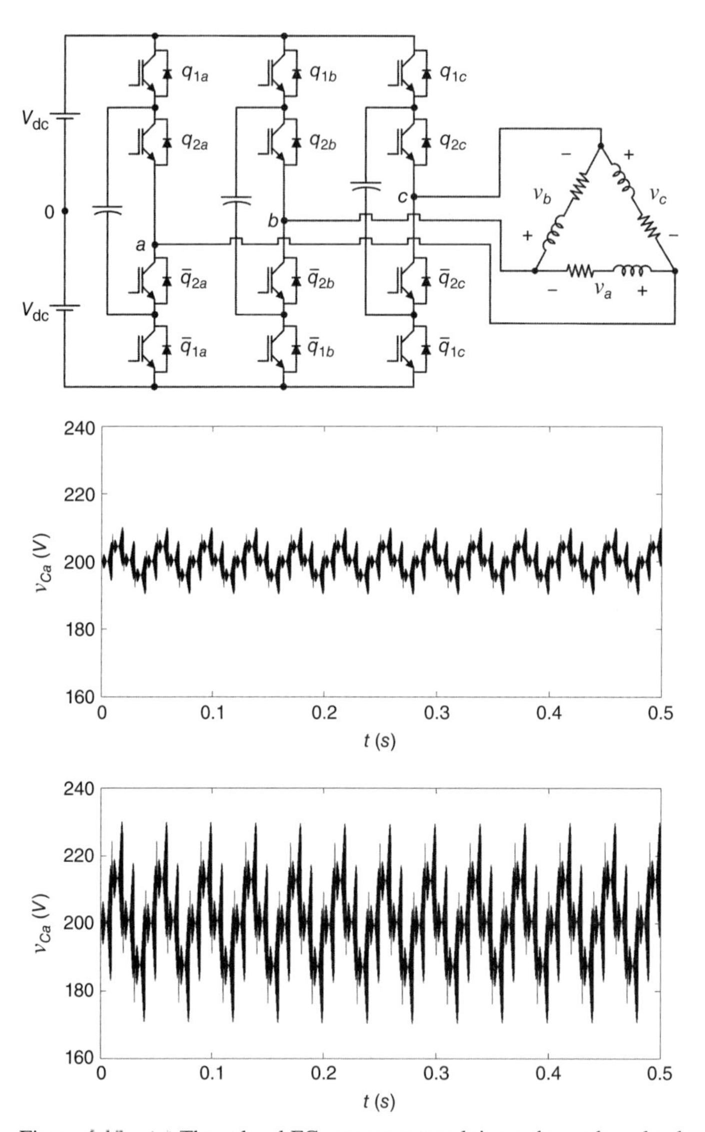

Figure 6.13 (a) Three-level FC converter supplying a three-phase load connected in Δ arrangement (b) FC voltage with (top) wye and (bottom) delta load connection.

6.7 NONCONVENTIONAL FC CONVERTERS WITH THREE-LEVEL LEGS

The goal in this section, as in other chapters in this book, is to present different power circuits obtained by changing the way the switches are arranged. Even assuming that those configurations may not be considered as traditional ones, they are useful in providing a learning intervention. In this sense, following the possibilities regarding the connection between the blocks that define a conventional two-level leg and those for FC three-level leg, it is possible that five- and four-level topologies can be obtained, as in Fig. 6.14(a) and 6.14(b), respectively.

In fact, despite using three dc voltage sources and having lesser levels at the output converter side, the converter in Fig. 6.14(b) has the advantage of operating with switches under the same blocking voltage. To guarantee that all power switches operate under the same blocking voltage and to guarantee a symmetrical output voltage, it is necessary to make $V_1 = V_3 = V_{dc}$ and $V_2 = 2V_{dc}$.

Considering all possibilities of switching states available (that is, eliminating the prohibited states), the output voltage is determined by Table 6.2. From this table and avoiding the undesired states, there are four levels for v_o. It is evident that Fig. 6.14(a) is a particular case of the converter with four levels with $V_1 = V_3 = 0$, which will lead necessarily to a converter with irregular blocking voltage among the switches.

As can be observed in Fig. 6.14(b), the single-phase load is connected between the points a and b of the converter, which leads to an output voltage given by

$$v_o = (2q_{1a} + 2q_{2a} - 2q_b - 1)V_{dc} \tag{6.15}$$

The modulation strategy for this converter can be decided, assuming a combination of the two-level and three-level PWM approaches, which means that one triangular carrier signal (v_{t2}^*) is employed for the two-level leg, and two triangular carrier signals (v_{t1}^* and v_{t3}^*) are used for a three- level leg. Since each leg is capable of synthesizing different values of voltages, that is, $2V_{dc}$ for the two-level leg and $4V_{dc}$ for the three-level leg, the sinusoidal waveforms employed to define PWM signals should follow the same ratio. Indeed, there are two requirements: the reference voltage for the three-level leg must be twice as big as the two-level leg and their difference must be the desired voltage (v_{sin}^*), which leads to

$$v_a^* = \left(\frac{2}{3}\right) v_{sin}^* \tag{6.16}$$

$$v_b^* = -\left(\frac{1}{3}\right) v_{sin}^* \tag{6.17}$$

Figure 6.14(c) shows the analog implementation of the PWM approach for the proposed converter. Note that the level-shift technique has been applied to the FC three-level leg. Figure 6.14(d) shows the PWM signals with three triangular waveforms (v_{t1}^* and v_{t3}^* employed for the three-level leg and v_{t2}^* used for the two-level leg) and two sinusoidal waveforms, as in equations (6.16) and (6.17).

Figure 6.15(a) shows the three-phase version of the FC four-level converter presented in Fig. 6.14(b) by using an open-end motor drive system. Again, this configuration presents all switches processing the voltage with the same blocking voltage. The model and PWM strategy for this topology can be adapted directly from Fig. 6.14(c). Figure 6.15(b) presents the machine phase voltage for phase 1 (v_a). The three-phase two-leg FC topology is depicted in Fig. 6.16.

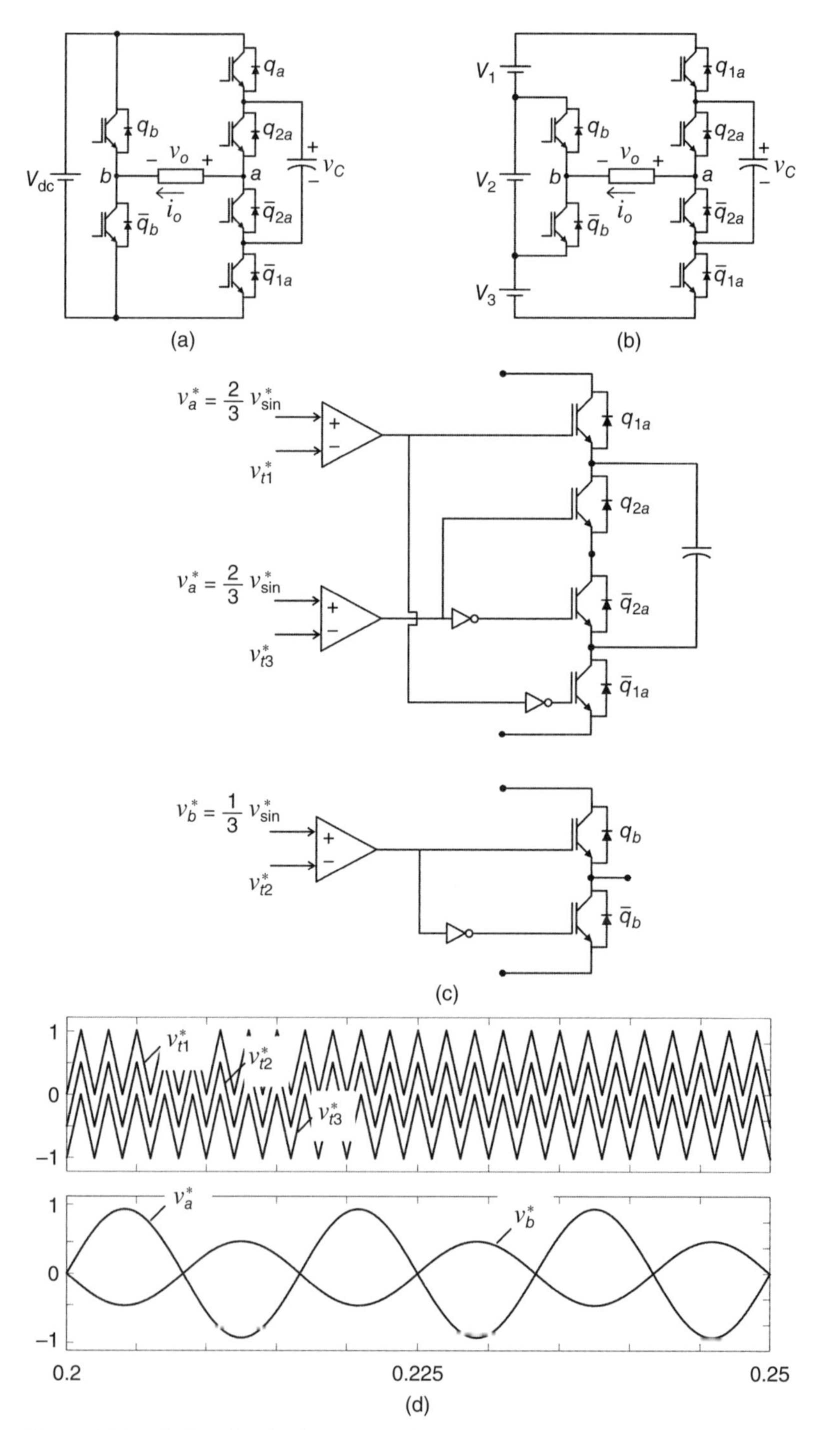

Figure 6.14 (a) Five-level FC topology. (b) Four-level FC topology. (c) PWM strategy for the four-level circuit and (d) its waveforms.

TABLE 6.2 Output Voltage Considering All Switching States Available

State	$\{q_{1a}q_{2a}q_b\}$	v_o
1	{0 0 0}	$-V_{dc}$
2	{0 0 1}	$-3V_{dc}$
3	{0 1 0}	$-V_{dc}$
4	{0 1 1}	V_{dc}
5	{1 0 0}	
6	{1 0 1}	
7	{1 1 0}	$3V_{dc}$
8	{1 1 1}	V_{dc}

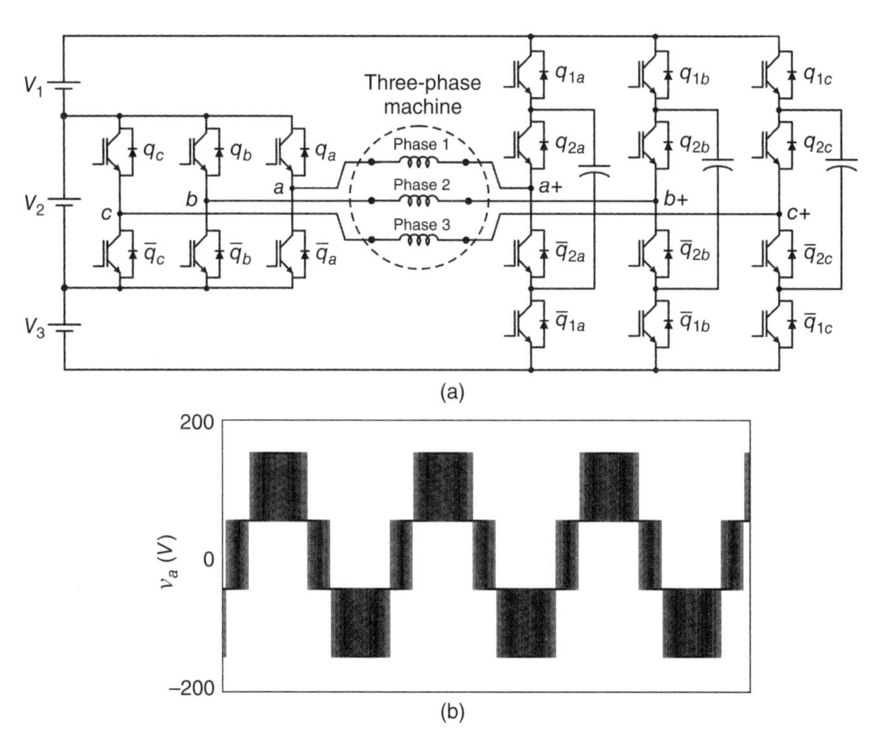

Figure 6.15 (a) Three-phase version of the four-level FC converter with open-end winding three-phase motor. (b) Machine phase voltage.

6.8 FOUR-LEVEL CONFIGURATION

Following the axioms and postulates involving the PBs, presented in the Chapter 3, the four-level FC leg can be obtained as seen in Fig. 6.17(a). Figure 6.17(b)–6.17(i) shows the methods used to obtain the levels at the output converter side, specified in Table 6.3. Figure 6.17(b) illustrates that it is necessary to activate the PBs 2 and 3 to obtain $v_o = -V_4$, while $v_o = V_1$ is obtained when PBs 1 and 2 are activated.

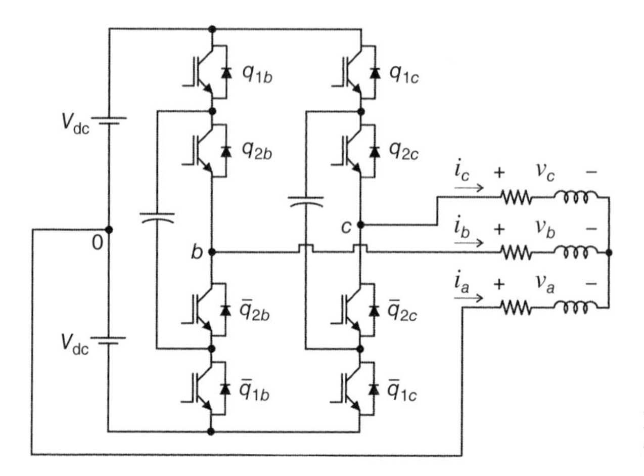

Figure 6.16 Three-phase two-leg FC converter.

TABLE 6.3 Table of Variables for the Converter in Fig. 6.17(a)

State	q_1	q_2	q_3	v_o
1	0	0	0	$-V_4$
2	0	0	1	$V_3 - V_4$
3	0	1	0	$-V_3 + V_2 - V_4$
4	0	1	1	$V_2 - V_4$
5	1	0	0	$-V_2 + V_1$
6	1	0	1	$V_3 - V_2 + V_1$
7	1	1	0	$-V_3 + V_1$
8	1	1	1	V_1

The other levels are: $v_o = V_3 - V_4$ [Fig. 6.17(c)], $v_o = -V_3 + V_2 - V_4$ [Fig. 6.17(d)], $v_o = V_2 - V_4$ [Fig. 6.17(e)], $v_o = -V_2 + V_1$ [Fig. 6.17(f)], $v_o = V_3 - V_2 + V_1$ [Fig. 6.17(g)], $v_o = -V_3 + V_1$ [Fig. 6.17(h)], and $v_o = V_1$ [Fig. 6.17(i)]. In general terms, to generate an output waveform with four symmetrical levels as in Fig. 6.18, it is necessary to activate the PBs in a sequence as observed in the same figure, with $V_1 = 1.5V_{dc}$, $V_2 = 2V_{dc}$, $V_3 = 1V_{dc}$, and $V_4 = 1.5V_{dc}$. Table 6.4 shows the symmetrical levels as a function of the switching states. The intermediate levels ($-0.5V_{dc}$ and $0.5V_{dc}$) can be obtained with three different circuits; such a characteristic can be employed to guarantee the FC voltage with reduced ripple.

The equation employed to describe the output voltage v_o as a function of both the input voltages and the binary variables (q_1, q_2, and q_3) is presented below. Theoretically, it is possible to guarantee each input voltage at v_o using only two binary variables, since $2^2 = 4$. However, it is necessary to associate each binary variable to a switching state of the four-level configuration, which means a minimum of three binary variables (q_1, q_2, and q_3).

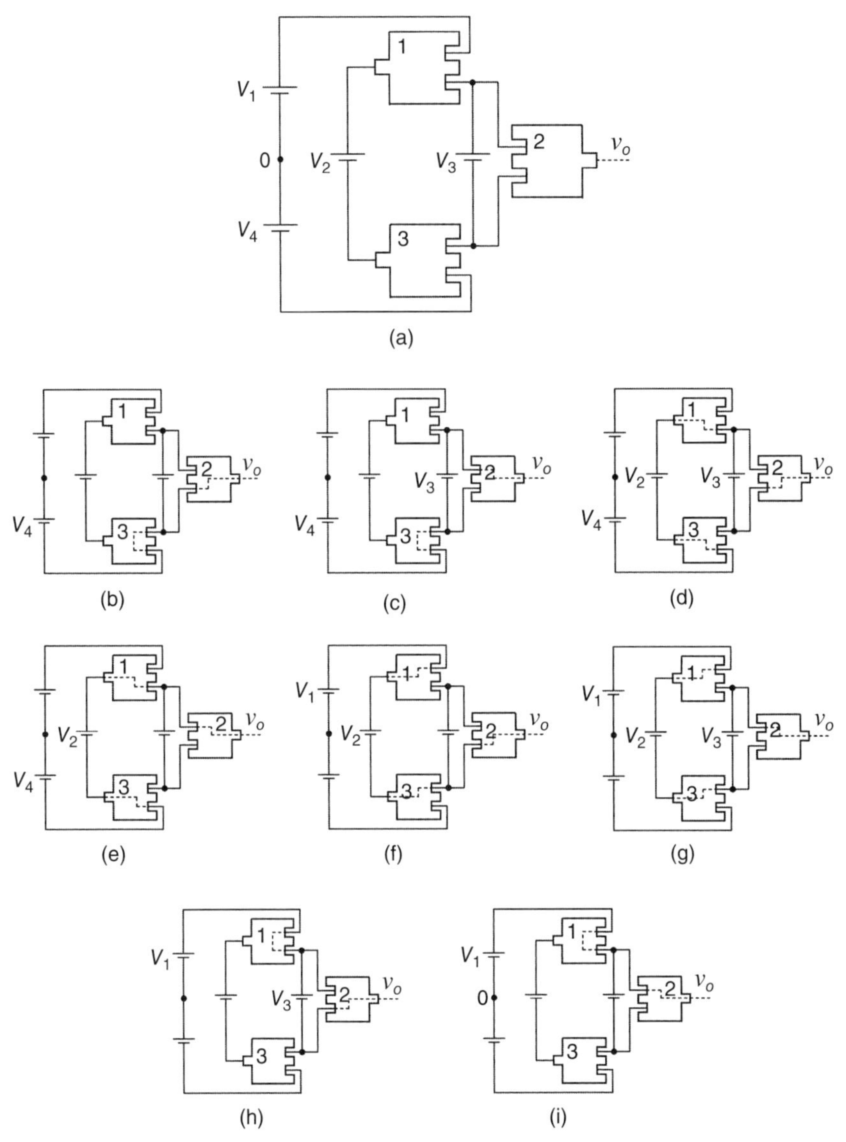

Figure 6.17 (a) Four-level FC configuration with PBs. (b) $v_o = -V_4$. (c) $v_o = V_3 - V_4$. (d) $v_o = -V_3 + V_2 - V_4$. (e) $v_o = V_2 - V_4$. (f) $v_o = -V_2 + V_1$. (g) $v_o = V_3 - V_2 + V_1$. (h) $v_o = -V_3 + V_1$. (i) $v_o = V_1$.

In this sense, the output pole voltage is given by

$$\begin{aligned} v_o = {} & \overline{q}_1\overline{q}_2\overline{q}_3(-1.5V_{\text{dc}}) + \overline{q}_1\overline{q}_2 q_3(-0.5V_{\text{dc}}) + \overline{q}_1 q_2\overline{q}_3(-0.5V_{\text{dc}}) \\ & + \overline{q}_1 q_2 q_3(0.5V_{\text{dc}}) + q_1\overline{q}_2\overline{q}_3(-0.5V_{\text{dc}}) + q_1\overline{q}q_3(0.5V_{\text{dc}}) \\ & + q_1 q_2\overline{q}_3(0.5V_{\text{dc}}) + q_1 q_2 q_3(1.5V_{\text{dc}}) \end{aligned} \tag{6.18}$$

where $\overline{q}_1 = 1 - q_1$, $\overline{q}_2 = 1 - q_2$, and $\overline{q}_3 = 1 - q_3$.

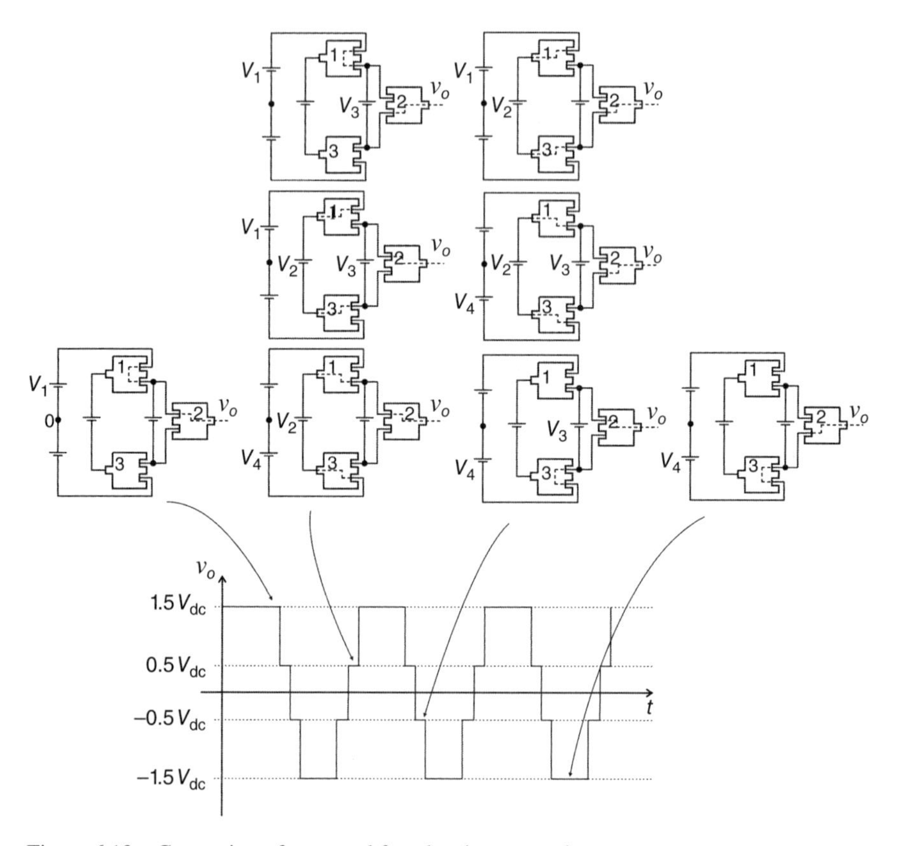

Figure 6.18 Generation of a general four-level output voltage.

TABLE 6.4 Table of Variables for the Converter in Fig. 6.18(a) with $1.5V_{dc}$, $0.5V_{dc}$, $-0.5V_{dc}$, or $-1.5V_{dc}$

State	q_1	q_2	q_3	v_o
1	0	0	0	$-1.5V_{dc}$
2	0	0	1	$-0.5V_{dc}$
3	0	1	0	$-0.5V_{dc}$
4	0	1	1	$0.5V_{dc}$
5	1	0	0	$0.5V_{dc}$
6	1	0	1	$0.5V_{dc}$
7	1	1	0	$0.5V_{dc}$
8	1	1	1	$1.5V_{dc}$

Developing (6.18) it is possible to write v_o as follows:

$$v_o = (q_1 + q_2 + q_3)V_{dc} - 1.5V_{dc} \tag{6.19}$$

As in the three-level FC configuration, the four-level one is able to generate the desired output voltage either for positive or negative voltage.

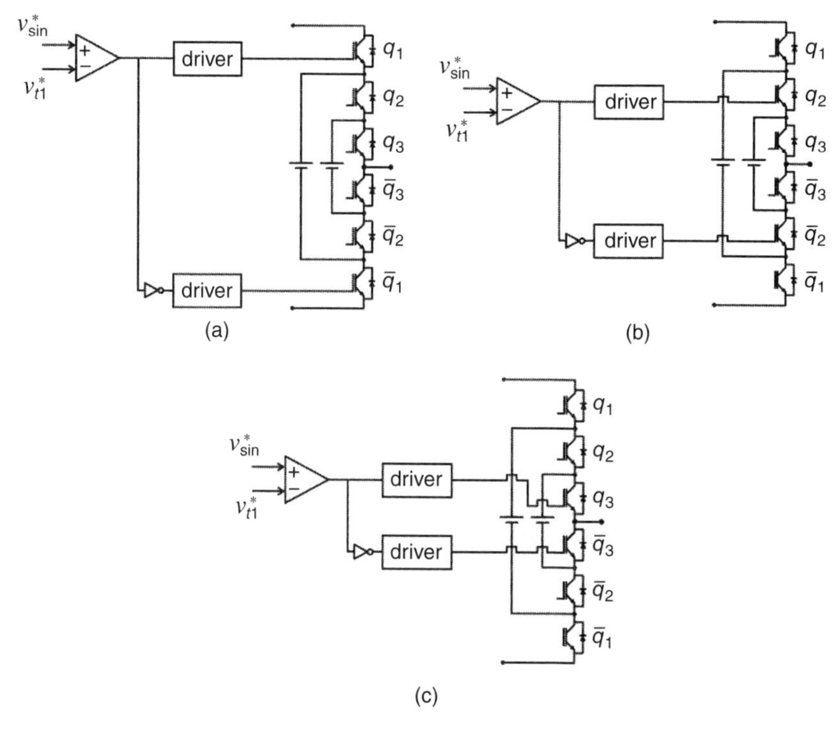

Figure 6.19 Modulation strategy for gating signal generation.

Considering the need for generation of an output waveform with reduced harmonic distortion, Fig. 6.19 shows the analog solution for the PWM signal generation. The states of the six switches are determined by a comparison among three high frequency triangular waveforms (v^*_{t1}, v^*_{t2}, and v^*_{t3}) and a sinusoidal waveform (v^*_{sin}). Such a comparison will allow the generation of the levels as in Table 6.4. Figure 6.20 shows the power converter and waveforms for a half-bridge four-level configuration.

Although the topologies in Figs 6.17–6.20 present the conventional way to obtain a four-level voltage by using FC configuration, it is also possible to obtain a four-level output voltage with only two PBs, that is, with four switches per leg, as seen in Fig. 6.21(a). In this case, it is necessary to guarantee the following relationship among the input voltage sources: $V_1 = V_3 = 2V_{dc}$ and $V_2 = V_{dc}$ and adapt the PWM generation to assure an output voltage with four levels. The pole voltage is given by

$$v_o = (3q_1 + q_2 - 2)V_{dc} \tag{6.20}$$

Figure 6.21(b) shows the desired sinusoidal voltage that must be synthesized by the studied converter. This waveform is intentionally separated into four sectors. To synthesize the desired voltage in each sector it is necessary to activate the switches q_1 and q_2 as described in Fig. 6.21(b) – bottom. While Fig. 6.22(a) shows the analog

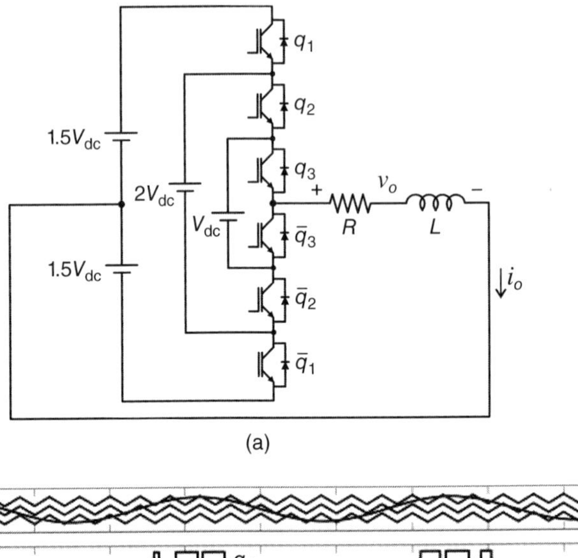

(a)

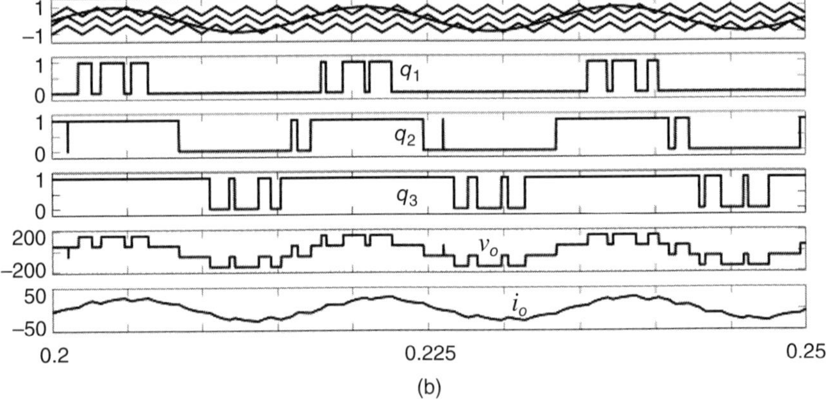

(b)

Figure 6.20 (a) Four-level FC half-bridge converter. (b) (from top to bottom) $v^*_{\sin}$, v^*_{t1}, v^*_{t2}, v^*_{t3}; gating signal of switches q_1, q_2, and q_3; load voltage and load current.

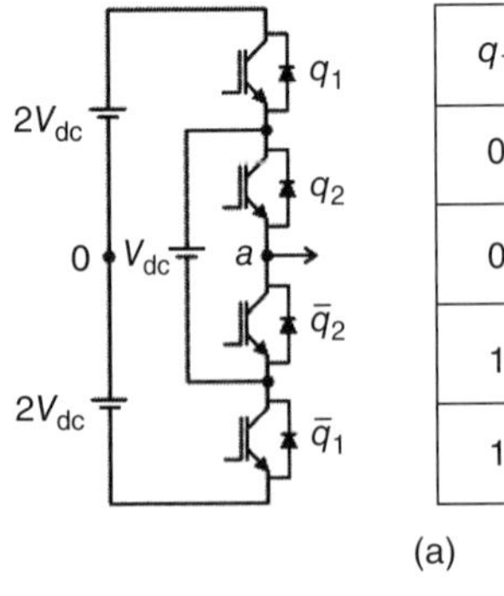

q_1	q_2	v_{a0}
0	0	$-2V_{dc}$
0	1	$-V_{dc}$
1	0	V_{dc}
1	1	$2V_{dc}$

(a)

Figure 6.21 (a) Four-level FC converter with four switches (b) Desired waveform divided in sectors.

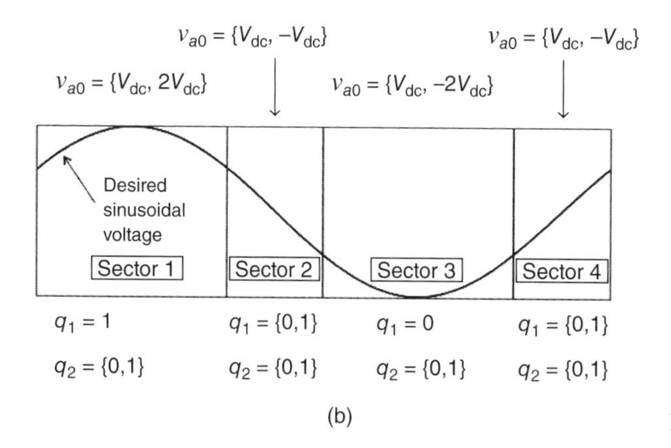

(b)

Figure 6.21 (*Continued*)

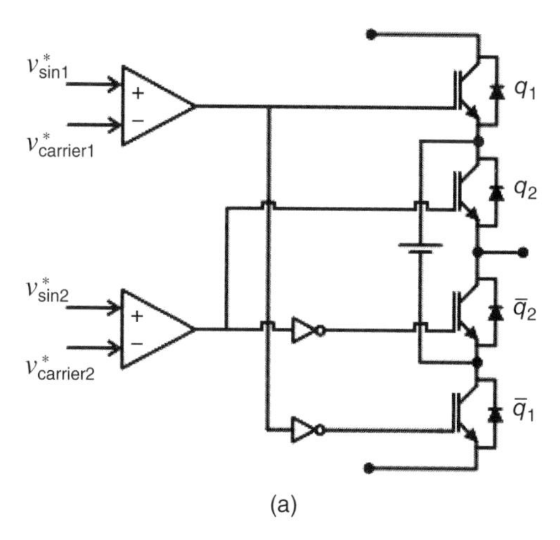

(a)

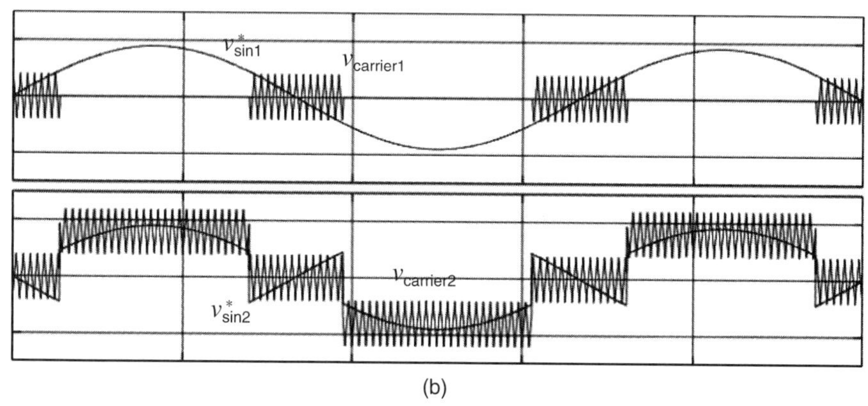

(b)

Figure 6.22 (a) Four-level FC converter with four switches (b) Desired waveform divided in sectors.

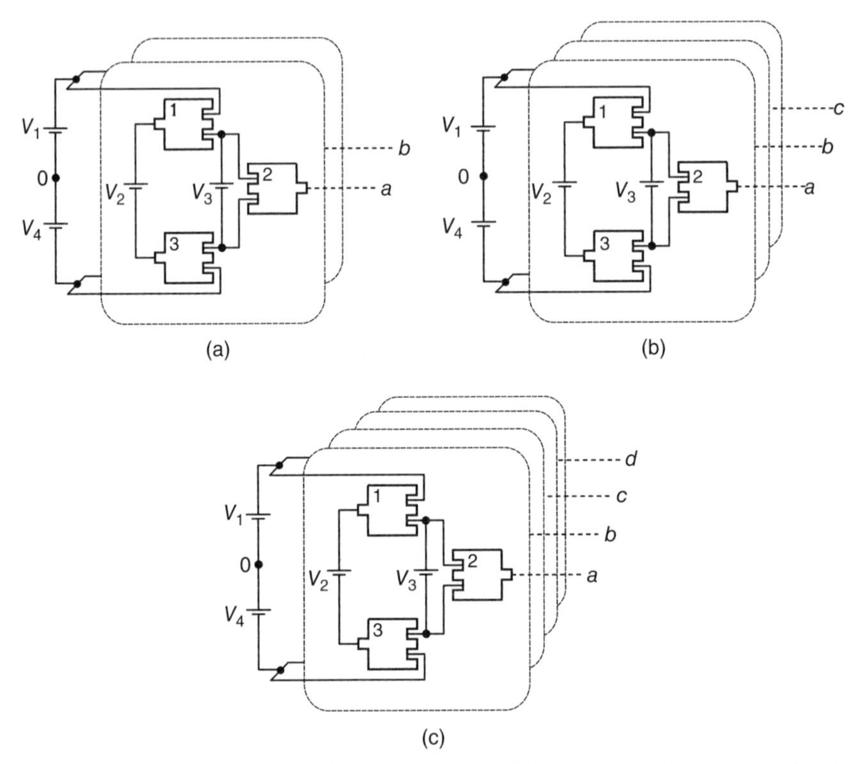

Figure 6.23 (a) Single-phase full-bridge four-level FC converter. (b) Three-phase four-level FC converter. (c) Three-phase four-wire four-level FC converter.

circuit for the four-level topology in Fig. 6.21(a), Fig. 6.22(b) shows its PWM signals. Notice that the voltages $v^*_{\sin 1}$, $v^*_{\sin 2}$, v_{carrier1}, and v_{carrier2} furnish the switching states necessary for a symmetrical output voltage with four levels.

Full-bridge four-level FC converters for both single-phase and three-phase loads are obtained by connecting the PBs as in Fig. 6.23(a) and 6.23(b), respectively. Figure 6.23(c) shows the option in which a three-phase four-wire system is required.

6.9 GENERALIZATION

Following the same approach as for the three- and four-level configurations, the N-level converter can be obtained as in Fig. 6.24(e) for an odd value for N, and in Fig. 6.24(f) for an even value for N. Theoretically, the number of levels in an FC topology can be increased as far as necessary, although practical limitations do appear when the number of levels is higher than five. Note that the evolution from three to N levels is presented in Fig. 6.24(a) to Fig. 6.24(f). It is worth mentioning that the PBG brings a simple and intuitive aspect of connecting blocks for creation of FC topologies from three to N levels. The model and PWM strategy applied for

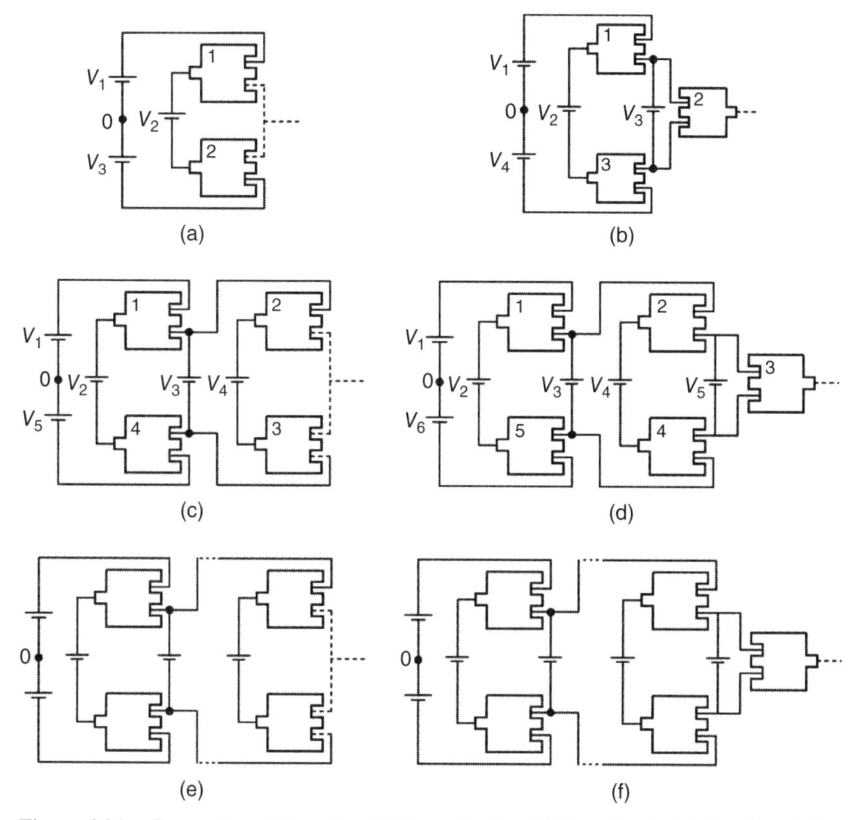

Figure 6.24 Generation of the n-level FC topologies: (a) three-level, (b) four-level, (c) five-level, (d) six-level, (e) N-level (with N odd), and (f) N-level (with N even).

the configurations with number of levels higher than four can be easily adapted from the previous discussion in this chapter.

6.10 SUMMARY

This chapter was dedicated to the multilevel topologies that are characterized by the presence of the FC employed to generate different levels. For the sake of simplification, first the FC was substituted by a DC source and then the circuit with capacitor was presented along with the analysis to keep v_c with minimum ripple. As in the case of other configurations in this book, such as neutral-point-clamped and cascade arrangements, FC topologies were presented in a systematic way following this sequence: (i) PB connections obtained from the rules presented in Chapter 3, (ii) once the topology was established the Table of Variables was obtained, (iii) modeling the converter by writing the output voltage to describe what was obtained in the Table of Variables, and (iv) developing the PWM approach from the Table of Variables. As a learning intervention, nonconventional converters were also employed in this chapter. There is an extensive technical literature on this topic some of which is given as References from 1 to 9.

REFERENCES

[1] Meynard TA, Foch H. Multilevel conversion: high voltage chopper and voltage source inverters. Proceedings of IEEE PESC'92; 1992. p 397–403.

[2] Meynard TA, Foch H, Thomas P, Courault J, Jakob R, Nahrstaedt M. Multicell converters: basic concepts and industry applications. IEEE Trans Ind Electron 2002;49(5):955–964.

[3] Jeon J-H, Kim T-J, Kang D-W, Hyun D-S. A symmetric carrier technique of CRPWM for voltage balance method of flying capacitor multi-level inverter. IEEE Trans Ind Electron 2005;52(3):879–888.

[4] Lin B-R, Hung T-L, Huang C-H. "Single-phase AC/AC converter with capacitor-clamped scheme". IEE Proc-Electr Power Appl 2003;150(4):464–470.

[5] Lin BR, Huang CH. Single-phase switching-mode rectifier with capacitor-clamped topology. IEE Proc-Electr Power Appl 2005;152(1):9–16.

[6] Lin B-R, Huang C-H. Three-phase capacitor-clamped converter with fewer switches for use in power factor correction. IEE Proc-Electr Power Appl 2005;152(3):596–604.

[7] Kou X, Corzine KA, Familiant Y. Full binary combination schema for floating voltage source multi-level inverters. IEEE Trans Power Electron 2002;17(6):891–897.

[8] Jing Huang, Keith Corzine. Extended operation of flying capacitor multilevel inverters. Proceedings of IEEE IAS' 04; 2004. p 813–819.

[9] Radermacher H, Schmidt B, De Doncker R. Determination and comparison of losses of single phase multi-level inverters with symmetric supply. Proceedings of IEEE PESC'04; vol. 6; 2004. p 4428–4433.

CHAPTER 7

OTHER MULTILEVEL CONFIGURATIONS

7.1 INTRODUCTION

After dealing with either conventional power circuits for multilevel topologies or nontraditional configurations based on the three most used circuits (neutral point clamped (NPC), cascade, and flying capacitor (FC)), this chapter presents some other topologies obtained independently. Such power circuits are also easily obtained from the axioms and postulates of the power blocks geometry (PBG) presented in Chapter 3.

Although many of the circuits presented in this chapter have not been considered in industrial applications, such converters have characteristics that bring up specific advantages as compared to the traditional circuits. Furthermore, the nonconventional multilevel configurations presented in this chapter have an important role in helping to understand how new topologies have been conceived in the literature. Such NPCs are presented, following the same procedure as before, with the following steps: (i) consider the PBs and the rules to determine the connection of blocks and consequently the connection of the power switches, (ii) determine a Table of Variables to associate the output voltage with binary variables (switching states), (iii) find an equation to describe the Table of Variables and consequently model the converter, and (iv) determine a simple pulse width modulation (PWM) approach to obtain the gating signals of the switches.

The sections in this chapter deal with different types of nontraditional converters and are organized as follows: Section 7.2 presents the nested configuration achieved by connecting only the first and second PB-ac in an appropriate way. Section 7.3 shows an alternate way to obtain a multilevel output voltage with a coupled inductor placed at the output of the converter, as well as a topology known as modular multilevel converter (MMC). Section 7.4 presents the ANPC topology, highlighting its thermal stress reduction among the power switches. A series of other topologies are included in Section 7.5, which brings up a review stressing how the multilevel circuits proposed in the technical literature can be in fact obtained easily from the Power Block Geometry approach. Finally, Section 7.6 summarizes this chapter.

Advanced Power Electronics Converters: PWM Converters Processing AC Voltages,
Forty Fifth Edition. Euzeli Cipriano dos Santos Jr. and Edison Roberto Cabral da Silva.

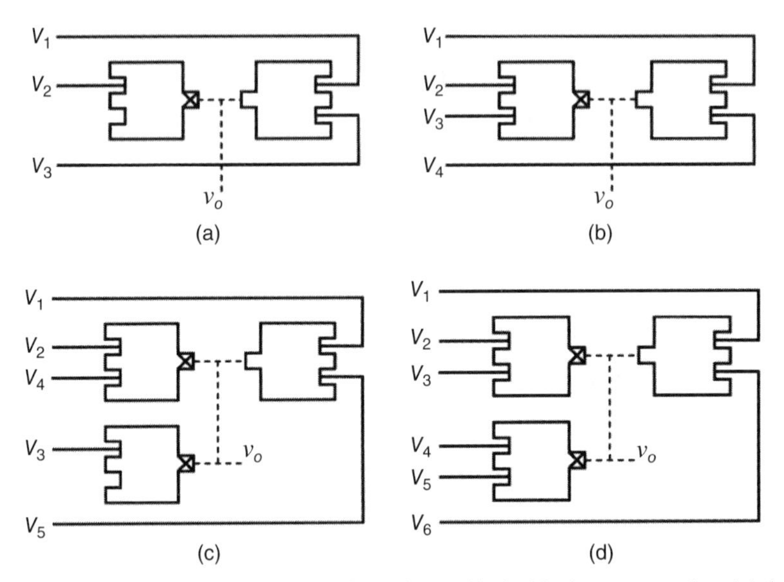

Figure 7.1 Nested multilevel configurations with the block representation: (a) three-level, (b) four-level, (c) five-level, and (d) six-level.

7.2 NESTED CONFIGURATION

This section presents multilevel topologies based on the concept of nested arrangement. Such topologies are called nested multilevel converters because the external connections of the blocks involve the internal ones. Figure 7.1(a) shows a three-level circuit obtained with the first PB-ac and second PB-ac. Figure 7.1(b), 7.1(c), and 7.1(d) depict a four-, five-, and six-level converters, respectively. Once the topologies are obtained via the power blocks, the circuits can be considered in their lower level (representation with switches), as presented in Fig. 7.2. Figure 7.2(a) and 7.2(b) show three- and four-level converters, respectively. The converters presented in this figure show both single-leg and three-phase versions of the nested configuration. Figure 7.2(c) shows the evolution from five to seven levels.

When compared to the NPC topologies, the nested configurations can be considered as an interesting option due to the reduced number of power diodes. A drawback would be different blocking voltage per switch.

Although Figs 7.1 and 7.2 present topologies able to generate a number of levels between 3 and 7, the analysis in this section will be focused on the four-level converter, which is enough to give the reader a comprehensive idea of how to deal with this family of multilevel converters.

The converter leg in Fig. 7.2(b) (left side) is constituted by two controlled switches (q_1 and q_4) and two bidirectional controlled switches (q_2 and q_3). For the version of the converter with three legs, required in three-phase systems, there are six controlled switches (q_{1x} and q_{4x} – $x = a, b$, and c) and six bidirectional controlled

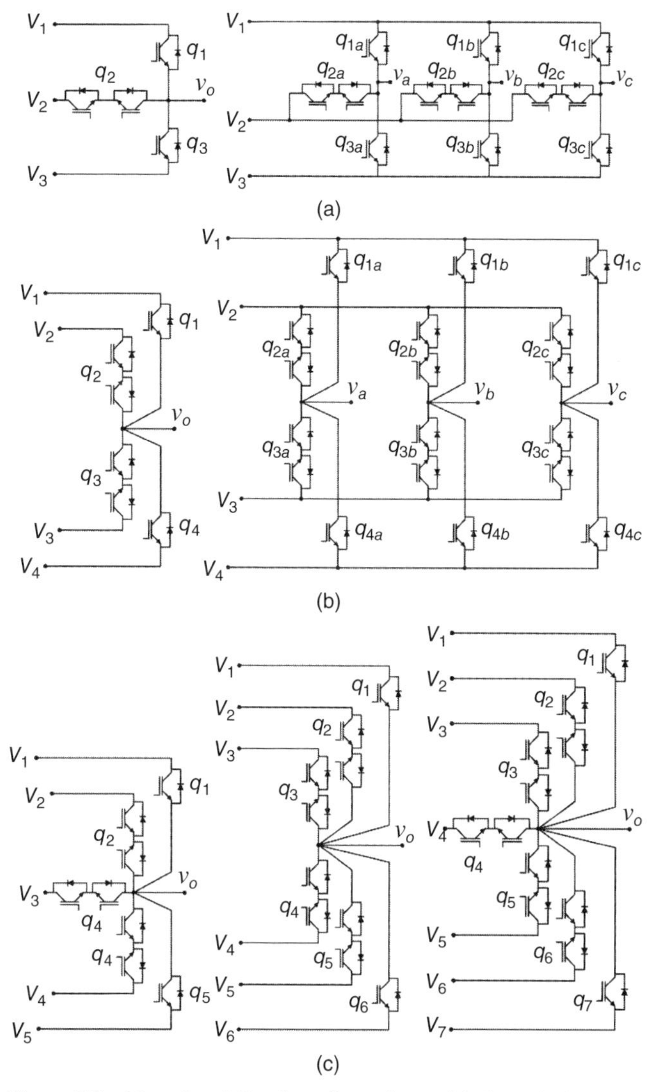

Figure 7.2 Nested multilevel configurations with the switches representation: (a) three-level, (b) four-level, and (c) evolution from five- to seven-level.

switches (q_{2x} and q_{3x}). The Table of Variables for the leg of the nested topology with four levels is presented in Table 7.1.

Such a table can be described by the following equation:

$$v_o = q_1\overline{q}_2\overline{q}_3\overline{q}_4V_1 + \overline{q}_1q_2\overline{q}_3\overline{q}_4V_2 + \overline{q}_1\overline{q}_2q_3\overline{q}_4V_3 + \overline{q}_1\overline{q}_2\overline{q}_3q_4V_4 \tag{7.1}$$

with $\overline{q}_y = 1 - q_y$ for $y = 1, 2, 3,$ and 4.

TABLE 7.1 Table of Variables for the Four-Level Nested Converter

State	q_1	q_2	q_3	q_4	v_o
1	1	0	0	0	V_1
2	0	1	0	0	V_2
3	0	0	1	0	V_3
4	0	0	0	1	V_4

For the generation of a symmetrical waveform with the maximum output voltage given by V_{dc}, the input voltage can be obtained as follows: $V_1 = 3V_{dc}/4$, $V_2 = V_{dc}/4$, $V_3 = -V_{dc}/4$, and $V_4 = -3V_{dc}/4$. In this case, it turns out that equation (7.1) becomes

$$v_o = \frac{(3q_1\overline{q}_2\overline{q}_3\overline{q}_4 + \overline{q}_1q_2\overline{q}_3\overline{q}_4 - \overline{q}_1\overline{q}_2q_3\overline{q}_4 - 3\overline{q}_1\overline{q}_2\overline{q}_3q_4)V_{dc}}{4} \tag{7.2}$$

Figure 7.3(a)–7.3(d) shows the positive and negative currents through the switches q_1, q_2, q_3, and q_4 when these switches are turned on, respectively. Notice that the bidirectional controlled switches (q_2 and q_3) have been employed on the inner leg, while the switches (q_1 and q_4) have been used on the outer leg.

Exercise 7.1

Explain what would happen if the positions of the switches placed on the inner and outer legs are switched, for example, if the figure below is used instead of Fig. 7.1(b). Justify your answer by using equivalent circuits such as in Fig. 7.3.

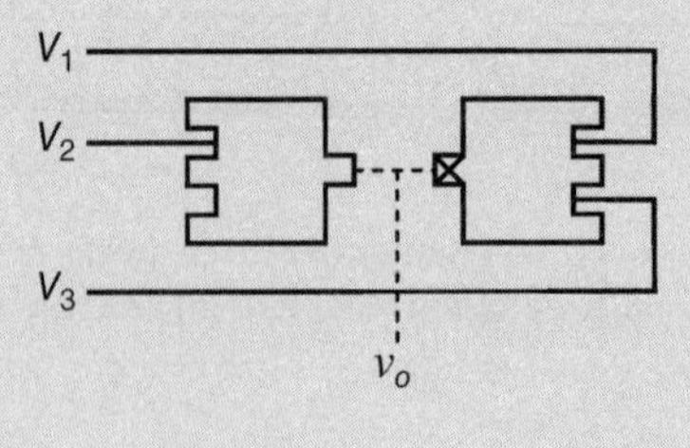

From the Table of Variables (Table 7.1), it is possible to determine a PWM solution as shown in Fig. 7.4. Notice that the PWM approach for the nested topology can be obtained with only one modulating signal and three carrier signals with a level shift technique.

Figure 7.5 shows the simulated results for the nested configuration with four levels for a three-phase application. The waveforms presented in Fig. 7.5(a) and 7.5(b)

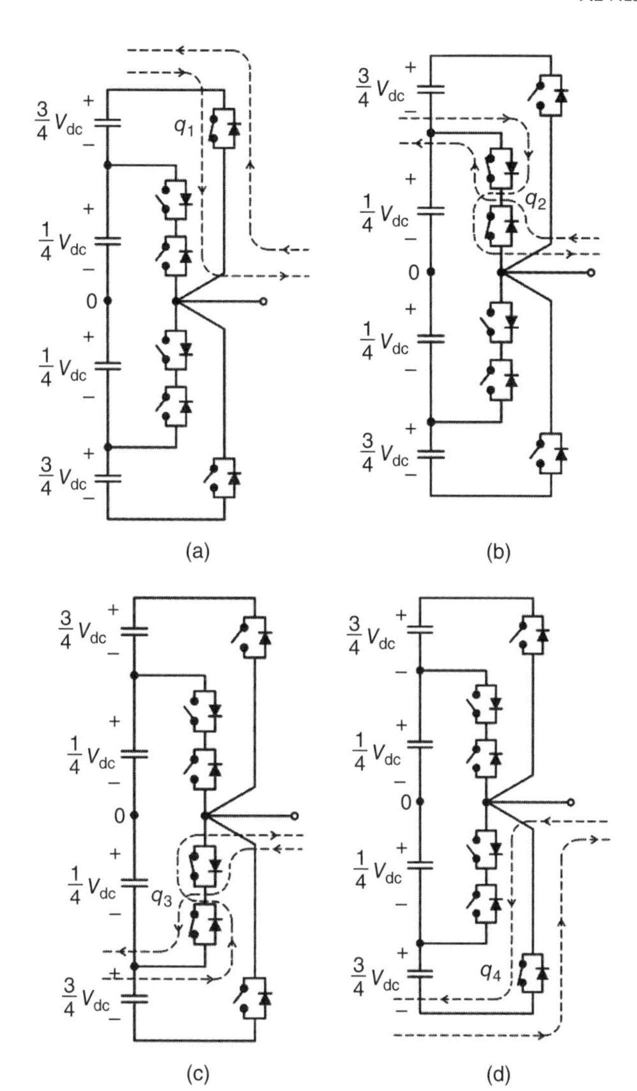

Figure 7.3 Current flows through the switches when (a) $q_1 = 1$, (b) $q_2 = 1$, (c) $q_3 = 1$, and (d) $q_4 = 1$.

are: (a) pole voltages; (b) from top to bottom: output line voltage, pole voltage, and phase current.

Exercise 7.2

Compare the nested and the NPC topologies in terms of the blocking voltage for the switches employed in one leg. Assume that both topologies have four levels and are generating the same output voltage. Disregard any safety margin while specifying the voltages for each power switch.

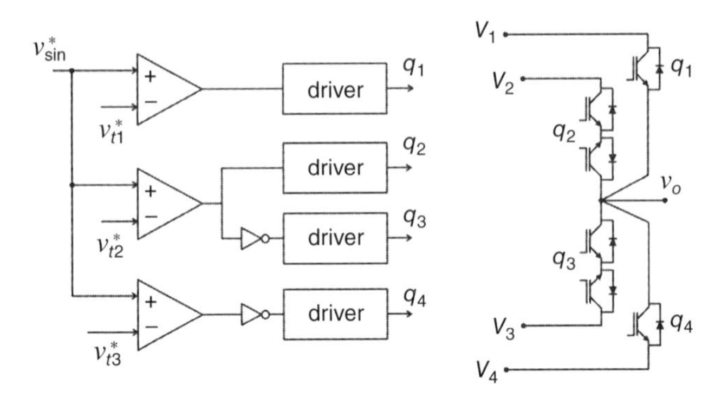

Figure 7.4 PWM approach for the nested configuration.

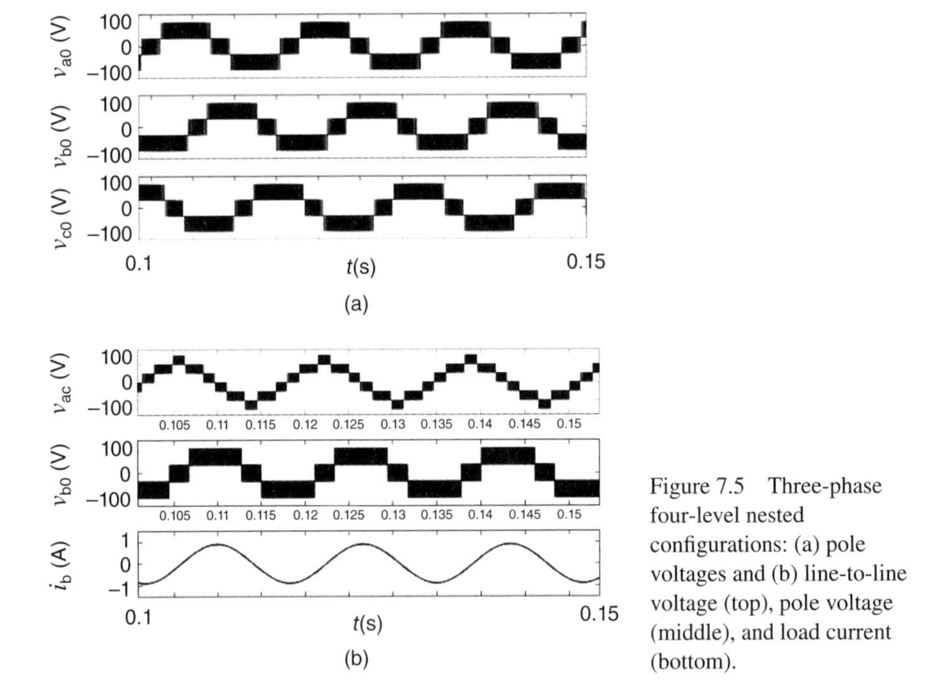

Figure 7.5 Three-phase four-level nested configurations: (a) pole voltages and (b) line-to-line voltage (top), pole voltage (middle), and load current (bottom).

Although sometimes there are practical limits for the converters with a higher number of levels, the nested topology allows its generation with N levels for both cases, N being an odd and an even number. Figure 7.6 presents a generalization of the nested multilevel converters for an odd [see Fig. 7.6(a)] and even [see Fig. 7.6(b)] number of levels.

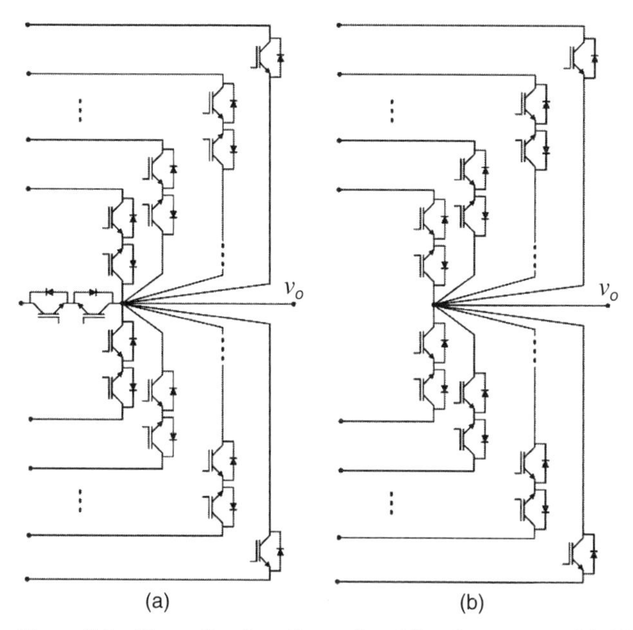

Figure 7.6 Generalization of nested multilevel converter with (a) odd and (b) even number of levels.

7.3 TOPOLOGY WITH MAGNETIC ELEMENT AT THE OUTPUT

A three-level voltage can be obtained either by using the nested topology (as described in Section 7.2) or by employing any of the conventional circuits (described in previous chapters). All circuits considered so far for generation of a three-level voltage at the output of the converter have been achieved by placing switches, sources, and/or capacitors appropriately, as seen in Fig. 7.7(a)–7.7(c).

The figures of the NPC, H-bridge, and FC configurations are repeated here in order to compare their principles with an alternate way to generate a three-level voltage. Observing such topologies, it is possible to use different strategies to generate the same number of levels at the output converter side. For instance, Fig. 7.7(a) shows an NPC topology that creates three different paths to connect the input voltages V_1, V_2, and V_3 to the output (v_o). The H-bridge circuit shown in Fig. 7.7(b) combines both inputs to obtain the desirable three-level output voltage, that is, $V_1 - V_2$, 0, $V_2 - V_1$. On the other hand, the FC topology in Fig. 7.7(c) associates sources (or sources and capacitors) in series to obtain the desirable levels at the output.

By employing a different approach, Fig. 7.7(d) shows another way to implement a three-level power converter. Such an arrangement, using a split-wound coupled inductor at the output converter side, guarantees a three-level output voltage with only two PBs-dc. The lower level representation of this converter is furnished

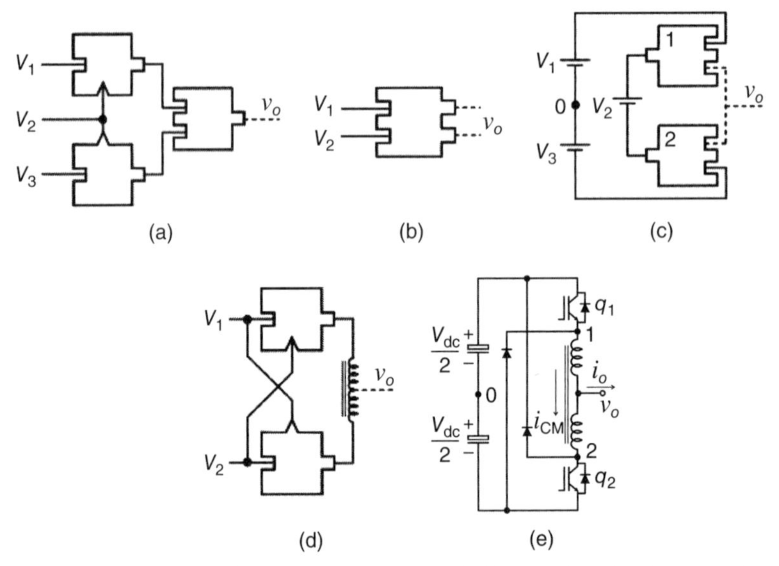

Figure 7.7 Three-level configurations: (a) NPC, (b) cascade, (c) flying capacitor, (d) configuration with two PB-dc and split-wound coupled inductors, and (e) representation of the converter in Fig. 7.7(d) with switches.

in Fig. 7.7(e), which highlights the connections of the power switches and coupled inductor.

It is assumed that the currents through the coupled inductor are in continuous conduction mode, that is, the current is always higher than zero. Figure 7.8(a), 7.8(b), 7.8(c), and 7.8(d) shows the equivalent circuit for the leg when (q_1,q_2) are given by (0,0), (0,1), (1,0), and (1,1), respectively. Table 7.2 shows the Table of Variables for this converter. Notice that the output voltage has three levels, as explained later.

The voltages v_{10} and v_{20} [voltages from the points 1 and 2 to the dc-link capacitor midpoint – "0" in Fig. 7.7(e)] can be expressed as a function of the state of the switches q_1 and q_2, as follows:

$$v_{10} = (2q_1 - 1)\frac{V_{\rm dc}}{2} \tag{7.3}$$

$$v_{20} = (2q_2 - 1)\frac{-V_{\rm dc}}{2} \tag{7.4}$$

The pole voltages as in equations (7.3) and (7.4) lead to the model presented in Fig. 7.9(a). Then the output voltage v_o with respect to "0" can be in turn written as

$$v_o = \frac{1}{2}(v_{10} + v_{20}) \tag{7.5}$$

Substituting (7.3) and (7.4) in (7.5) it is possible to write v_o as a function only of the state of the switches and dc-link voltage

$$v_o = (q_1 - q_2)\frac{V_{\rm dc}}{2} \tag{7.6}$$

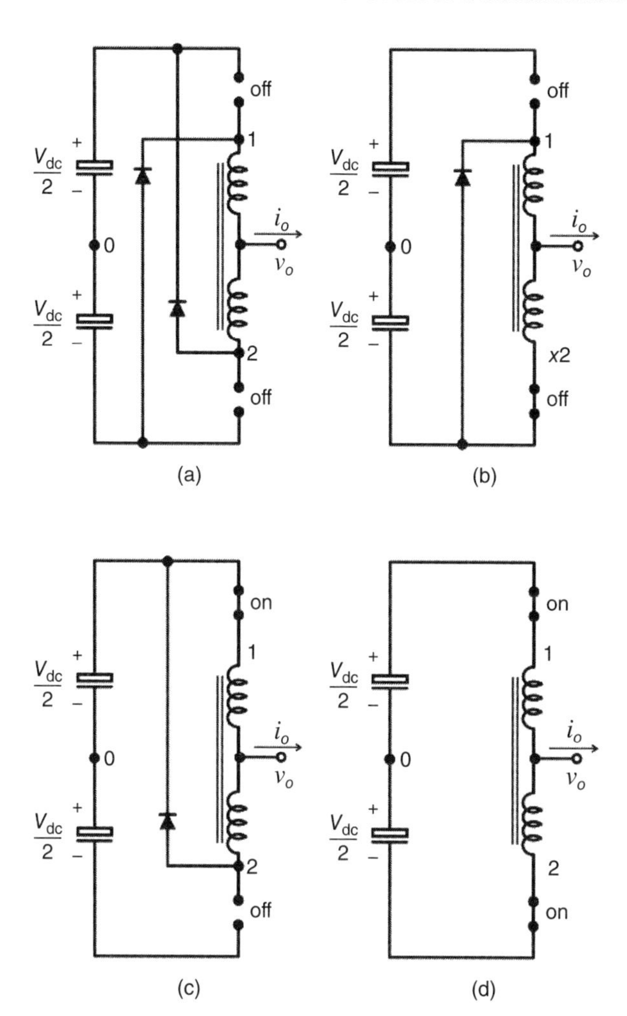

Figure 7.8 Equivalent circuits for the topology presented in Fig. 7.7(d): (a) $q_1 = q_2 = 0$, (b) $q_1 = 0$ and $q_2 = 1$, (c) $q_1 = 1$ and $q_2 = 0$, and (d) $q_1 = q_2 = 1$.

TABLE 7.2 Pole Voltage as a Function of the Switching States

q_1	q_2	v_{10}	v_{20}	v_o
0	0	$-V_{dc}/2$	$V_{dc}/2$	0
0	1	$-V_{dc}/2$	$-V_{dc}/2$	$-V_{dc}/2$
1	0	$V_{dc}/2$	$V_{dc}/2$	$V_{dc}/2$
1	1	$V_{dc}/2$	$-V_{dc}/2$	0

From the Table of Variables presented in Table 7.2 it is possible to define the PWM strategies as in Fig. 7.9(b). Figure 7.9(c) shows the main waveforms associated with this circuit.

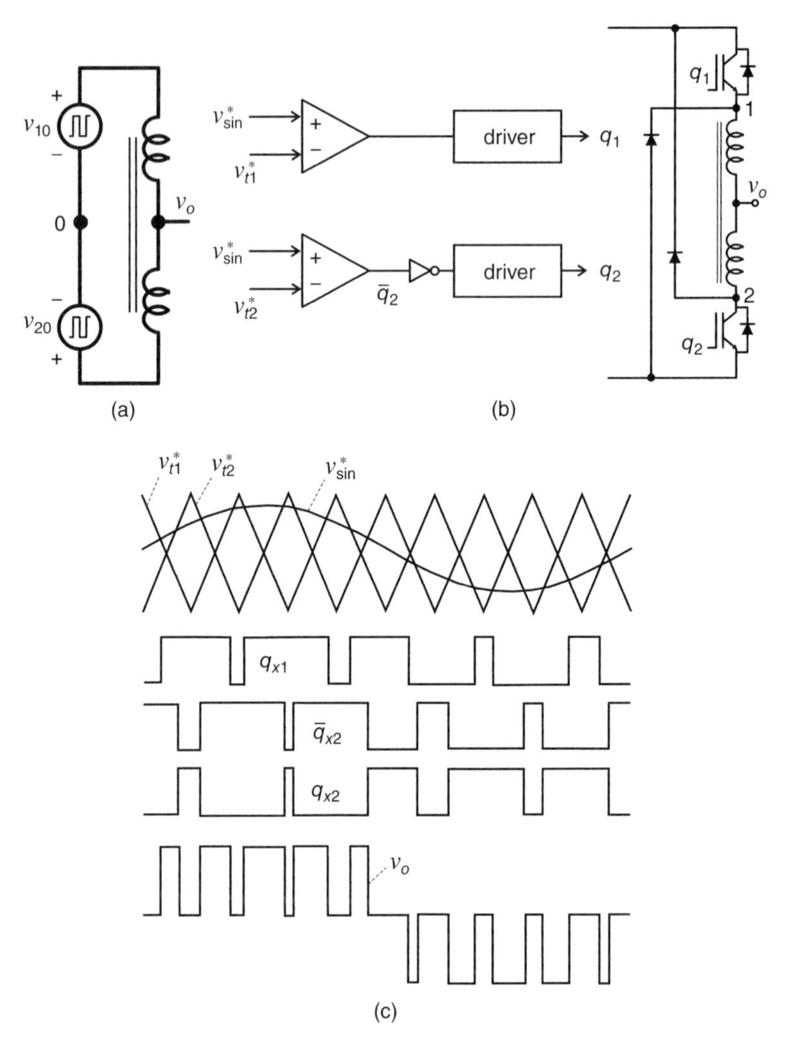

Figure 7.9 (a) Model of converter in Fig. 7.7(e). (b) PWM circuit. (c) Waveforms for generation of a three-level voltage.

Example 7.1

Deduce equation (7.5) for the three-level topology presented in Fig. 7.7(e).

Solution

The output voltage v_o can be written with respect to the point "0" in two different ways, $v_o = -v_{x1} + v_{10}$ and $v_o = v_{x2} + v_{20}$, where v_{x1} and v_{x2} are the drop voltages on the coupled inductor for the upper and lower parts of the inductor, respectively. Adding both equations v_o is easily obtained as $(v_{10} + v_{20})/2$ for the symmetrical case ($v_{x1} = v_{x2}$).

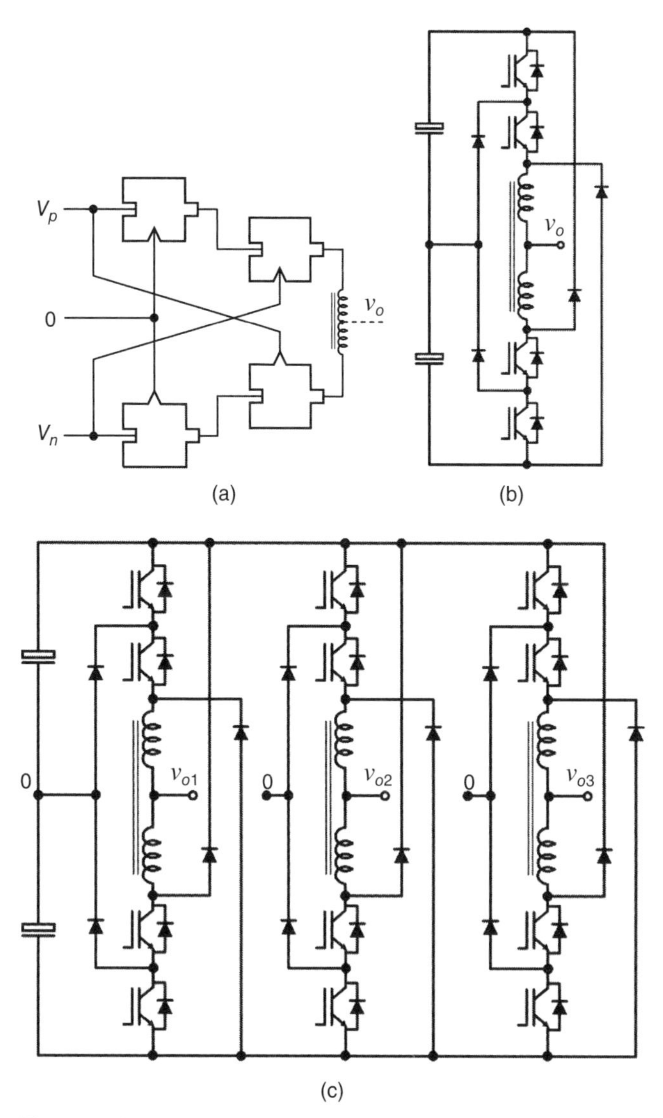

Figure 7.10 Five-level topology with coupled inductor placed at the output converter side: (a) PBG representation, (b) representation with switches, and (c) three-phase version.

More levels can be obtained by combining the solution with coupled inductor and NPC converter. The circuit depicted in Fig. 7.10 is a five-level topology with a PBG representation [see Fig. 7.10(a)] and with a representation using switches [see Fig. 7.10(b)]. Its three-phase version is presented in Fig. 7.10(c). Notice that in this converter the number of levels is higher than the input voltages available (V_p, 0, V_n), which follows the same characteristics of the three-level version in Fig. 7.7(e) – in the three-level case there are three levels obtained from only two voltages.

Although many of the converters presented so far in this and previous chapters are suitable for high voltage high power applications, the MMC combines excellent

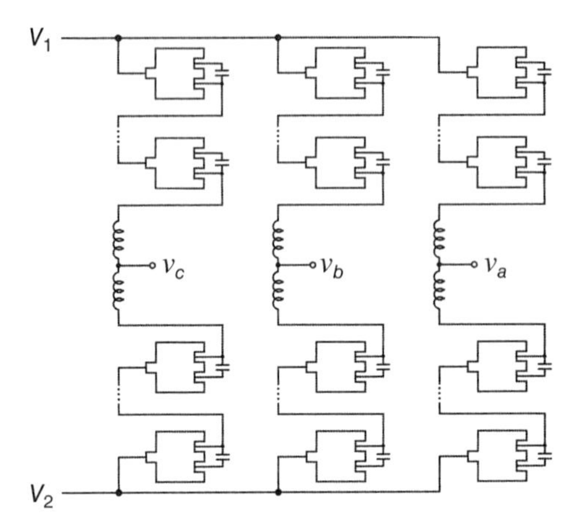

Figure 7.11 Modular multilevel converter with PBs.

output voltage waveforms, very high efficiencies and an overall high blocking voltage capability with reduced stress on the switching devices of each submodule. Those characteristics are desirable in applications such as high power motor drives. Figures 7.11 and 7.12 show this configuration with power blocks and switches, respectively. There are some challenges related to the operation point of view of this converter. They are capacitor voltage ripple control for each module as well as circulating current regulation.

Each phase of the converter shown in Figs. 7.11 and 7.12 constitutes two branches of half-bridge converters (also known as submodules), series connected to dc terminals V_1 and V_2. Those branches are connected through inductors. The ac terminals (v_a, v_b, and v_c) are connected to the middle point of the inductors. In general terms, the ac voltage is controlled by changing the number of half-bridge converters connected to each leg. From Fig. 7.12 i_a and the circulating current (i_{cc}) can be written, respectively, by

$$i_a = i_{a1} - i_{a2} \tag{7.7}$$

$$i_{cc} = \frac{i_{a1} + i_{a2}}{2} \tag{7.8}$$

Consequently the currents i_{a1} and i_{a2} can be expressed as a function of both circulating current and output ac current as follows:

$$i_{a1} = \frac{i_{cc} + i_a}{2} \tag{7.9}$$

$$i_{a2} = \frac{i_{cc} - i_a}{2} \tag{7.10}$$

The circulating current is associated to the energy exchange between the phase leg and the dc-link source. In fact, the control strategy applied for this converter should guarantee the capacitor voltage control for each submodule as well as desired output voltages v_a, v_b, and v_c.

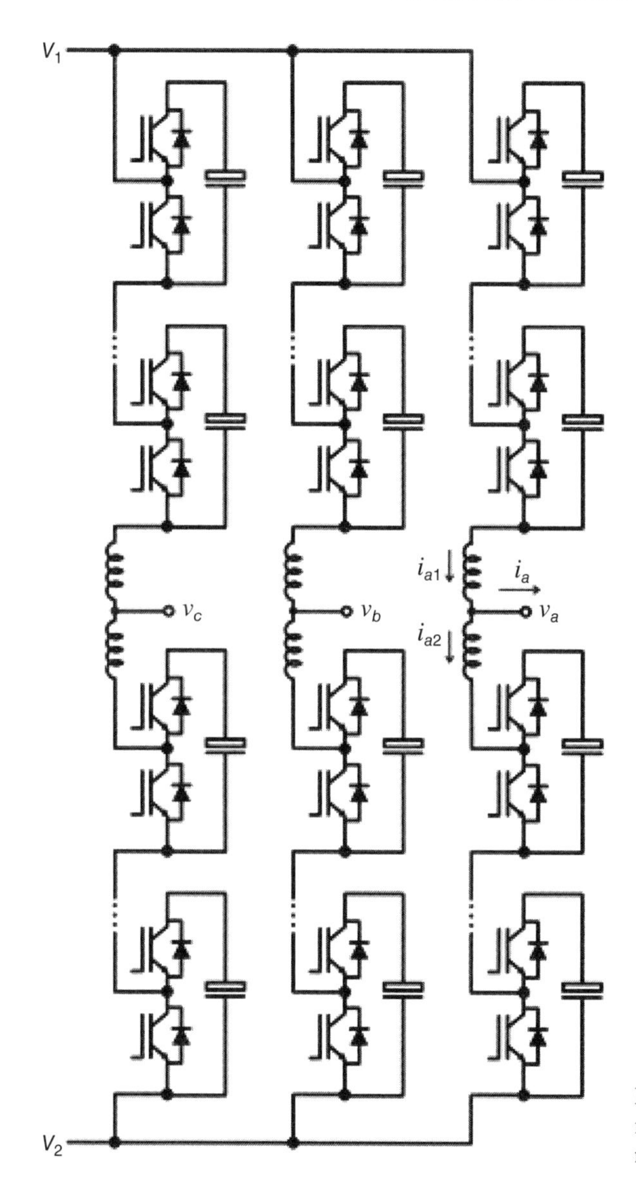

Figure 7.12 Modular multilevel converter – representation with switches.

7.4 ACTIVE-NEUTRAL-POINT-CLAMPED CONVERTERS

The converter presented in Fig. 7.13 shows another arrangement using only the first PB-ac, which is called ANPC configuration. Figure 7.13(a) shows the three-level topology, while Fig. 7.13(b) presents the four-level version of the ANPC circuit. Yet more levels can be reached by adding more blocks as done in Chapter 4 for the NPC converter. Notice that for this family of multilevel converters controlled switches are employed instead of power diodes (as in NPC configurations) to obtain zero voltage

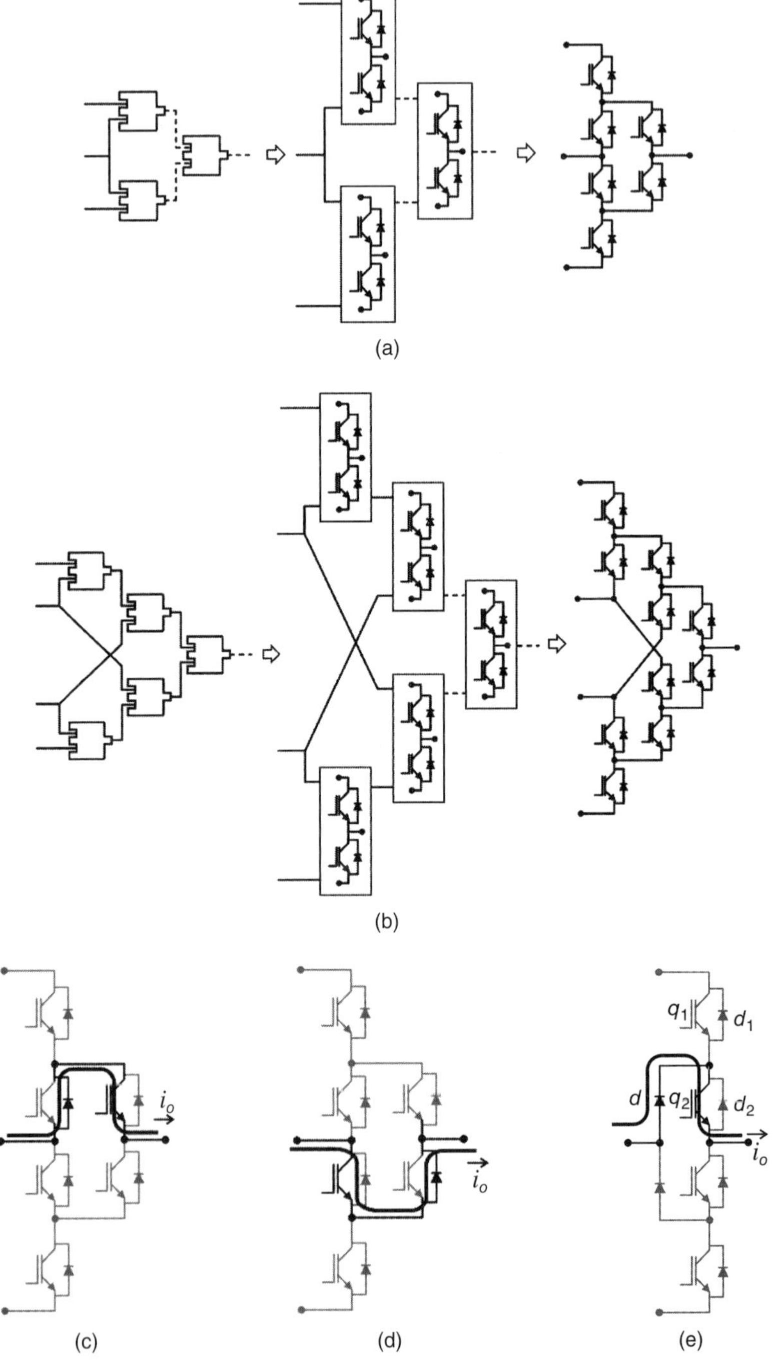

Figure 7.13 ANPC configuration: (a) three-level circuit, (b) four-level circuit, (c) zero output voltage with positive current, (d) second path for zero output voltage with positive current, and (e) zero output voltage with positive current for NPC converter.

at the output of the converter. The main advantages as compared to the conventional NPC topology [see Fig. 7.7(a)] is a better balance distribution of losses, which allows an increase in the maximum output power.

In general terms, the design of a multilevel power electronics converter can be established by limiting the switching frequency in order to meet the maximum allowable losses. In this sense and assuming a high power density scenario, a topology such as NPC, with irregular power losses distribution, presents unequal junction temperature distribution on the power converter circuit. As a direct consequence, some power devices become hotter than others, restricting the switching frequency even more.

On the other hand, due to the presence of controlled power switches, the ANPC topology can allow alternative paths to obtain the same zero voltage at the output converter side. It turns out that it is possible to manage the losses distribution of the switches by changing these paths. For example, Fig. 7.13(c) depicts the ANPC topology for the case in which a zero voltage is obtained for positive output current. On the other hand, Fig. 7.13(d) shows the same case as before (zero voltage obtained at the output with positive current) reached by using an alternative circuit. Finally, Fig. 7.13(e) shows that for the NPC converter there is only one path to obtain zero level voltage at the output of the converter with a positive current.

Since the heat sink is essential for the thermal stability and lifetime of the semiconductor devices and for operation of the power converter as a whole, its specification must guarantee heat transfer capability for the switches, especially those under the highest thermal stress. Indeed, such a characteristic of the ANPC topology presented in Fig. 7.13(c) and 7.13(d) plays an important role in the thermal design of this converter. The irregular junction temperature among diodes and power switches of the NPC topology is presented in Fig. 7.14(a). The components d, q_1, q_2, d_1, and d_2 are the semiconductor devices in Fig. 7.13(e). To avoid such irregular junction temperature among the power switches while synthesizing the desired voltage, Fig. 7.14(b) shows one leg of the ANPC topology with a PWM defined by the temperature of the devices, which means that the hot points can be avoided by using alternative paths highlighted in Fig. 7.13(c) and 7.13(d).

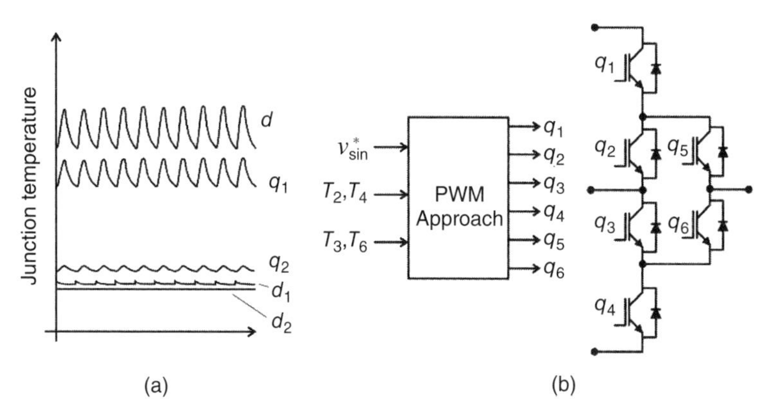

Figure 7.14 (a) Junction temperature among the semiconductor devices. (b) PWM strategy for the ANPC topology.

TABLE 7.3 Table of Variables for the ANPC Topology

q_1	q_2	q_3	q_4	q_5	q_6	v_o
1	1	0	0	0	0	V_{dc}
0	1	0	0	1	0	0
0	0	1	0	0	1	0
0	0	0	1	0	1	$-V_{dc}$

As mentioned previously, such irregular losses distribution can be avoided with the ANPC topology and a regulated loss-balancing system. This loss control strategy can be achieved by either measuring or estimating the junction temperature. From Table 7.3 it is possible to establish a PWM strategy, depending on the temperature either measured or estimated from each semiconductor device.

7.5 MORE MULTILEVEL CONVERTERS

Since the goal of this section is to demonstrate that many topologies can be represented by PBG, their presentation is limited to the description of the circuits with both blocks and switches. All analysis in terms of Table of Variables, model, and PWM can be obtained by the reader following a similar approach as in the previous chapters.

Also, this section aims to reiterate that the power blocks and the rules and postulates presented in Chapter 3 are an interesting way to understand how the multilevel configurations have been presented in the technical literature. By following the rules and using specific blocks, it is possible to, for example, obtain the configuration depicted in Fig. 7.15, which has been named by the authors as active switch NPC (ASNPC). Figure 7.15(a) shows this converter under PBG representation, which employs four PBs-ac. The equivalent lower-level representation using power switches is presented in Fig. 7.15(b). The improvement brought up by this converter lies on the reduction of the average switching frequency for all power devices.

The three-level configuration obtained using just the second PB-ac comes up with two different implementations as presented in Fig. 7.16(a) and 7.16(c).

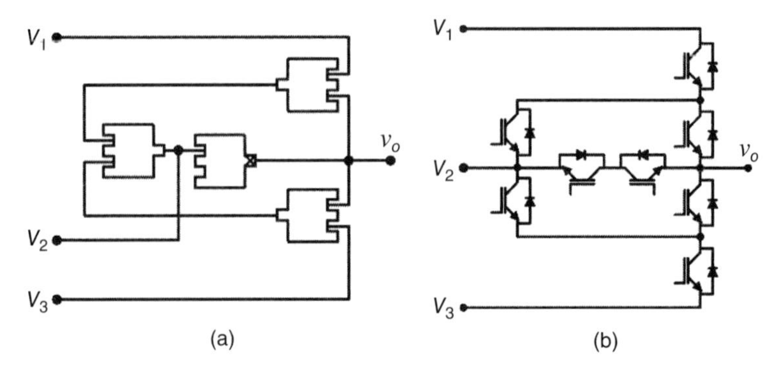

Figure 7.15 Three-level configuration ASNPC (a) blocks and (b) switches.

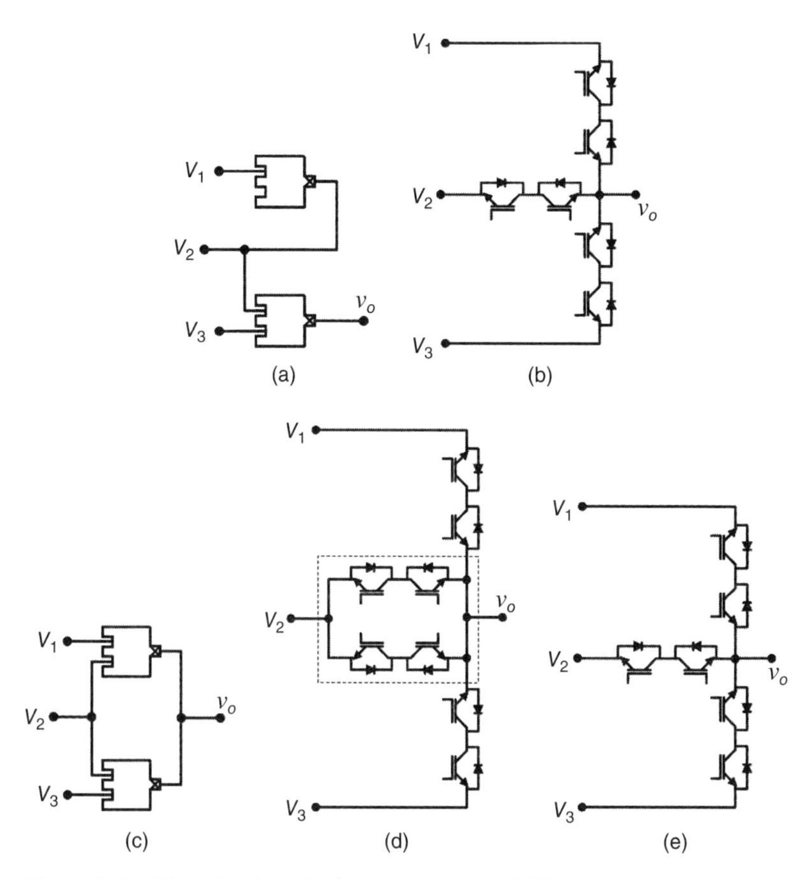

Figure 7.16 Three-level topologies using the second PB-ac only.

Figure 7.16(b) shows the same topology with switches. Considering the representation with the power switches as in Fig. 7.16(d), it is evident that the four switches between the v_o and V_2 can be simplified by only two antiseries switches as in Fig. 7.16(e). Indeed, Fig. 7.16(a) and 7.16(c) end up with the same topology, as presented in Fig. 7.16(b) and 7.16(e).

Figure 7.17 shows a family of multilevel topologies known as mixed NPC/FC converters. Figure 7.17(a) depicts such a mixed three-level converter using both PB-dc and PB-ac, while Fig. 7.17(b) shows it with only PBs-ac. The representation with switches for those converters is presented in Fig. 7.17(c) and 7.17(d), respectively. Higher levels can be easily obtained by stacking more elements as in Fig. 7.17(e)–7.17(h).

Another way to merge a FC with the first PB-ac is presented in Fig. 7.18. Figure 7.19 depicts a similar configuration using a second PB-ac instead of the first one. Finally, Fig. 7.20(a) shows that two PBs-ac can be vertically stacked, which creates a differential converter, and Fig. 7.20(b) and 7.20(c) present different ways to generate a four-level converter.

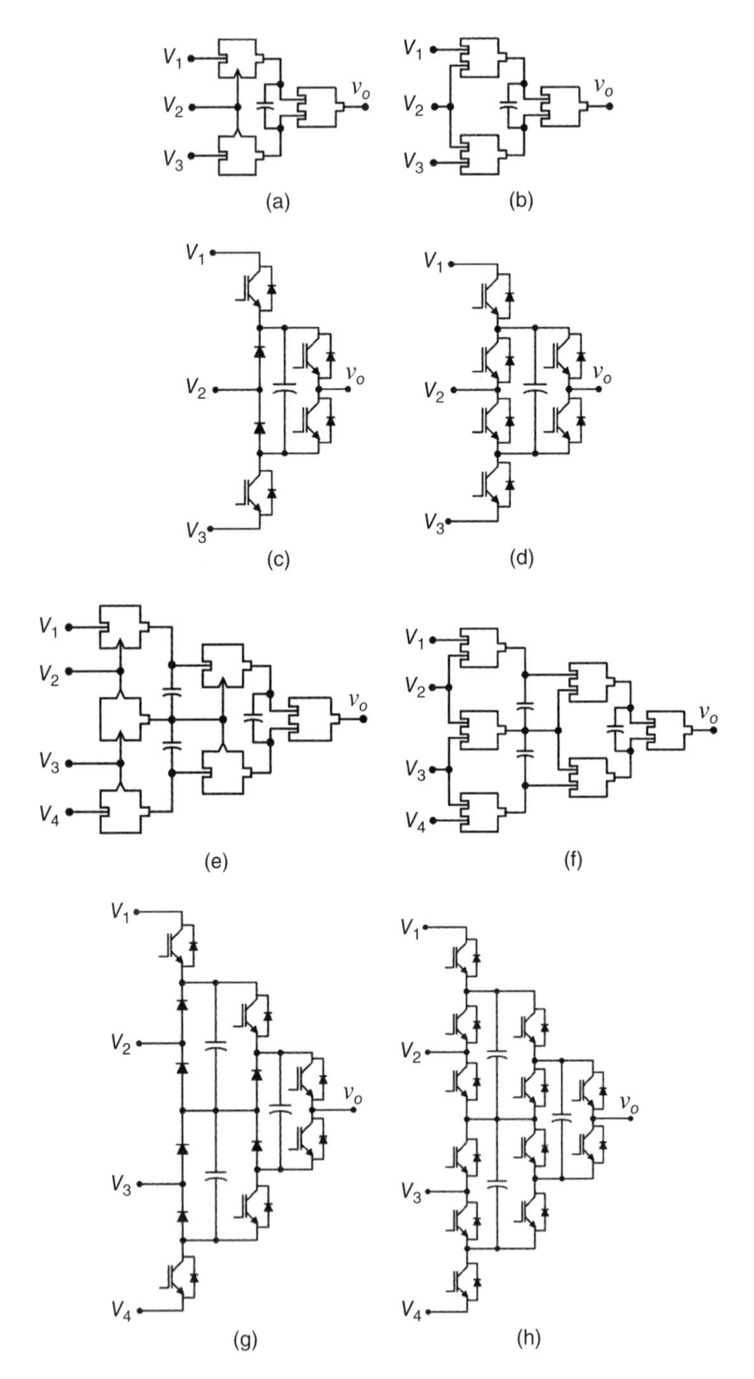

Figure 7.17 Mixed NPC/FC converter.

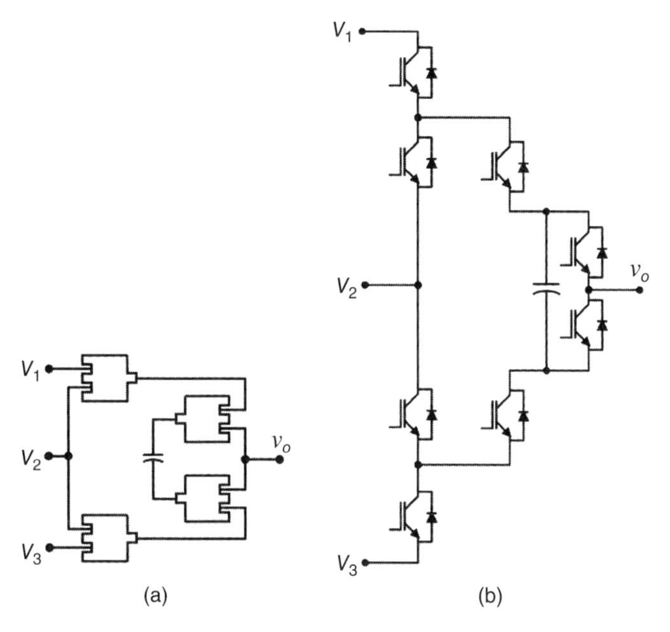

Figure 7.18 Mixed NPC/FC converter using flying capacitor and first PB-ac.

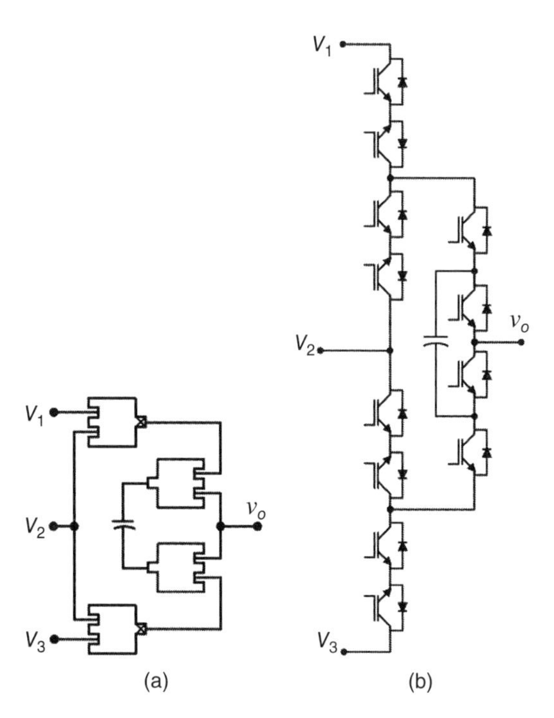

Figure 7.19 Mixed NPC/FC converter using flying capacitor and first/second PB-ac.

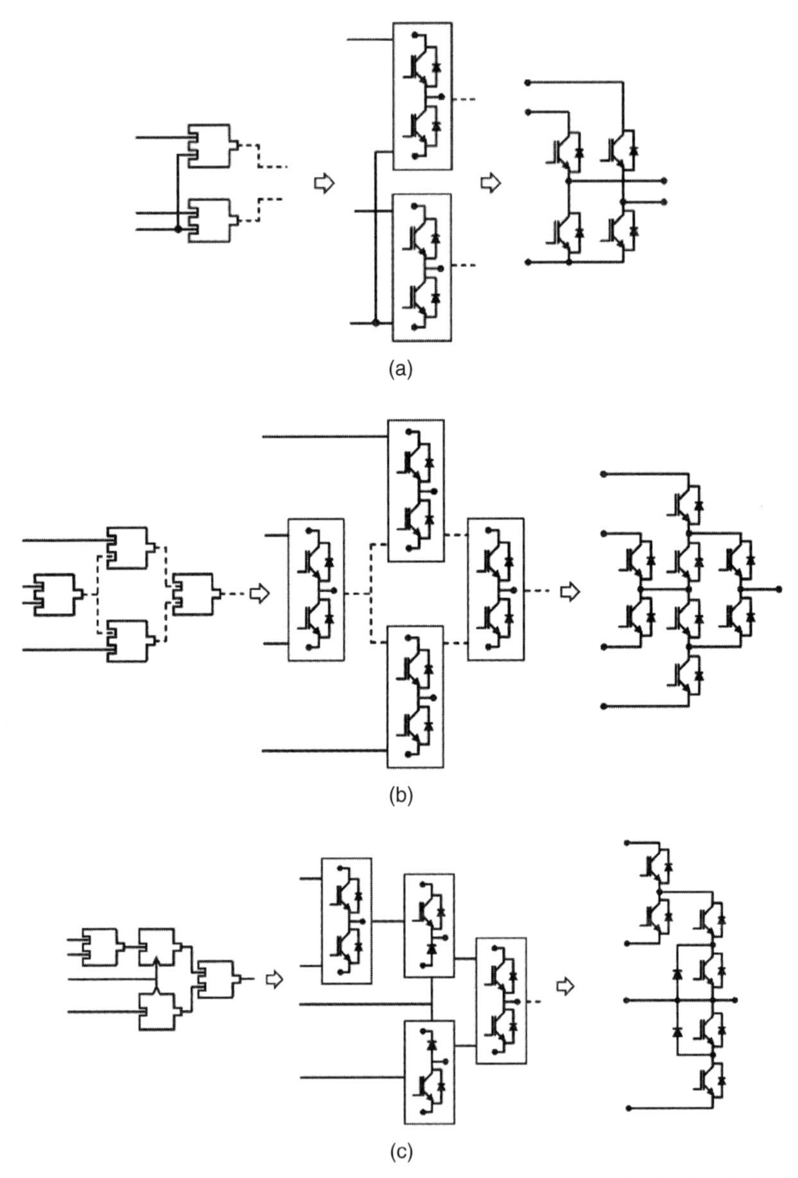

Figure 7.20 (a) Differential converter. (b) Four-level converter obtained with the first PB-ac. (c) Asymmetrical four-level topology.

7.6 SUMMARY

This chapter dealt with of nontraditional multilevel converters. The approach used, with the PBG, shows that there is a high number of possibilities for generation of multilevel topologies and such an approach helps to understand how many of these topologies have been desscribed in the technical literature. Although not all possibilities have been deeply examined, three of them have been analyzed, considering the PBs and the rules to determine the connection of blocks, Table of Variables and model of the converter. There is an extensive number of possibilities of multilevel topologies to which this technique can be used. Some of them, indicated in References from 1 to 26 have been used in this chapter.

REFERENCES

[1] Bhagwat PM, Stefanovic VR. Generalized structure of a multilevel PWM inverter. IEEE Trans Ind Appl 1983;IA-19(6):1057–1069.

[2] Holmes D, Lipo T. *Pulse Width Modulation for Power Converters: Principles and Practice*. Wiley-IEEE Press; 2003.

[3] dos Santos Jr. EC, Muniz JHG, da Silva ERC. 2L3L inverter. Proceedings of COBEP; 2011. p 924–929.

[4] dos Santos Junior EC, Muniz JHG, da Silva ERC, Jacobina CB. Nested multilevel configurations. IEEE Energy Conversion Congress and Exposition (ECCE); 2012. p 324–329.

[5] Matsui K, Murai Y, Watanabe M, Kaneko M, Ueda F. A pulsewidth modulated inverter with parallel-connected transistors by using current sharing reactors. IAS Annual Meeting Conference 1985; 1985. p 1015–1019.

[6] Fratta A, Griffero G, Nieddu S, Pellegrino G Villata F. Inductive three-level V.-supplied conversion cell by new hybrid coupling reactor family. Proceedings of IAS'02; 2002. p 2378–2385.

[7] Salmon J, Ewanchuk J, Knight AM. PWM inverters using split-wound coupled inductors. IEEE Trans Ind Appl 2009;45(6):2001–2009.

[8] Teixeira CA, McGrath BP, Holmes DG, Topologically reduced multilevel converters using complementary unidirectional phase-legs. IEEE International Symposium on Industrial Electronics (ISIE); 2012. p 2007–2012.

[9] Brückner T, Bernet S, Güldner H. The active NPC converter and its loss-balancing control. IEEE Trans Ind Electron 2005;52(3):855–868.

[10] Bruckner T, Bernet S, Steimer PK. Feedforward loss control of three-level active NPC converters. IEEE Trans on Ind Appl 2007;43(6):1588–1596.

[11] Andrejak JM, Lescure M. High voltage converters promising technological developments. Proc. Rec. EPE Conf.; 1987. p 1.159–1.162.

[12] Choi NS, Cho JG, Cho GH. A general circuit topology of multilevel inverter. Proc. Rec. IEEE PESC; 1991. p 96–103.

[13] Carpita M, S Tenconi. A novel multilevel structure for voltage source inverter. Proc. Rec. EPE Conf.; 1991. p 1.90–1.94.

[14] Floricau D, Gateau G, Leredde A. New active stacked NPC multilevel converter: operation and features. IEEE Trans Ind Electron 2010;57(7):2272–2278.

[15] Gateau G, Meynard TA, Foch H. Stacked multicell converter (SMC): properties and design. IEEE 32nd Annual Power Electronics Specialists Conference, PESC 2001; 2001 Jun 17–22; Vancouver, Canada; 2001.

[16] Suh B-S, Hyun D-S. A new n-level high voltage inversion system. IEEE Trans Ind Electron 1997;44(1):107–115.

[17] Peng F. A generalized multilevel inverter topology with self voltage balancing. IEEE Trans Ind Appl 2001;37(2).

[18] Barbosa P, Steimer P, Steinke J, Winkelnkemper M, Celanovic N. Active-neutral-point-clamped multilevel converters. Proc. IEEE Power Electron. Spec. Conf.; 2005. p 2296–2301.

[19] da Silva ERC, Muniz JHG, dos Santos EC, Silva RNA, Barreto LH. Capacitor balance in a five-level based halfbridge converter by use of a mixed active-cell. IEEE Energy Conversion Congress and Exposition (ECCE); 2013. p 414–419.

[20] Meilil J, Ponnaluri S, Serpa L, Steimer PK, Kolar JW. Optimized pulse patterns for the 5-level ANPC converter for high speed high power applications. IEEE IECON 06; 2006. p 2597–2592.

[21] Meynard T, Lienhardt AM, Gateau G, Haederli Ch, Barbosa P. Flying capacitor multiCell converters with reduced stored energy. Proceedings of IEEE ISIE; 2006. p 914–918.

[22] Ramos RR, Ruiz-Caballero D, Ortmann MS, Mussa SA. New symmetrical hybrid multilevel dc-ac converters. Proceedings of Power Electronics Specialists Conference (PESC2008); 2008. p 1916–1922.

[23] Ratnayake KR, Ishikawa H, Wang D, Yoshida M, Watanabe T. Novel hybrid five level inverter. IPEC; 2000. p 2025–2028.

[24] Chen A, Hu L, and He X. A novel type of combined multilevel converter topologies. The 30th Annual Conference of the IEEE Industrial Electronics Society; 2004. p 2290–2294.

[25] Haque MT. Series sub-multilevel voltage source inverters (MLVSIs) as a high quality MLVSI. Proceedings of SPEEDAM; 2004. p F1B-1–F1B-4.

[26] Babaei E. A cascade multilevel converter topology with reduced number of switches. IEEE Trans Power Electron 2008;23(6):2657–2664.

CHAPTER 8

OPTIMIZED PWM APPROACH

8.1 INTRODUCTION

Many pulse width modulation (PWM) approaches have been proposed by authors throughout the years for the systems studied in this book, that is, dc–ac and ac–dc–ac power conversions. In general terms, it is possible to sort the PWM strategies into two main categories: (i) sinusoidal pulse width modulation (SPWM) and (ii) space vector modulation (SVPWM). In SPWM, introduced by Schonnung in 1964 [1], to produce the output voltage waveform, a sinusoidal control signal (modulating signal) is compared to a triangular signal (carrier signal). SVPWM uses a complex voltage vector to define the pulse widths. Although one of the first suggestions for employing complex voltage vectors in a PWM control was made by Jardan [2], the SVPWM technique was first published by Busse and Holtz [3] followed by Pfaff et al. [4], in the same year.

The majority of the PWM approaches presented in the previous chapters of this book can be classified as SPWM. For each of the configurations studied so far, intensive research has developed PWM strategies, bringing different levels of optimization in terms of the THD of the waveforms generated and also in terms of the efficiency of the converter. As a direct consequence of the high volume of research devoted to this topic, several improvements have been observed since the first studies. For instance, modification of the modulating signal has introduced many improvements to the SPWM technique, resulting in nonsinusoidal carrier-based PWM (CPWM) techniques [5–8]. After the reports by Buja and Indri [9] it has been gradually recognized that the addition of an adequate third-harmonic zero-sequence component to each of the reference pole voltage increases 15.5% of the fundamental component of the output voltages in a three-phase converter. As concluded by Depenbrock [10], more than introducing the benefit of 15.5%, the zero sequence component can be used to reduce the number of times the switches are turned on and off, by clamping each pole voltage during 60° of the sinusoidal waveform.

This chapter presents some techniques for optimization of the PWM approach considering the fact that the number of pole voltages is higher than the number of voltages demanded by the load, and then there is a degree of freedom to be explored. For instance, in Fig. 8.1 there is one degree of freedom for the H-bridge converter supplying a single-phase load. There are two pole voltages available (v_{10} and v_{20})

Advanced Power Electronics Converters: PWM Converters Processing AC Voltages,
Forty Fifth Edition. Euzeli Cipriano dos Santos Jr. and Edison Roberto Cabral da Silva.

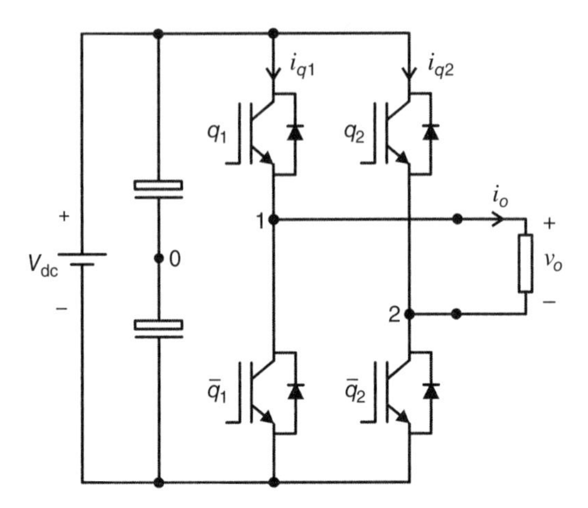

Figure 8.1 Converter with two legs supplying a single-phase load.

at the converter side and just one voltage (v_o) must be controlled. Following this introduction, this chapter is organized in seven sections. Sections 8.2 and 8.3 show an optimized PWM strategy for converters with two and three legs, respectively. For both sections, a PWM approach is presented with both analog and digital implementations, and finally the influence of the parameter μ is addressed for the quality of the waveform generated as well as for the losses of the converter. Section 8.4 deals with space vector modulation for a three-phase converter, while Section 8.5 presents the CPWM for both a three-leg converter feeding a two-phase symmetrical machine and a four-leg converter supplying a three-phase four-wire load. The CPWM is also employed in non-conventional converters (Section 8.6), such as split-wound coupled inductors, Z-source converter, and open-end winding motor.

8.2 TWO-LEG CONVERTER

8.2.1 Model

A two-leg converter, as seen in Fig. 8.1, consists of four power switches q_1, $\overline{q}_1$, q_2, and $\overline{q}_2$. Note that the circuitry necessary to generate the gating signals is not presented in this figure. The upper (q_1 and q_2) and bottom ($\overline{q}_1$ and $\overline{q}_2$) switches are complementary, which means that $q_1 = 1 - \overline{q}_1$ and $q_2 = 1 - \overline{q}_2$. As considered in the previous chapters, the conduction state of all switches can be represented by the binary variable with $q_x = 1$ indicating a closed switch and $q_x = 0$ indicating an open one (with $x = 1$ and 2).

The pole voltages v_{10} and v_{20} can be defined as a function of the switching states, as follows:

$$v_{10} = (2q_1 - 1)\frac{V_{dc}}{2} \tag{8.1}$$

$$v_{20} = (2q_2 - 1)\frac{V_{dc}}{2} \tag{8.2}$$

TABLE 8.1 Voltages and Currents of the Switches as a Function of the Switching States

Switching States		Voltages Across the Switches				Currents through the Switches			
q_1	q_2	v_{q1}	$v_{\bar{q}1}$	v_{q2}	$v_{\bar{q}2}$	i_{q1}	$i_{\bar{q}1}$	i_{q2}	$i_{\bar{q}2}$
0	0	V_{dc}	0	V_{dc}	0	0	i_o	0	i_o
0	1	V_{dc}	0	0	V_{dc}	0	i_o	i_o	0
1	0	0	V_{dc}	V_{dc}	0	i_o	0	0	i_o
1	1	0	V_{dc}	0	V_{dc}	i_o	0	i_o	0

The output voltage is given by

$$v_o = v_{10} - v_{20} \tag{8.3}$$

$$v_o = (q_1 - q_2)V_{dc} \tag{8.4}$$

Note that there are two control variables (switching states q_1 and q_2) to define just one output voltage (v_o). While the load voltage is generated by the converter, the load current (i_o) depends on the load connected to the converter ($i_o = v_o/Z$, where Z is the impedance for a linear load). The voltages across the switches and currents through the switches are presented in Table 8.1.

From Table 8.1, the following equations can be used to define the voltage and currents of the switches:

$$v_{qx} = (1 - q_x)V_{dc} \tag{8.5}$$

$$v_{\bar{q}x} = q_x V_{dc} \tag{8.6}$$

$$i_{qx} = q_x i_o \tag{8.7}$$

$$i_{\bar{q}x} = (1 - q_x)i_o \tag{8.8}$$

with $x = 1, 2$.

8.2.2 PWM Implementation

From equation (8.3) it is possible to define the desired output voltage as a function of the reference pole voltages, as follows:

$$v_o^* = v_{10}^* - v_{20}^* \tag{8.9}$$

A reference voltage [$v_o^* = V_o^* \cos(\omega_o t)$] must be synthesized through two reference pole voltages (v_{10}^* and v_{20}^*), which means one degree of freedom to be explored. Then, an auxiliary variable (v_h^*) can be defined to allow some level of improvement. In this case it is possible to write:

$$v_{10}^* = v_o^* + v_h^* \tag{8.10}$$

$$v_{20}^* = v_h^* \tag{8.11}$$

Equation (8.9), in principle, is satisfied for any value of v_h^*. However, the auxiliary voltage must be obtained in order to guarantee that the reference pole voltages are limited as follows:

$$\frac{-V_{dc}}{2} \leq v_{10}^* \leq \frac{V_{dc}}{2} \tag{8.12}$$

$$\frac{-V_{dc}}{2} \leq v_{20}^* \leq \frac{V_{dc}}{2} \tag{8.13}$$

When equations (8.12)–(8.13) are satisfied, it is said that the converter is operating in its linear region. The following development in this subsection aims to determine an analytical expression for v_h^*, and consequently defining v_{10}^* and v_{20}^* for PWM implementation.

Substituting (8.10) and (8.11) into (8.12) and (8.13), it is possible to write:

(a) *Inequality* 1. $v_h^* \geq \frac{-V_{dc}}{2} - v_o^*$;

(b) *Inequality* 2. $v_h^* \geq \frac{-V_{dc}}{2}$;

(c) *Inequality* 3. $v_h^* \leq \frac{V_{dc}}{2} - v_o^*$;

(d) *Inequality* 4. $v_h^* \leq \frac{V_{dc}}{2}$.

As long as v_o^* is a sinusoidal waveform, it can assume either positive or negative values; then we have two cases, Case 1: when $v_o^* > 0$ and Case 2: for $v_o^* < 0$. Figure 8.2 shows the hatched area (feasible region) relative to inequalities 1 and 2 for Case 1, while Fig. 8.3 shows the same inequalities for Case 2. Even with v_o^* being a sinusoidal waveform, it appears in the following analysis (Figs 8.2–8.5) as a constant due to the short interval of time considered.

Note that in Case 1 (Fig. 8.2) the inequality 1 is predominant over inequality 2 (since satisfying inequality 1, the inequality 2 will be necessarily satisfied), while in Case 2 inequality 2 is predominant.

Then, equation (8.14) is enough to succinctly describe both scenarios presented in Figs 8.2 and 8.3:

$$v_h^* \geq \frac{-V_{dc}}{2} - \text{MAX}\{v_o^*, 0\} \tag{8.14}$$

Equation (8.14) is a generic way to present inequalities 1 and 2, and it is equivalent to the graphical analysis depicted in Figs 8.2 and 8.3.

A similar approach can be used for inequalities 3 and 4. Figure 8.4 shows the hatched area (feasible region) related to inequalities 3 and 4 for Case 1, while Fig. 8.5 shows the same inequalities for Case 2. Note that in Case 1 inequality 4 is predominant, while in Case 2 inequality 3 must be satisfied. Then, it is possible to write v_h^* in a short way as presented in

$$v_h^* \leq \frac{V_{dc}}{2} - \text{MIN}\{v_o^*, 0\} \tag{8.15}$$

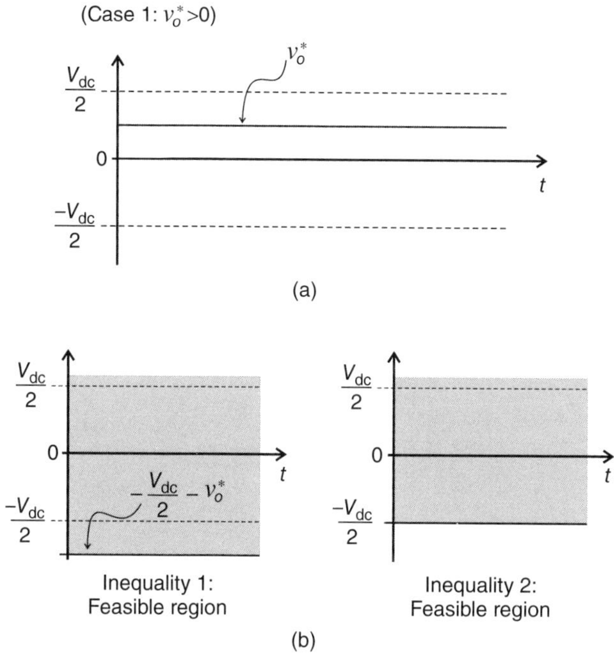

Figure 8.2 (a) Limits of v_h^* when $v_o^* > 0$ (Case 1). (b) Feasible region for inequality 1 (left) and inequality 2 (right).

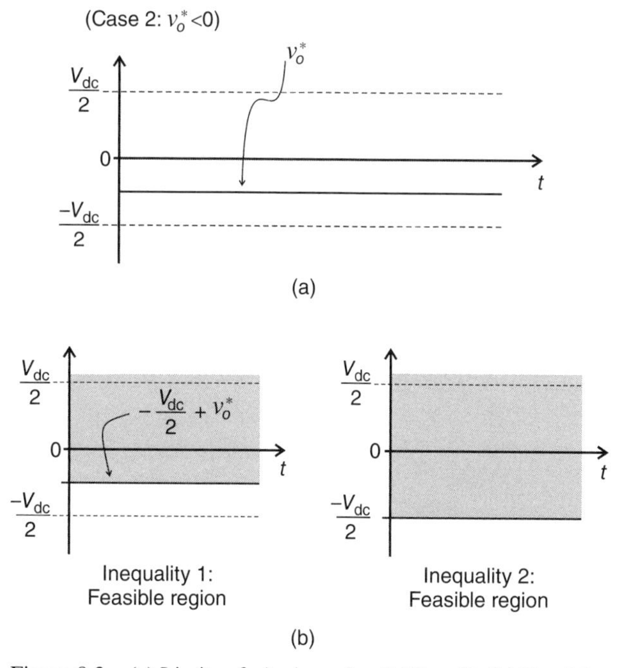

Figure 8.3 (a) Limits of v_h^* when $v_o^* < 0$ (Case 2). (b) Feasible region for inequality 1 (left) and inequality 2 (right).

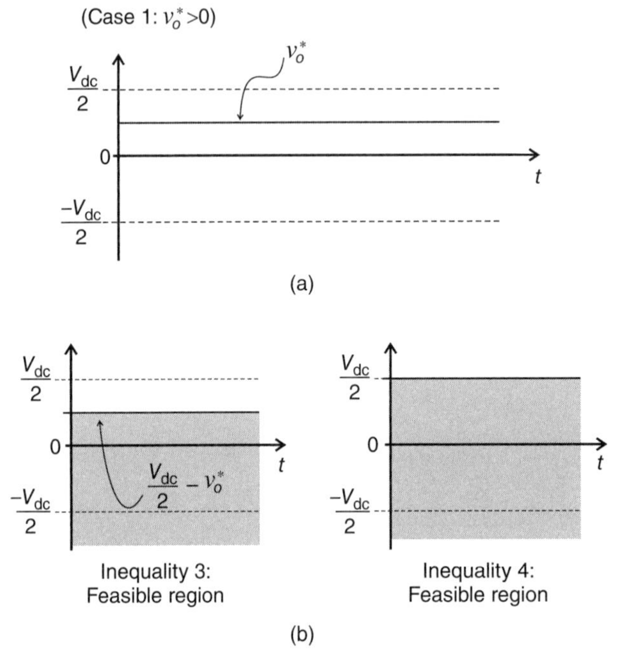

Figure 8.4 (a) Limits of v_h^* when $v_o^* < 0$. (b) Feasible region for inequality 3 (left) and inequality 4 (right).

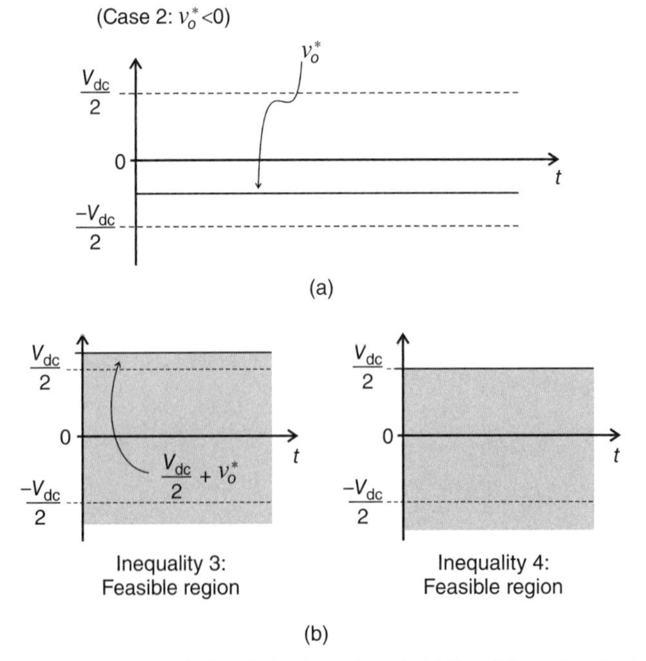

Figure 8.5 (a) Limits of v_h^* when $v_o^* < 0$. (b) Feasible region for inequality 3 (left) and inequality 4 (right).

As in equation (8.14), equation (8.15) can be used instead of employing the graphical analysis depicted in Figs 8.4 and 8.5. Hence, it is possible to write equations (8.14) and (8.15) together, as:

$$\frac{-V_{dc}}{2} - \mathrm{MAX}\{v_o^*, 0\} \leq v_h^* \leq \frac{V_{dc}}{2} - \mathrm{MIN}\{v_o^*, 0\} \tag{8.16}$$

or

$$\frac{-V_{dc}}{2} - V_{max} \leq v_h^* \leq \frac{V_{dc}}{2} - V_{min} \tag{8.17}$$

where $V_{max} = \mathrm{MAX}\{v_o^*, 0\}$ and $V_{min} = \mathrm{MIN}\{v_o^*, 0\}$.

For the operation inside a linear region [see equations (8.12) and (8.13)] v_h^* must be considered between the limits presented in equation (8.17) and observed in Fig. 8.6(a). Note from this figure that the lower and upper values for v_h^* are V_1 and V_2, respectively. These voltages are defined as follows: $V_1 = (-V_{dc}/2) - V_{max}$ and $V_2 = (V_{dc}/2) - V_{min}$. It is possible to create an index μ (distribution factor) in such a way that the voltage v_h^* can be changed proportionally as a function of μ, as observed in Fig. 8.6(b), with $0 \leq \mu \leq 1$. Therefore, the equation of the line in Fig. 8.6(b) can be defined as follows:

$$v_h^*(\mu) = V_1 + (V_2 - V_1)\mu \tag{8.18}$$

By developing equation (8.18),

$$v_h^* = V_{dc}\left(\mu - \frac{1}{2}\right) + (\mu - 1)V_{max} - \mu V_{min} \tag{8.19}$$

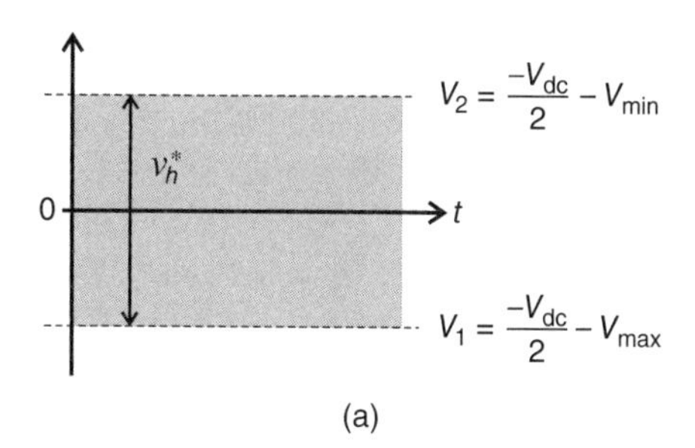

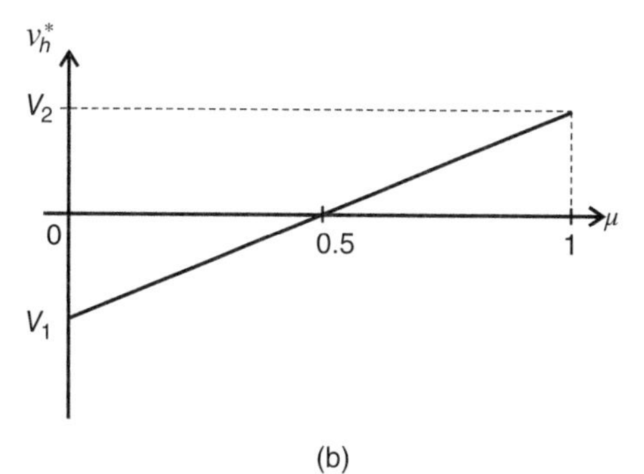

Figure 8.6 (a) Limits of v_h^*. (b) v_h^* as a function of the distribution factor.

With v_h^* defined as in (8.19), the reference pole voltages (v_{10}^* and v_{20}^*) presented in (8.10), (8.11) are used either to compare with a high frequency triangular waveform (in the analog solution) or to generate timer counters (in the digital solution).

Example 8.1

Determine the waveforms for v_{10}^* and v_{20}^* when $\mu = 0.5$ (distribution factor) and assuming $v_o^* = V_o^* \cos(\omega_o t)$.

Solution

To determine the waveforms for v_{10}^* and v_{20}^*, from (8.10) and (8.11), it is necessary to find v_h^*, as in equation (8.19). Equation (8.19) with $\mu = 0.5$ becomes $v_h^* = -1/2(V_{\min}^* + V_{\max}^*)$. As defined earlier, $V_{\max} = \mathrm{MAX}\{v_o^*, 0\}$ and $V_{\min} = \mathrm{MIN}\{v_o^*, 0\}$, which means that:

(a) When $v_o^* > 0$, $V_{\max} = v_o^*$ and $V_{\min} = 0$

(b) When $v_o^* \leq 0$, $V_{\max} = 0$ and $V_{\min} = v_o^*$.

Hence, v_h^* is equal to $-1/2(v_o^*)$ for both cases $v_o^* > 0$ and $v_o^* \leq 0$. Therefore, $v_{10}^* = \frac{1}{2}v_o^*$ and $v_{20}^* = -\frac{1}{2}v_o^*$. The waveforms for v_{10}^* and v_{20}^* are two sinusoidal waves with the same amplitude but 180° apart.

8.2.3 Analog and Digital Implementation

Figure 8.7 shows the analog implementation of the PWM solution for the two-leg converter as depicted in Fig. 8.1. Note that the reference pole voltages, obtained from equations (8.10)–(8.11) are compared to the carrier signal v_t^* with amplitude equal to V_t. The relationship between the amplitude of the modified voltages v_{j0}^* (i.e., V_{j0}^* –with $j = 1, 2$) and V_t defines the amplitude modulation ratio ($m_a = V_{j0}^*/V_t$). On the other hand, the relationship between the switching frequency (f_s – frequency of the carrier signal) and the frequency of the sinusoidal waveform (f_o) defines the frequency modulation ratio ($m_f = f_s/f_o$).

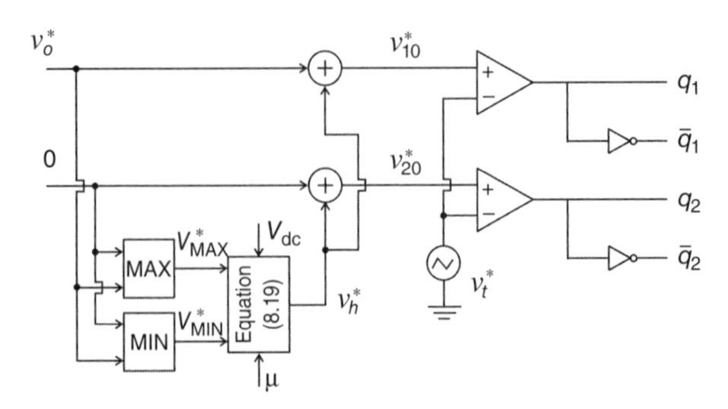

Figure 8.7 Analog PWM implementation.

Example 8.2

Considering $\mu = 0.5$, determine the load voltage v_o assuming the amplitude modulation ratio (m_a) equal to infinite.

Solution

By definition m_a is given by V_{j0}^*/V_t, which means that for a modulation ratio equal to infinite the amplitude of the sinusoidal waveform is always higher than the carrier signal amplitude. Then, $q_j = 1$ if $v_{j0}^* > 0$ and $q_j = 0$ if $v_{j0}^* < 0$ with $j = 1, 2$. Then the load voltage will be a square waveform with the same frequency of v_o^* and with amplitude equal to V_{dc}.

The equivalent digital solution can be obtained as an alternative for the analog PWM implementation by defining the time in which the switches (q_1 and q_2) are on. For determination of the pulse widths, used to define when the switches are either on or off (i.e., emulating the OpAmp operation in Fig. 8.7), it is assumed that the modified reference voltages v_{10}^* and v_{20}^* are constants over the switching period T_s. From Fig. 8.8, this hypothesis (i.e., v_{10}^* and v_{20}^* constants over T_s) becomes more accurate as the switching frequency becomes much higher than the frequency of the modulating signal (i.e., $f_s \gg f_o$).

Then, assuming that the triangular frequency is high enough to guarantee the condition in Fig. 8.8(b), the average values of v_{10} and v_{20} over T_s should be equal to the medium values of v_{10}^* and v_{20}^*, respectively. Figure 8.9 shows the waveforms for the first leg highlighting that the medium values of v_{10}^* and v_{10} must be equal. Hence, it is possible to write

$$\frac{1}{T_s}\int_t^{t+T_s} v_{10}^*(t)dt = \frac{1}{T_s}\int_t^{t+T_s} v_{10}(t)dt \tag{8.20}$$

$$\frac{1}{T_s}\int_t^{t+T_s} v_{20}^*(t)dt = \frac{1}{T_s}\int_t^{t+T_s} v_{20}(t)dt \tag{8.21}$$

Since v_{10}^* and v_{20}^* are assumed to be constant over T_s,

$$v_{10}^* = \left[\frac{V_{\mathrm{dc}}}{2}\tau_1 - \frac{V_{\mathrm{dc}}}{2}\left(T_s - \tau_1\right)\right]\frac{1}{T_s} \tag{8.22}$$

$$v_{20}^* = \left[\frac{V_{\mathrm{dc}}}{2}\tau_2 - \frac{V_{\mathrm{dc}}}{2}\left(T_s - \tau_2\right)\right]\frac{1}{T_s} \tag{8.23}$$

where τ_1 and τ_2 are the time in which the switches q_1 and q_2 are on.

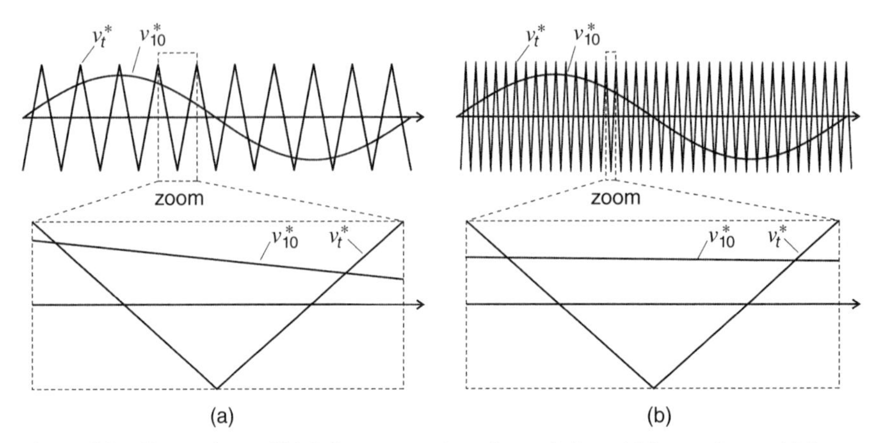

Figure 8.8 Comparison of high frequency triangular and sinusoidal waveforms: (a) low frequency modulation ratio and (b) high frequency modulation ratio.

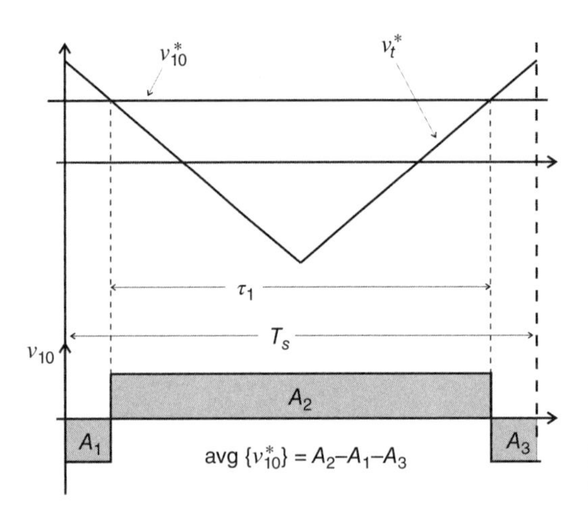

Figure 8.9 Average value of v_{10}^* equal to the average value v_{10}.

From expressions (8.22)–(8.23) it is possible to calculate the time intervals in which the switches q_1 and q_2 are on, that is

$$\tau_1 = \left(\frac{v_{10}^*}{V_{dc}} + \frac{1}{2} \right) T_s \tag{8.24}$$

$$\tau_2 = \left(\frac{v_{20}^*}{V_{dc}} + \frac{1}{2} \right) T_s \tag{8.25}$$

After the definition of the pulse widths (τ_1 and τ_2), it is necessary to program timers for implementation with either a digital signal processing (DSP) or a field-programmable gate array (FPGA) device, for instance.

Exercise 8.1

Assume a single-phase dc–ac converter as in Fig. 8.1 with $V_{dc} = 100\,\text{V}$. Also consider two ways to generate the PWM signals: (i) analog implementation with OpAmps with maximum voltage allowed equal to 15 V (due to operation restrictions of the OpAmps), that is, maximum voltage for v_{10}^*, v_{20}^*, and v_t^*, must be equal to 15 V; and (ii) digital implementation with the code written in C++ for a DSP, with $v_{10}^* = 50\sin(\omega t)$ and $v_{20}^* = 50\sin(\omega t + 180°)$. Question: Is it possible to generate a desirable ac output voltage with amplitude equal to 100 V for both analog and digital implementations as described before?

8.2.4 Influence of μ for PWM Implementation

The parameter μ and consequently v_h^* play an important role in the PWM implementation described already. As mentioned earlier, the distribution factor (μ) can be employed to optimize the converter operation. For instance, to emphasize such an

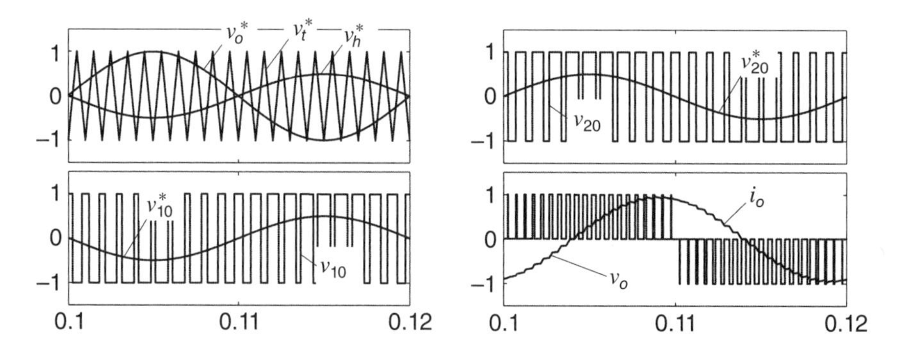

Figure 8.10 Waveforms obtained with $\mu = 0.5$.

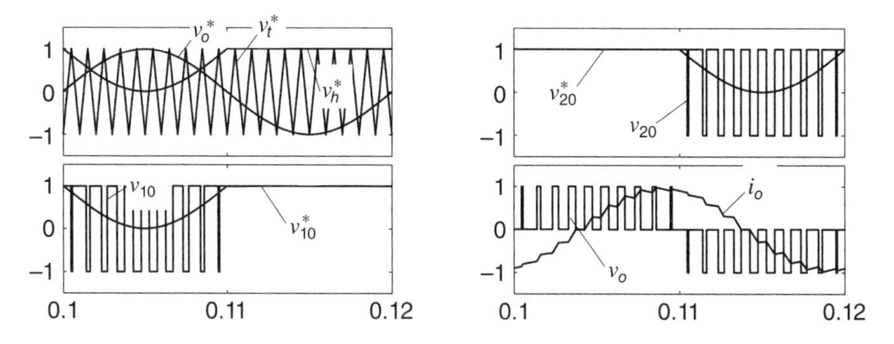

Figure 8.11 Waveforms obtained with $\mu = 0.0$.

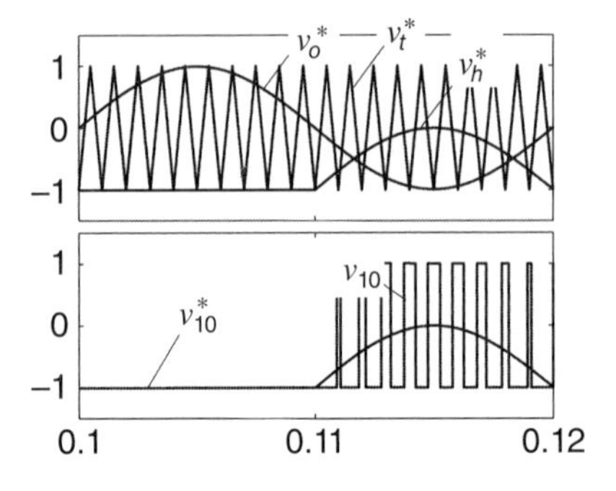

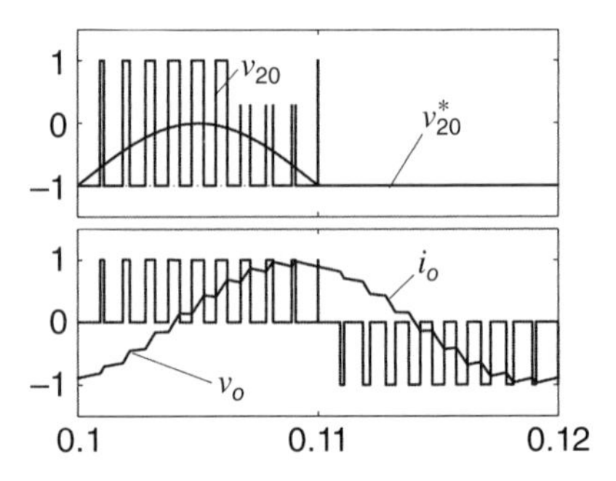

Figure 8.12 Waveforms obtained with $\mu = 1.0$.

optimization, a set of the main converter's variables are shown in Figs 8.10, 8.11, and 8.12 for μ equal to 0.5, 0.0, and 1.0, respectively. Notice that those results are normalized and the maximum value is always one for both power and control waveforms. Figure 8.10 shows the case with symmetrical waveforms (v_{10}^* and v_{20}^*) obtained with $\mu = 0.5$, which represents the best option in terms of the total harmonic distortion (THD).

Note that the ripple of the current in this result is lower than that shown in Figs 8.11 and 8.12. On the other hand, $\mu = 0$ and $\mu = 1$ generate asymmetrical waveforms (see Figs 8.11 and 8.12) with the main advantage related to the switching losses reduction. There is a large period of time (180° of the modulating signal) in which there is no switching for either v_{10} or v_{20}.

It is important to highlight that although μ plays an important role creating waveforms with distortions to deal with either THD improvement or loss reduction, these parameters (μ and consequently v_h^*) will not change the lengths of the active states, as observed in Fig. 8.13.

Example 8.3

Why does the load current (i_o) for all values of μ have the same average value? Does it change if another value for μ is chosen rather than 0.5, 0.0, or 1.0, for example $\mu = 0.3$?

Solution

Even with different instantaneous load voltages due to different values of v_h^*, as observed in Figs 8.10–8.12, the average values of v_o should be the same since the common term v_h^* is cancelled [see equation (8.9)]. That is why the amplitude value of the load current is kept constant. Another explanation can be obtained from Fig. 8.13, which shows the unchanged active vector length by either using v_h^* or not, that is, if the same active vectors are applied, the same amplitude for the currents is expected.

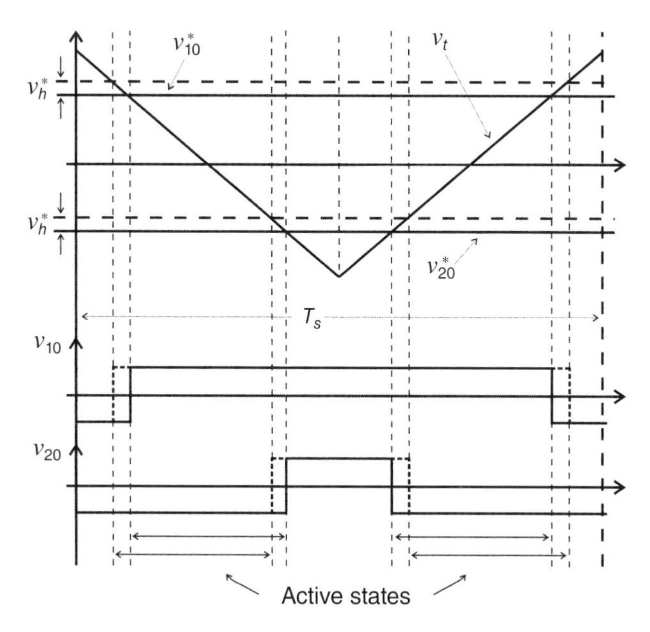

Figure 8.13 Influence of v_h^* on the active states generation.

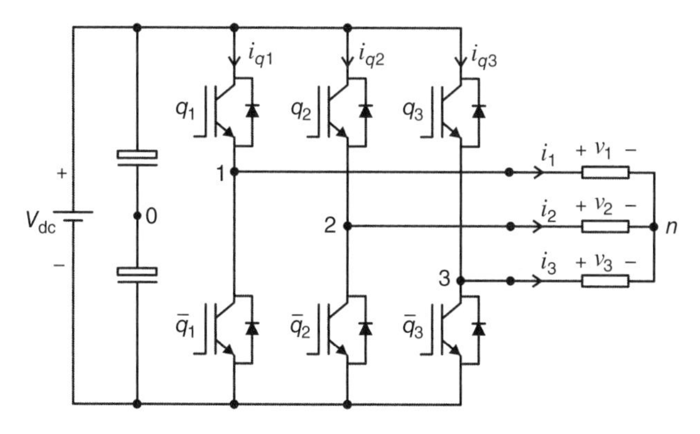

Figure 8.14 Three-leg converter supplying a three-phase load.

8.3 THREE-LEG CONVERTER AND THREE-PHASE LOAD

8.3.1 Model

A three-leg converter, as seen in Fig. 8.14, is constituted by six power switches q_1, $\overline{q}_1$, q_2, $\overline{q}_2$, q_3, and $\overline{q}_3$. For simplification purposes, the circuitry necessary to generate the gating signals is not presented in this figure; instead just the power part of the circuit has been presented. The upper (q_1, q_2, and q_3) and bottom ($\overline{q}_1$, $\overline{q}_2$, and $\overline{q}_3$) switches are complementary, which means that $q_x = 1 - \overline{q}_x$ (with $x = 1, 2, 3$). The conduction state of all switches can also be represented by a binary variable with $q_x = 1$ indicating a closed switch and $q_x = 0$ indicating an open one.

The pole voltages v_{10}, v_{20}, and v_{30} can be defined as a function of the switching states, as follows:

$$v_{10} = (2q_1 - 1)\frac{V_{dc}}{2} \tag{8.26}$$

$$v_{20} = (2q_2 - 1)\frac{V_{dc}}{2} \tag{8.27}$$

$$v_{30} = (2q_3 - 1)\frac{V_{dc}}{2} \tag{8.28}$$

The output voltages (v_1, v_2, and v_3) are given by

$$v_1 = v_{10} - v_{n0} \tag{8.29}$$

$$v_2 = v_{20} - v_{n0} \tag{8.30}$$

$$v_3 = v_{30} - v_{n0} \tag{8.31}$$

Considering a balanced three-phase system (i.e., $v_1 + v_2 + v_3 = 0$) and from (8.29)–(8.31), it is possible to define v_{n0} as in equation (8.32):

$$v_{n0} = \frac{1}{3}(v_{10} + v_{20} + v_{30}) \tag{8.32}$$

Substituting (8.32) into (8.29)–(8.31), we get

$$v_1 = \frac{2}{3}v_{10} - \frac{1}{3}v_{20} - \frac{1}{3}v_{30} \tag{8.33}$$

$$v_2 = -\frac{1}{3}v_{10} + \frac{2}{3}v_{20} - \frac{1}{3}v_{30} \tag{8.34}$$

$$v_3 = -\frac{1}{3}v_{10} - \frac{1}{3}v_{20} + \frac{2}{3}v_{30} \tag{8.35}$$

Note that it is possible to explore the redundancy inherent in a three-phase system. When two voltages are controlled, the third is indirectly controlled, such as: $v_3 = -(v_1 + v_2)$.

Substituting (8.26)–(8.28) into (8.33)–(8.35), it is possible to write the output voltages only as a function of the state of the switches:

$$v_1 = \left[\frac{2}{3}q_1 - \frac{1}{3}q_2 - \frac{1}{3}q_3\right] V_{dc} \tag{8.36}$$

$$v_2 = \left[-\frac{1}{3}q_1 + \frac{2}{3}q_2 - \frac{1}{3}q_3\right] V_{dc} \tag{8.37}$$

$$v_3 = -v_1 - v_2 \tag{8.38}$$

The three-leg converter supplying a three-phase load (Fig. 8.14) also presents a degree of freedom. Note that there are only two output voltages to be controlled (v_1 and v_2 – or any other combination of two voltages among v_1, v_2, and v_3) throughout three controlled variables q_1, q_2, and q_3.

While the load voltages are generated by the converter, the load currents (i_1, i_2, and i_3) depend on the load connected to the converter. In the case of a linear load, it can be written as $i_x = v_x/Z_x$, where Z_x is the load impedance for phase x; with $x = 1, 2, 3$. The voltages and currents through the switches are presented in Table 8.2.

TABLE 8.2 Voltages and Currents of the Switches as a Function of the Switching States

Switching States	Voltages of the Switches		Currents of the Switches	
q_x	v_{qx}	$v_{\bar{q}x}$	i_{qx}	$i_{\bar{q}x}$
0	V_{dc}	0	0	i_l
1	V_{dc}	0	i_l	0

From Table 8.2, the following equations can be used to define the voltage and currents of the switches, respectively.

$$v_{qx} = (1 - q_x)V_{dc} \tag{8.39}$$

$$v_{\bar{q}x} = q_x V_{dc} \tag{8.40}$$

$$i_{qx} = q_x i_x \tag{8.41}$$

$$i_{\bar{q}x} = (1 - q_x) i_x \tag{8.42}$$

with $x = 1, 2, 3$.

8.3.2 PWM Implementation

From equations (8.29)–(8.31) it is possible to define the desired output voltages as a function of both reference pole voltages and voltage v_{n0}^*, as follows:

$$v_1^* = v_{10}^* - v_{n0}^* \tag{8.43}$$

$$v_2^* = v_{20}^* - v_{n0}^* \tag{8.44}$$

$$v_3^* = v_{30}^* - v_{n0}^* \tag{8.45}$$

The voltage v_{n0}^* can be considered as an auxiliary variable (v_h^*) which is expected to bring a level of optimization for the converter, as for the single-phase converter presented previously. If the desired balanced three-phase voltages are given by $v_1^* = V_o^* \cos(\omega_o t)$, $v_2^* = V_o^* \cos(\omega_o t + 120°)$, and $v_3^* = V_o^* \cos(\omega_o t - 120°)$, the reference pole voltages can be defined by:

$$v_{10}^* = v_1^* + v_h^* \tag{8.46}$$

$$v_{20}^* = v_2^* + v_h^* \tag{8.47}$$

$$v_{30}^* = v_3^* + v_h^* \tag{8.48}$$

with

$$v_h^* = V_{dc}\left(\mu - \frac{1}{2}\right) + (\mu - 1)V_{max} - \mu V_{min} \tag{8.49}$$

where $V_{max} = \text{MAX}\{v_1^*, v_2^*, v_3^*\}$ and $V_{min} = \text{MIN}\{v_1^*, v_2^*, v_3^*\}$; μ is the distribution factor and determines how the free-wheeling time interval can be distributed throughout the switching period.

Equations (8.19) and (8.49) are identical equations, just changing the terms $V_{\max}$ and $V_{\min}$. In fact, v_h^* is a generalized expression that can be employed in different power converters processing ac voltage, as discussed later in this chapter.

8.3.3 Analog and Digital Implementation

Figure 8.15 shows the analog implementation of the PWM solution for the three-leg converter depicted in Fig. 8.14. The reference pole voltages are obtained as in equations (8.46)–(8.48) and compared to the carrier signal v_t^*.

Following the same approach employed for the single-phase converter, it is possible to calculate the time intervals in which the switches q_1, q_2, and q_3 are on, that is

$$\tau_1 = \left(\frac{v_{10}^*}{V_{dc}} + \frac{1}{2}\right) T_s \tag{8.50}$$

$$\tau_2 = \left(\frac{v_{20}^*}{V_{dc}} + \frac{1}{2}\right) T_s \tag{8.51}$$

$$\tau_3 = \left(\frac{v_{20}^*}{V_{dc}} + \frac{1}{2}\right) T_s \tag{8.52}$$

After defining the pulse widths (τ_1, τ_2, and τ_3), it is necessary to program three timers associated with each leg.

8.3.4 Influence of μ for PWM Implementation in a Three-Leg Converter

As considered for the two-leg converter, the voltage v_h^* can be used to optimize specific parameters such as weighted total harmonic distortion (WTHD) and losses.

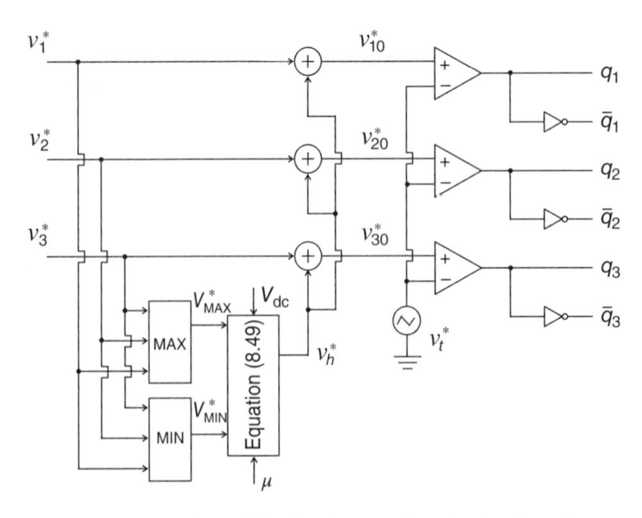

Figure 8.15 Analog PWM implementation for the three-leg converter.

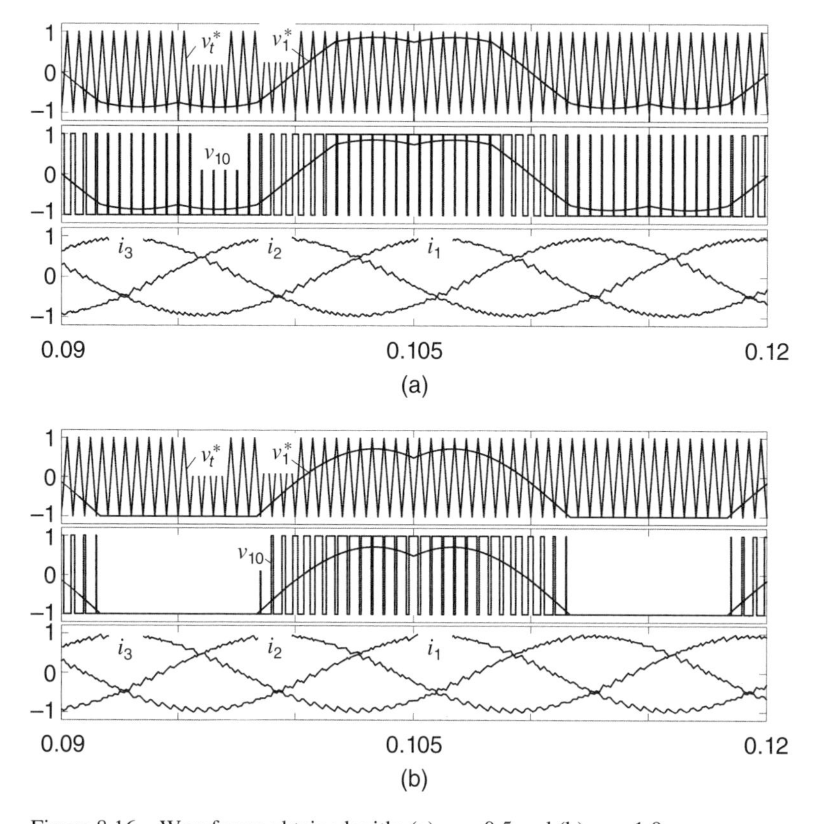

Figure 8.16 Waveforms obtained with: (a) $\mu = 0.5$ and (b) $\mu = 1.0$.

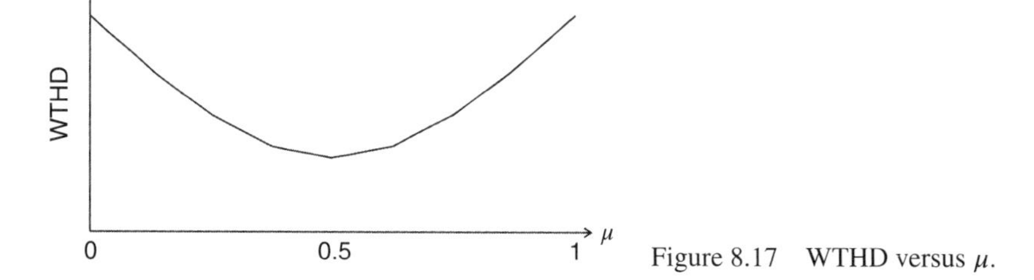

Figure 8.17 WTHD versus μ.

Figure 8.16(a) and 8.16(b) shows the main waveforms for μ equal to 0.5 and 0, respectively. While Fig. 8.17 shows the WTHD of the load phase voltage as a function of μ, Fig. 8.18 shows the same waveforms as in Fig. 8.16 with μ changing from 0 to 1 six times per period. Note that $\mu = 0.5$ generates symmetrical waveforms with the benefit of better WTHD. On the other hand, $\mu = 0$ leads to benefits in terms of loss reduction due to the clamped pole voltages. One way to bring up the benefits of both $\mu = 0.5$ and $\mu = 1.0$ is to change the parameter μ from 0 to 1 many times inside the sine period. This will guarantee an average value of 0.5 while keeping the pole voltages clamped.

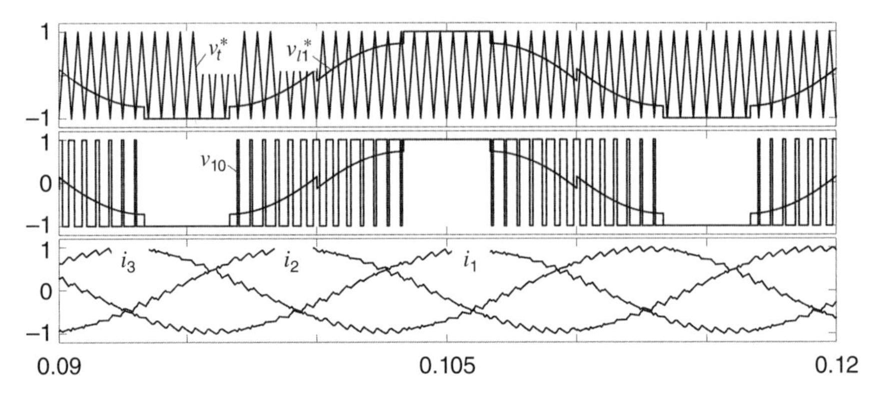

Figure 8.18 Waveforms with μ changing from 0 to 1.

8.3.5 Influence of the Three-Phase Machine Connection over Inverter Variables

Traditionally, a three-phase machine is designed to be connected in a wye (Y) or delta (Δ) arrangement depending on the voltage level available from the utility grid. In a motor drive system, the connections of the three-phase machine can still be defined as a function of the voltage level available on the dc-link capacitor. If the dc-link voltage for a given application is not restrictive, the choice between both possibilities of machine connections [Y – see Fig. 8.19(a) or Δ – see Fig. 8.19(b)] favors the Y connection, as shown. The following analysis is obtained with the assumption that the same power switches and dc-link capacitors are employed for both motor

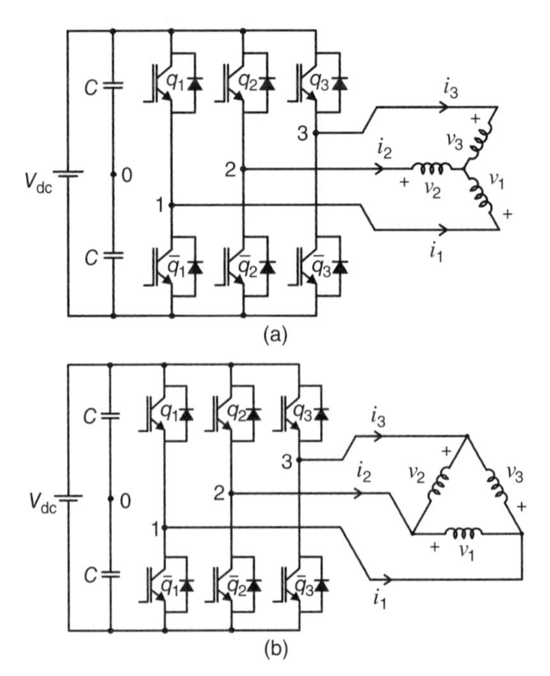

Figure 8.19 Ac motor drive system with a three-phase machine in a: (a) Y connection and (b) Δ connection.

drive systems (Y and Δ), which is reasonable due to the range of IGBTs available in the market. For instance, there is a class of IGBTs that operates at 600 V, the following IGBT available deals with 1200 V. As a consequence, the same IGBT can be employed for a three-phase system with $220V_{\text{rms}}$ (phase voltage) for both Y and Δ, since a Y arrangement would require power switches with 539 V (peak), while a Δ arrangement 311 V (peak).

This section aims to show the advantages and disadvantages of the Y and Δ connections of the three-phase machine in a motor drive system, highlighting the influence of these connections on the converter variables. Some parameters of the inverter will be considered in this analysis, that is, (i) dc-link voltage requirement, (ii) RMS current of the dc-link capacitor, (iii) WTHD of the motor voltages, (iv) converter losses, and (v) fault tolerance capability.

The dc-link voltage of a voltage source inverter (VSI) can be defined as a function of the rated voltage demanded by the motor, as given in (8.53) and (8.54) for wye and delta connection, respectively

$$V_{\text{dc}} = \sqrt{3}V_{\text{ph}} \tag{8.53}$$

$$V_{\text{dc}} = V_{\text{ph}} \tag{8.54}$$

where V_{ph} is the amplitude of the motor phase voltage.

The dc-link voltage requirement of the motor drive system in Fig. 8.19(a) is $\sqrt{3}$ times higher than that in the Fig 8.19(b), considering the motor phase voltage is the same.

On the other hand, the dc-link capacitor current of the system with Y and Δ connections are given by (8.55) and (8.56), respectively:

$$i_c = q_1 i_1 + q_2 i_2 + q_3 i_3 \tag{8.55}$$

$$i_c = q_1(i_3 - i_1) + q_2(i_1 - i_2) + q_3(i_2 - i_3) \tag{8.56}$$

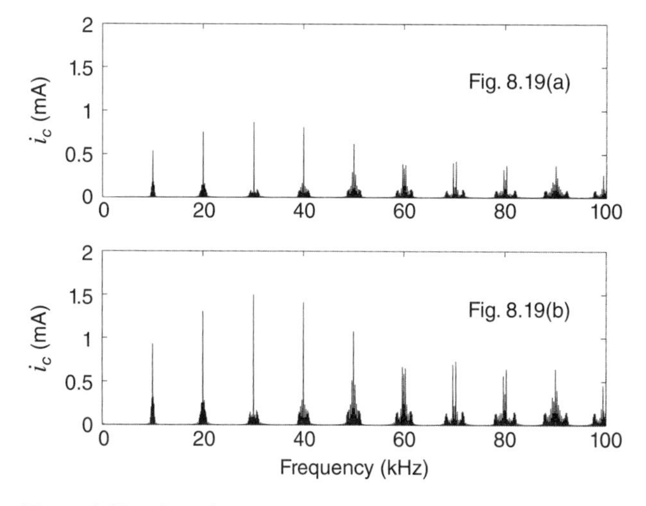

Figure 8.20 Capacitor current spectrum: (a) Y connection and (b) Δ connection.

Figure 8.20 shows the spectra of the dc-link capacitor current. As expected in (8.55) and (8.56), the amplitude of the harmonic components is higher in Δ, which means that the dc-link capacitor losses will be also higher and consequently the lifespan is expected to be lower for a Δ connection.

The quality of the waveform generated at the output converter side can be measured by WTHD, which can be calculated by

$$\mathrm{WTHD} = \frac{100}{a_1}\sqrt{\sum_{i=2}^{p}\left(\frac{a_i}{i}\right)^2} \tag{8.57}$$

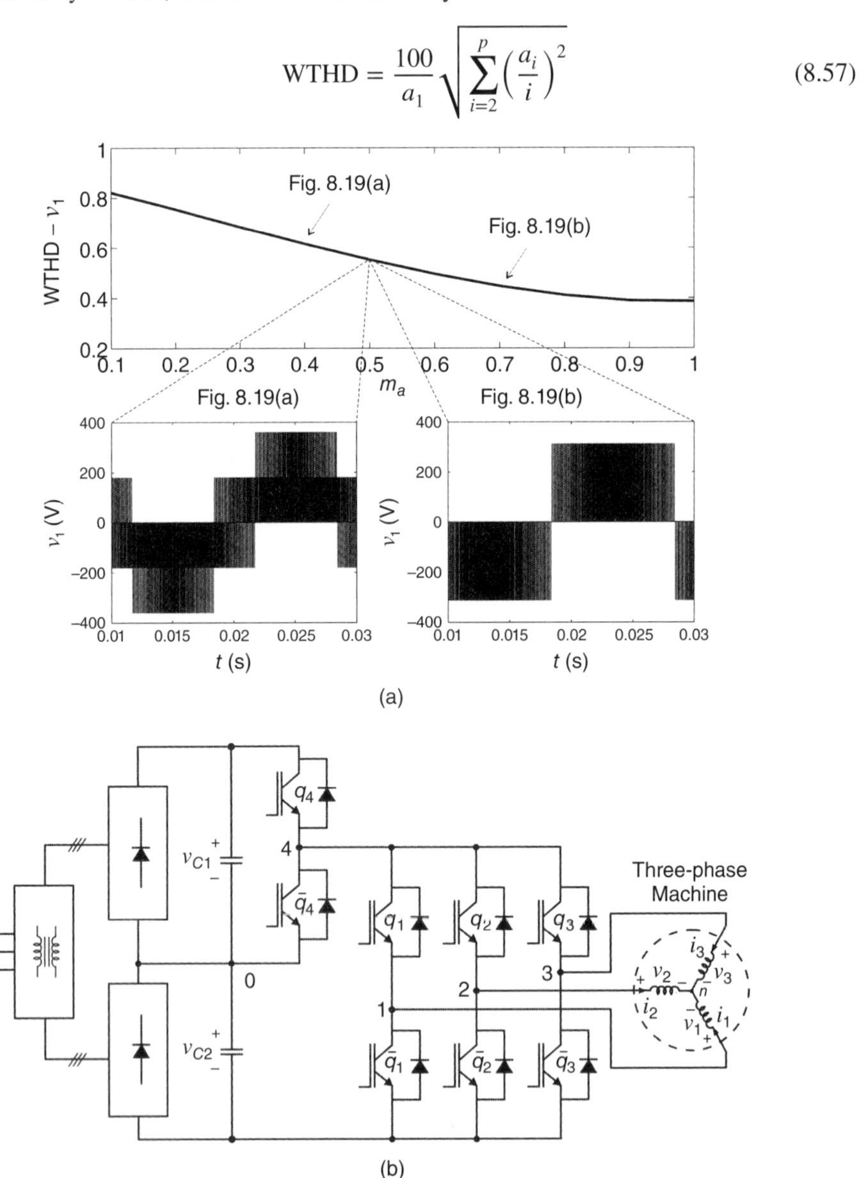

Figure 8.21 (a) (top) WTHD of the phase voltage (v_1) as a function modulation index. (bottom) Phase voltages with m_a = 0.5. (b) Solution with two dc voltages available. (c) THD of the motor current as a function of the modulation index.

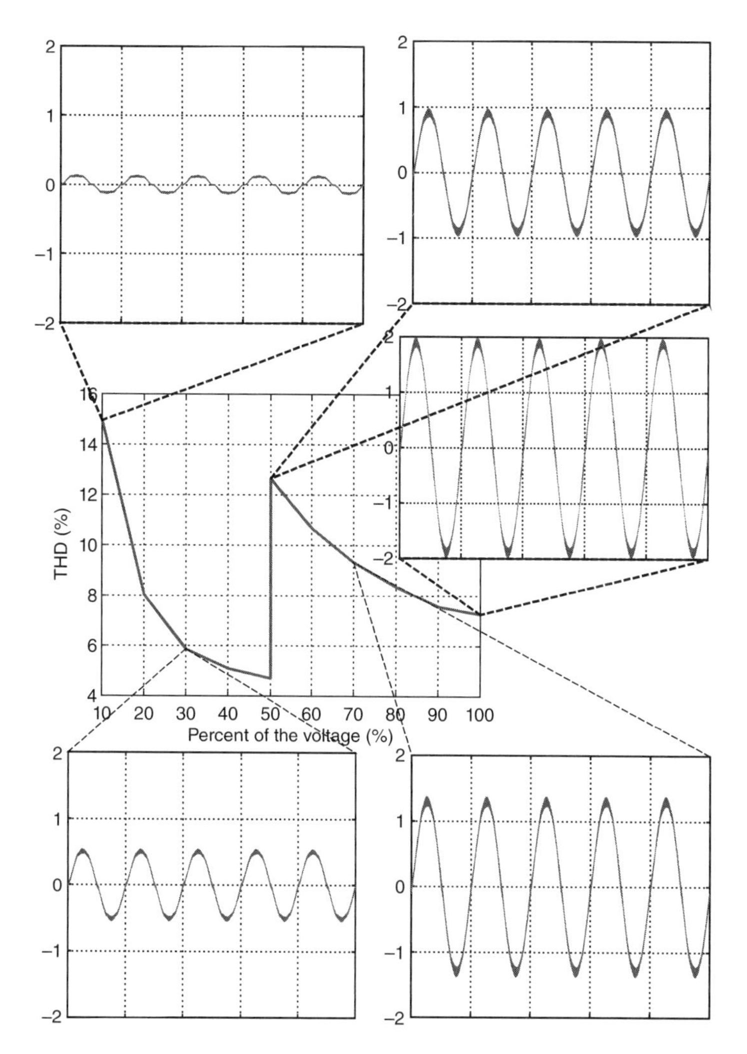

Figure 8.21 (*Continued*)

where a_1 is the amplitude of the fundamental voltage component, a_i is the amplitude of the ith harmonic, and p is the highest harmonic taken into consideration.

An interesting result is shown in Fig. 8.21(a) (top). This figure shows the WTHD of the phase voltage (v_1) as a function of the modulation index (m_a) for both systems shown in Fig. 8.19. Note that independent of the three-phase motor connections (wye or delta) the WTHD of the phase voltages are exactly the same, although the waveforms of the motor phase voltages are different, as seen in Fig. 8.21(b) (bottom). Even with the phase voltage of a three-phase machine, Y connection presents five levels instead of three in a Δ connection; such levels have higher dv/dt in their peaks and valleys as compared to the three-level voltage, which reduces the WTHD. Therefore, even with higher levels, both voltages are equal under the WTHD point of view. A VSI able to drive a three-phase induction motor with reduced THD for a wide range of m_a is shown in Fig. 8.21(b). When the range of the motor voltage is from 100% to 50%, the switch q_4 must be turned on to guarantee the dc-link voltage equal to $v_{C1} + v_{C2}$ and the modulation index goes from 1 to 0.5 ($1 \leq m_a \leq 0.5$). On the other hand, when the range of the motor voltage is from 50% to 0%, the switch q_4 must be turned off to guarantee the dc-link voltage equal to v_{c2} and the modulation index range goes back to $1 \leq m_a \leq 0.5$. It is worth mentioning that, even considering the operation at reduced output voltage range, the power converter in Fig. 8.21(b) operates with a high modulation index, which means reduced THD. Figure 8.21(c), in turn, shows the THD behavior of the motor current from 10% to 100% of rated voltage applied to the machine.

Another important feature affected by the connection of the machine is the converter's losses. Figure 8.22 shows the comparison of conduction, switching, and total (conduction + switching) losses of the VSI. In this figure the switching frequency of the converter is 10 kHz. The converter power losses are higher when the machine is connected in a Δ arrangement, due to the higher level of currents. Figures 8.20 and 8.22 indicate why a Y connection is the primary option for a three-phase machine connection in a motor drive system.

Exercise 8.2

When the machine is Δ connected, the reference pole voltages for the PWM generation can be defined as follows:

$$v_{10}^* = -v_1^* + v_h^*$$
$$v_{20}^* = v_h^*$$
$$v_{30}^* = v_2^* + v_h^*$$

with $v_1^* = V_o^* \cos(\omega_o t)$ and $v_2^* = V_o^* \cos(\omega_o t + 120°)$ and v_h^* given by equation (8.49) with $V_{\text{MAX}} = \{-v_1^*, 0, v_2^*\}$ and $V_{\text{MIN}} = \{-v_1^*, 0, v_2^*\}$. Explain why these equations guarantee a balanced three-phase voltage applied to the machine. What would happen if equations (8.46)–(8.48) are applied for the three-leg converter with a three-phase machine in a Δ arrangement?

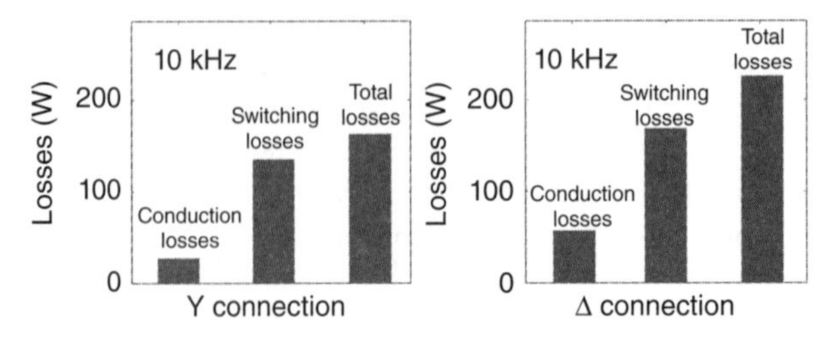

Figure 8.22 Converter losses of the ac motor drive system with a switching frequency equal to 10 kHz.

8.4 SPACE VECTOR MODULATION (SVPWM)

The nonsinusoidal CPWM techniques described earlier in this chapter could be implemented in either an analog or a digital way. On the other hand, the space vector modulation is an inherent digital way to generate the gating signals for q_1, q_2, and q_3. These three independent gating signals lead to eight possible switching combinations for a three-phase inverter.

From the electrical machine and reference-frame theory analyses, it is possible to define a transformation to represent the original three-phase variables in a simple system with variables in quadrature. Such a transformation changes the variables with no physical connotation. Hence, a three-phase system can be considered to form stationary vectors in the dq complex plane. The transformation of variables can be defined as

$$\overline{v}_{odq}^{s} = \overline{P}\overline{v}_{123}^{s} \tag{8.58}$$

where $\overline{v}_{123}^{s}$ are the voltages in the original three-phase system, $\overline{v}_{odq}^{s}$ are the voltages in the dq complex plane plus zero sequence component represented by "o"; the superscript "s" stands for stationary reference-frame, as the stator variables in a three-phase ac machine. $\overline{P}$ is the transformation of variables with

$$\overline{v}_{odq}^{s} = \begin{bmatrix} v_o^s \\ v_d^s \\ v_q^s \end{bmatrix}, \quad \overline{v}_{123}^{s} = \begin{bmatrix} v_1^s \\ v_2^s \\ v_3^s \end{bmatrix}, \quad \text{and} \quad \overline{P} = \sqrt{\frac{2}{3}} \begin{bmatrix} \frac{1}{\sqrt{2}} & \frac{1}{\sqrt{2}} & \frac{1}{\sqrt{2}} \\ 1 & -\frac{1}{2} & -\frac{1}{2} \\ 0 & \frac{\sqrt{3}}{2} & -\frac{\sqrt{3}}{2} \end{bmatrix}.$$

The dq voltages in the stationary reference-frame can be obtained from v_1^s, v_2^s and v_3^s, by using the matrix $\overline{P}$, as follows:

$$v_d^s = \sqrt{\frac{2}{3}}\left(v_1^s - \frac{1}{2}v_2^s - \frac{1}{2}v_3^s\right) \tag{8.59}$$

$$v_q^s = \frac{1}{\sqrt{2}}(v_2^s - v_3^s) \tag{8.60}$$

Since it is assumed a balanced case ($v_1^s + v_2^s + v_3^s = 0$) the term v_o^s is null. Equations (8.59)–(8.60) can be written as functions of the states of the switches [see equations (8.36)–(8.38)], as follows:

$$v_d^s = \sqrt{\frac{2}{3}}\left(q_1 - \frac{1}{2}q_2 - \frac{1}{2}q_3\right) V_{\text{dc}} \tag{8.61}$$

$$v_q^s = \frac{1}{\sqrt{2}}(q_2 - q_3)V_{\text{dc}} \tag{8.62}$$

Table 8.3 shows the vectors in the dq plane $V_x^s = V_d^s + jV_q^s$ $(x = 1, 2, \dots, 8)$ associated with each switching state. Each vector means a topological circuit for the converter, as shown in Fig. 8.23.

Note from Table 8.3 that there are two null (0 and 7) and six non-null (1, 2, 3, 4, 5, and 6) vectors – active vectors, which will be used to synthesize the desired

TABLE 8.3 PWM Vectors as a Function of the Switching States

Vector	q_1	q_2	q_3	V_x^s
0	0	0	0	0
1	1	0	0	$\sqrt{2/3}V_{dc}$
2	1	1	0	$V_{dc}/\sqrt{6}+jV_{dc}/\sqrt{2}$
3	0	1	0	$-V_{dc}/\sqrt{6}+jV_{dc}/\sqrt{2}$
4	0	1	1	$-\sqrt{2/3}V_{dc}$
5	0	0	1	$-V_{dc}/\sqrt{6}-jV_{dc}/\sqrt{2}$
6	1	0	1	$V_{dc}/\sqrt{6}-jV_{dc}/\sqrt{2}$
7	1	1	1	0

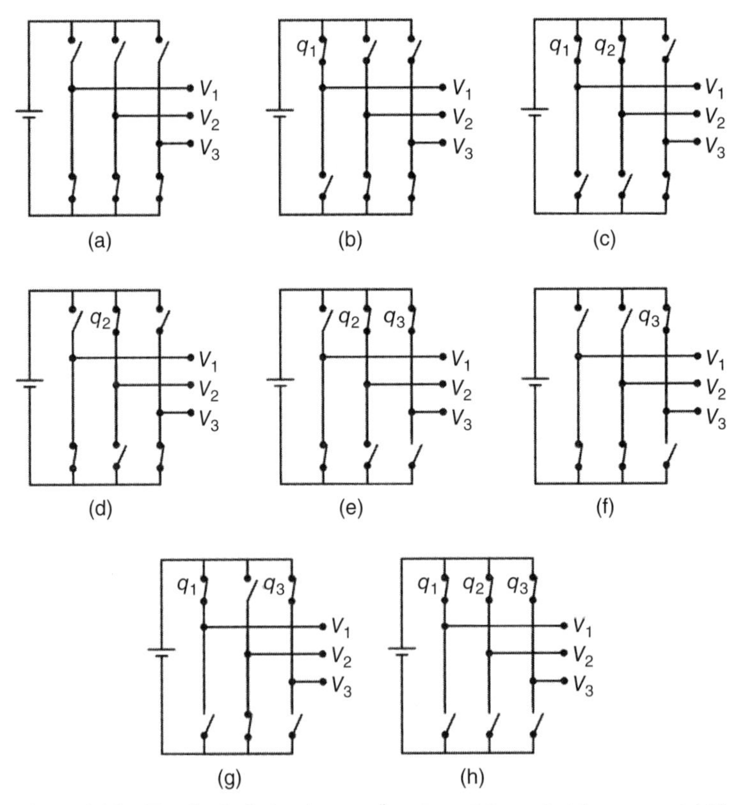

Figure 8.23 Topological circuits as a function of the switching states: (a) Vector 0, (b) Vector 1, (c) Vector 2, (d) Vector 3, (e) Vector 4, (f) Vector 5, (g) Vector 6, (h) Vector 7.

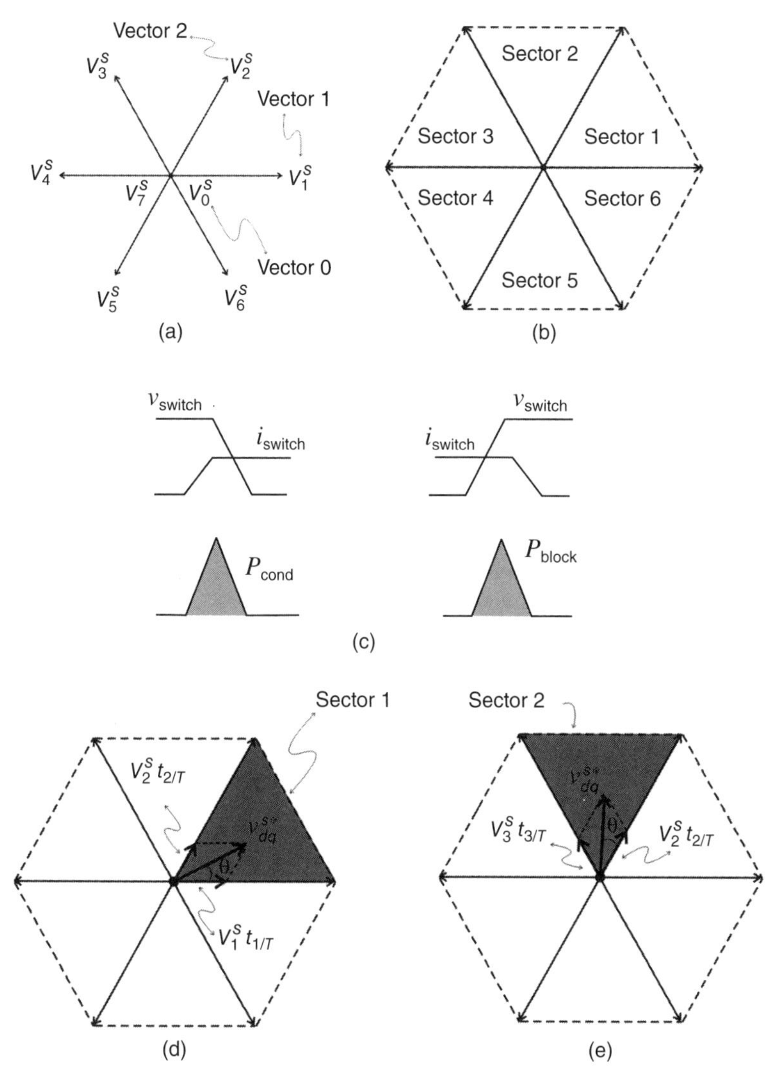

Figure 8.24 Graphical analysis for the space vector PWM showing the (a) vectors and (b) sectors (c) Losses associated to the switching states change A reference voltage in (d) Sector 1, and (e) Sector 2.

load voltage. All vectors V_x^s $(x = 0, 2, \ldots, 7)$ are shown in Fig. 8.24(a) as a graphical representation of Table 8.3. The area between two adjacent non-null vectors defines a sector as depicted in Fig. 8.24(b), which leads to six vectors.

It is possible to go from one vector to its adjacent one with just a single change for the switching state. For instance, from Vector 1 to Vector 2, q_2 is changed from 0 to 1, while q_1 and q_3 are kept constant. In the same way, from Vector 2 to Vector 3, q_1 is changed from 1 to 0, while q_2 and q_3 are kept constant. It means that it is possible to reduce the number of changes in the switching states by using adjacent vectors to synthesize a specific value of the desired voltage. The strategy of applying adjacent vectors plays an important role for converter losses reduction, since losses appear each time it is necessary to change the switching states, as seen in Fig. 8.24(c).

To reduce the switching losses, if the desired voltage is inside Sector 1, the Vectors 1 and 2 should be chosen [see Fig. 8.24(d)]. In the same way, if the desired voltage is inside Sector 2, Vectors 2 and 3 should be chosen [see Fig. 8.24(e)], and so on.

All pairs of adjacent vectors available in a three-leg converter are: (V_1^s and V_2^s), (V_2^s and V_3^s), (V_3^s and V_4^s), (V_4^s and V_5^s), (V_5^s and V_6^s), and (V_6^s and V_1^s). Notice that it is possible to define such a set of adjacent vectors in a generic way by using (V_k^s and V_l^s), with $k = 1, \ldots, 6$, and $l = k + 1$ if $k \leq 5$; and $l = 1$ if $k = 6$. The vectors V_k^s and V_l^s have their dq components (see Table 8.3), that is, $V_k^s = V_{dk}^s + jV_{qk}^s$ and $V_l^s = V_{dl}^s + jV_{ql}^s$. The reference voltage v_{dq}^{s*} can be obtained from a set of three-phase reference voltages (v_{123}^{s*}) by using the transformation $\overline{P}$ given in (8.58). As mentioned earlier, a switching sequence for q_1, q_2, and q_3 should be chosen to allow the converter to synthesize a given reference voltage v_{dq}^{s*}, which is constant in the switching period (T_s). In terms of average values, it is possible to write

$$\frac{1}{T_s}\int_0^{T_s} v_{dq}^{s*} dt = \frac{1}{T_s}\int_0^{t_k} V_k^s dt + \frac{1}{T_s}\int_0^{t_l} V_l^s dt \tag{8.63}$$

$$v_{dq}^{s*} = \frac{t_k}{T_s} V_k^s + \frac{t_l}{T_s} V_l^s \tag{8.64}$$

where t_k and t_l are intervals of time for application of the vectors V_k^s and V_l^s, respectively. Rewriting (8.64) in terms of dq components, we get

$$v_d^{s*} = \frac{t_k}{T_s} V_{kd}^s + \frac{t_l}{T_s} V_{ld}^s \tag{8.65}$$

$$v_q^{s*} = \frac{t_k}{T_s} V_{kq}^s + \frac{t_l}{T_s} V_{lq}^s \tag{8.66}$$

Equations (8.65) and (8.66) show that the reference voltages can be obtained from the interval of time for application of two adjacent vectors. From (8.65) and (8.66) it is possible to write

$$t_k = \frac{v_d^{s*} V_{lq}^s - v_q^{s*} V_{ld}^s}{V_{kd}^s V_{lq}^s - V_{ld}^s V_{qk}^s} T_s \tag{8.67}$$

$$t_l = \frac{v_q^{s*} V_{kd}^s - v_d^{s*} V_{kq}^s}{V_{kd}^s V_{lq}^s - V_{ld}^s V_{qk}^s} T_s \tag{8.68}$$

To keep the switching frequency of the converter constant, it is necessary to guarantee that the sum of the time for the application of the vectors is equal to T_s. Then, the null vectors (Vectors 0 and 7 in Table 8.3) can be used to keep the switching frequency constant, that is, $t_k + t_l + t_o = T_s$, and then $t_o = T_s - t_k - t_l$. Note that

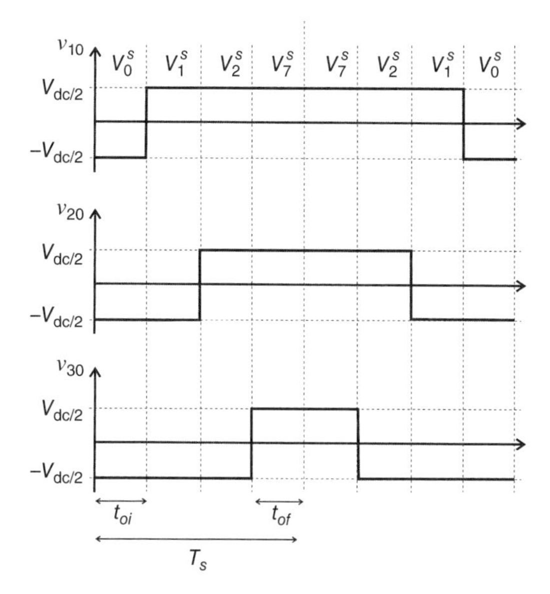

Figure 8.25 Pulses defining how the null vectors are divided at the beginning and end of the switching period.

there is a degree of freedom for t_o, meaning that this time could be divided between the beginning ($t_{oi} = \mu t_o$) and the end $[t_{of} = (1 - \mu)t_o,\ 0 \leq \mu \leq 1]$ of the sampling interval, as presented in Fig. 8.25. Note that this approach brings benefits in terms of THD reduction, since the pulse can be centered inside the sampling period.

A summary of the SPWM is presented in Fig. 8.26, which also shows the algorithm for the implementation of the space vector modulation by using a MATLAB code.

8.5 OTHER CONFIGURATIONS WITH CPWM

Although the SPWM approach represents one of the most popular options in industrial applications, the CPWM is able to generate PWM signals with the same benefits as in the SPWM. Furthermore, the expression derived in (8.19) is a generic way to represent the auxiliary voltage v_h^*. Such expressions can be employed in different converters with at least one degree of freedom, that is, the number of voltages available at the converter side higher than the number of voltage(s) demanded by the load. In the following sections, a symmetrical two-phase motor drive system and a converter for application in three-phase four-wire systems are employed as examples of the comprehensiveness of the CPWM.

8.5.1 Three-Leg Converter — Two-Phase Machine

Two-phase motors can be found in home appliances, industrial tools, or small power applications. Such a motor is often used in fixed speed drives. However, a dc–ac power converter can be used when a variable motor speed operation is required. The

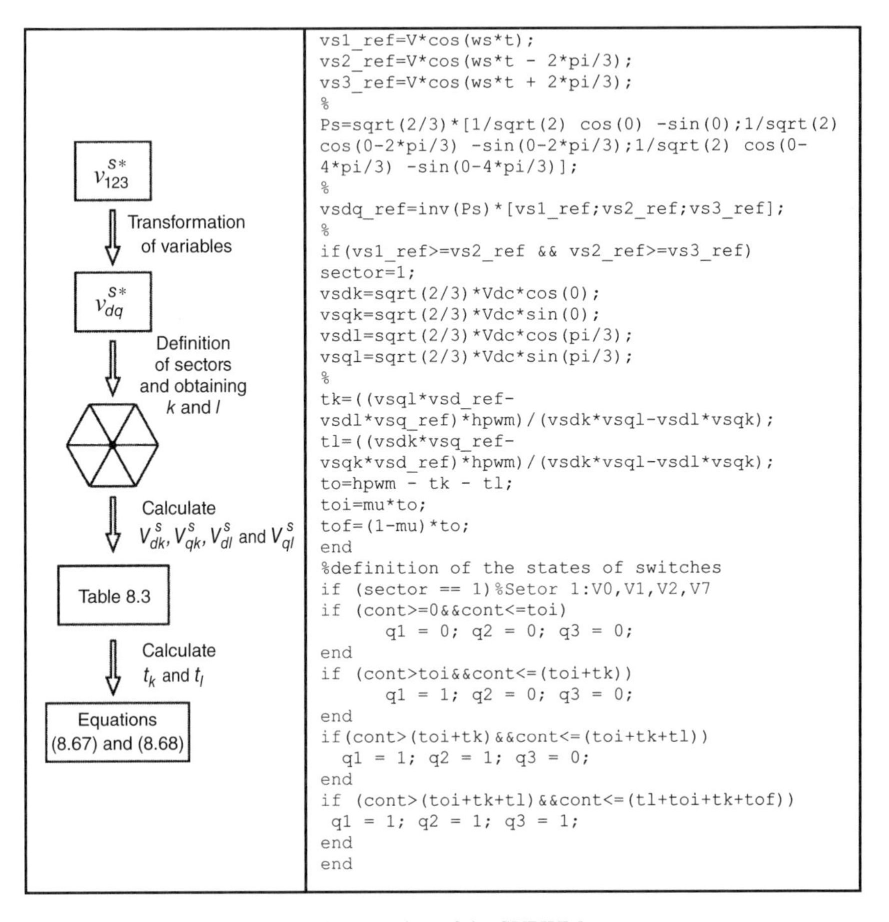

Figure 8.26 Algorithm for implementation of the SVPWM.

symmetrical two-phase motor can be considered as an interesting alternative for fractional horsepower variable-speed drive, due to its characteristic of no pulsating torque produced as observed in single-phase motors.

Since a symmetrical two-phase motor is a three-terminal electrical apparatus, a three-leg converter can be used to supply it. Figure 8.27 shows a three-leg converter supplying a two-phase symmetrical machine. Since this type of machine demands two voltages in quadrature (v_α^* and v_β^*), the reference pole voltages are as follows:

$$v_{10}^* = v_\alpha^* + v_h^* \tag{8.69}$$

$$v_{20}^* = v_\beta^* + v_h^* \tag{8.70}$$

$$v_{30}^* = v_h^* \tag{8.71}$$

where v_h^* can be obtained directly from equation (8.49) with $V_{\max} = \mathrm{MAX}\{v_\alpha^*, v_\beta^*, 0\}$ and $V_{\min} = \mathrm{MIN}\{v_\alpha^*, v_\beta^*, 0\}$.

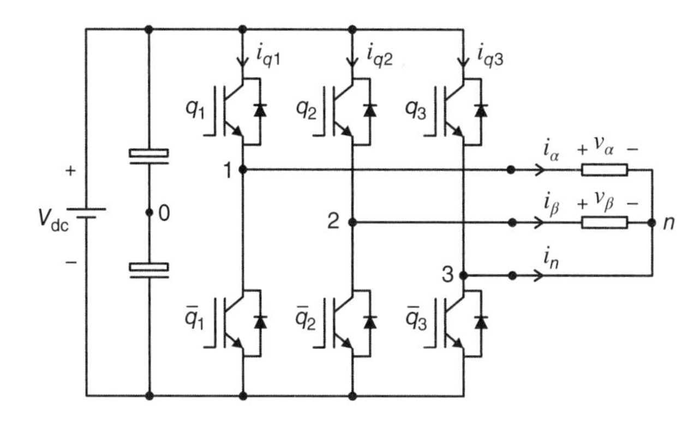

Figure 8.27 Three-leg converter supplying a symmetrical two-phase machine.

Exercise 8.3

Comparing the three-leg converter supplying a three-phase load (as in Fig. 8.14) with the same converter supplying a two-phase load (as in Fig. 8.27), determine the changes in the specification of the power switches assuming that both machines have the same phase voltage, active power, and power factor. Disregard any safety margin while specifying the power switches.

8.5.2 Four-Leg Converter

Note that the three-phase and two-phase motor drive systems require the same topology in terms of the number of switches employed. The difference is related to the load (e.g., type of machine) connected to the converter and also in terms of control considered for PWM and feedback strategies (if needed).

Similarly, the same CPWM approach can be employed for dc-ac converters with four legs considered in different applications. Figure 8.28 shows a four-leg converter feeding a three-phase four-wire system. In this type of application, there

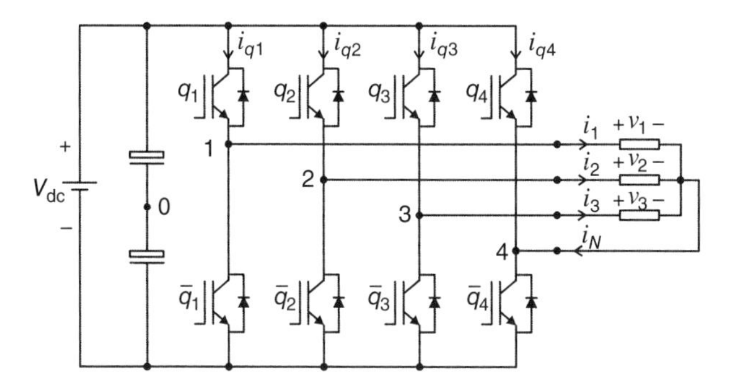

Figure 8.28 Four-leg converter supplying a three-phase four-wire load.

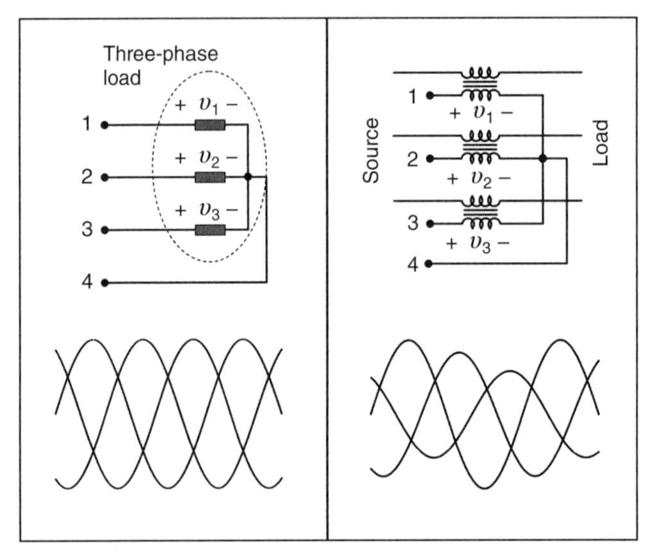

Figure 8.29 Three-phase four-wire systems with: (left) balanced and (right) unbalanced waveforms.

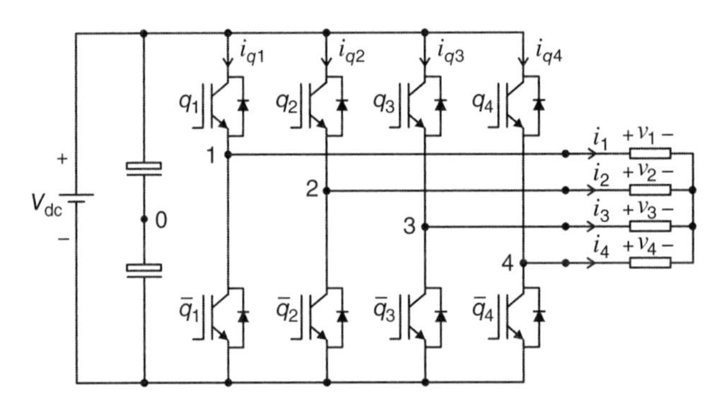

Figure 8.30 Four-leg converter supplying four-phase machine.

are two different scenarios: balanced three-phase voltages, as in Fig. 8.29 (left-hand side), and unbalanced three-phase voltages, as in Fig. 8.29 (right-hand side). The latter case is employed for applications such as dynamic-voltage-restorer and series-active-power-filter.

On the other hand, Fig. 8.30 shows the same four-leg converter supplying a four-phase machine. The PWM strategy with the inclusion of the voltage v_h^* is still valid for the converter with four legs, even considering the different applications highlighted previously. The use of a four-phase machine drive system brings some advantages such as eliminating the common mode, reducing the ratings of the power switches, and providing fault tolerance characteristics for the drive system.

Exercise 8.4

(a) Write the PWM equations (reference pole voltages) for the four-leg converter presented in Fig. 8.28 assuming both cases with balanced and unbalanced requirement. (b) Write the PWM equations for the four-leg converter supplying a four-phase machine with balanced voltages. (c) Also, draw a block diagram as in Fig. 8.15 for the (a) and (b) questions.

Application (Dynamic Voltage Restorer)

Dynamic voltage restorer (DVR) can be used to minimize the effect of operational malfunction of the electric power grid, which can lead to sags or swells. The DVR's benefits of sustaining and restoring the voltage supply are especially important in applications with critical loads requiring a regulated voltage. The operation principle of this electrical apparatus is as follows: during the normal operation of the grid, the static switch Q_3 is turned on, while during sags or swells the switches Q_1 and Q_2 are turned on to either add or subtract the grid voltage, as compensation voltage from the DVR.

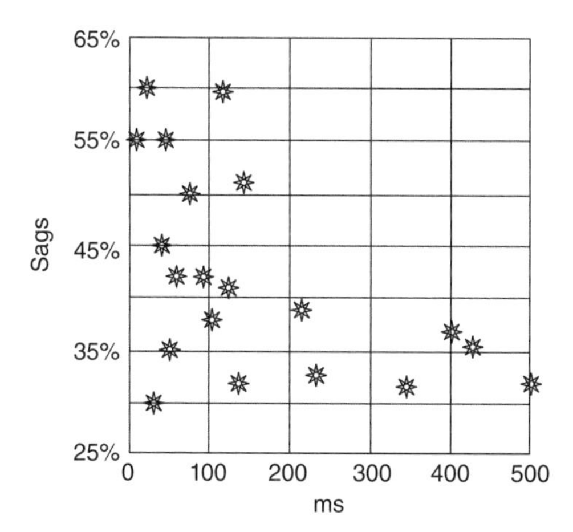

The figure below shows the compensation requirement for a typical situation that shows a trend of deeper voltage sags with shorter duration, while moderate sags

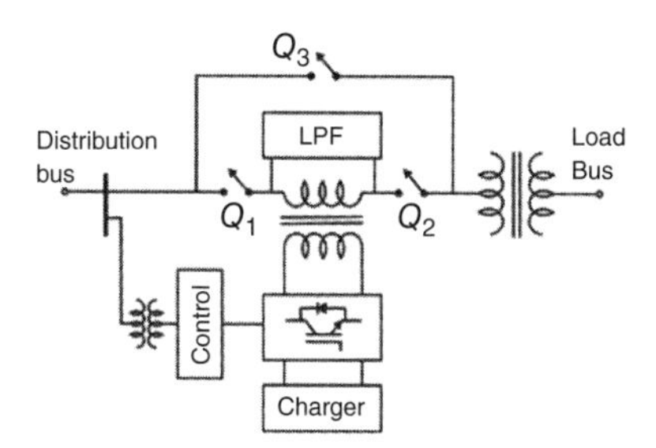

occur for a longer period of time. The information provided in the graph below plays an important role for designing the charger of the DVR.

The figure below presents the dynamic operation of the DVR highlighting the grid voltage with a 100 ms *sag duration. Note that this figure shows a sag with balanced voltages, but it is quite common to have an unsymmetrical sag, which requires a DVR implemented by a three-phase four-wire dc-ac converter.*

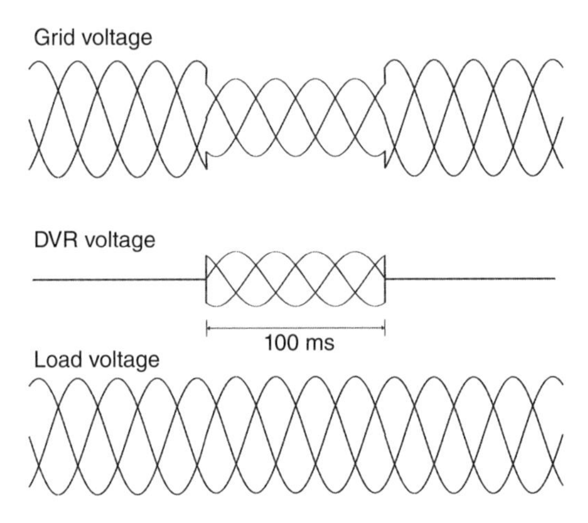

8.6 NONCONVENTIONAL TOPOLOGIES WITH CPWM

8.6.1 Inverter with Split-Wound Coupled Inductors

A converter having the same number of controlled power devices (six IGBTs) as in the conventional three-leg converter (Fig. 8.14) but with higher number of levels is presented in Fig. 8.31. Such a converter was presented in Section 7.3 with another PWM approach, consisting of two carrier and one modulating signal per phase. The

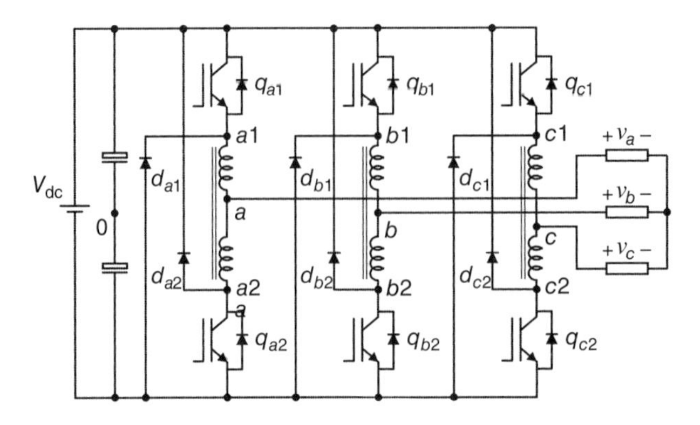

Figure 8.31 Three-phase PWM inverter with split-wound coupled inductors.

CPWM is presented in this section for the converter in Fig. 8.31 as an alternative option. The main advantages of this converter when compared to the solutions presented in the technical literature are higher number of levels for the pole voltages and no dead-time requirement for the converter's operation.

Such a PWM inverter with split-wound coupled inductors is constituted by switching power devices ($q_{x1} - q_{x2}$, with $x = a, b, c$), six diodes ($d_{x1} - d_{x2}$), and three split-wound coupled inductors. The characteristics and the operation modes of this converter are presented in Section 7.3.

The voltages v_{x10} and v_{x20} (voltages from the points $x1$ and $x2$ to the dc-link capacitor midpoint) can be expressed as a function of the state of the switches q_{x1} and q_{x2}, respectively as follows:

$$v_{x10} = (2q_{x1} - 1)\frac{V_{dc}}{2} \tag{8.72}$$

$$v_{x20} = (2q_{x2} - 1)\frac{-V_{dc}}{2} \tag{8.73}$$

It means that each leg can be modeled as in Fig. 8.32(a). Then the pole voltages v_{x0} will be given by

$$v_{x0} = \frac{1}{2}(v_{x10} - v_{x20}) \tag{8.74}$$

Hence from (8.72)–(8.74) it is possible to write v_{x0} as a function only of the state of the switches and dc-link voltage, as follows:

$$v_{x0} = (q_{x1} - q_{x2})\frac{V_{dc}}{2} \tag{8.75}$$

or

$$v_{x0} = v_{x1} - v_{x2} \tag{8.76}$$

where $v_{x1} = q_{x1}V_{dc}/2$ and $v_{x2} = q_{x2}V_{dc}/2$.

From (8.76) it is possible to model the three-phase converter as depicted in Fig. 8.32(b).

If the desired voltage in the three-phase load is given by v_a^*, v_b^*, and v_c^*, then the reference pole voltage can be written as

$$v_{a0}^* = v_a^* + v_h^* \tag{8.77}$$

$$v_{b0}^* = v_b^* + v_h^* \tag{8.78}$$

$$v_{c0}^* = v_c^* + v_h^* \tag{8.79}$$

where v_h^* can be obtained directly from equation (8.49). Once the reference pole voltages v_{a0}^*, v_{b0}^*, and v_{c0}^* are defined from (8.77)–(8.79), the voltages from the points x_1 and x_2 (with $x = a, b, c$) with respect to 0 are given by

$$v_{x1}^* = \frac{v_{x0}^*}{2} \tag{8.80}$$

$$v_{x2}^* = -\frac{v_{x0}^*}{2} \tag{8.81}$$

The PWM signals generation from equations (8.80) and (8.81) can be done as presented in Fig. 8.32(c).

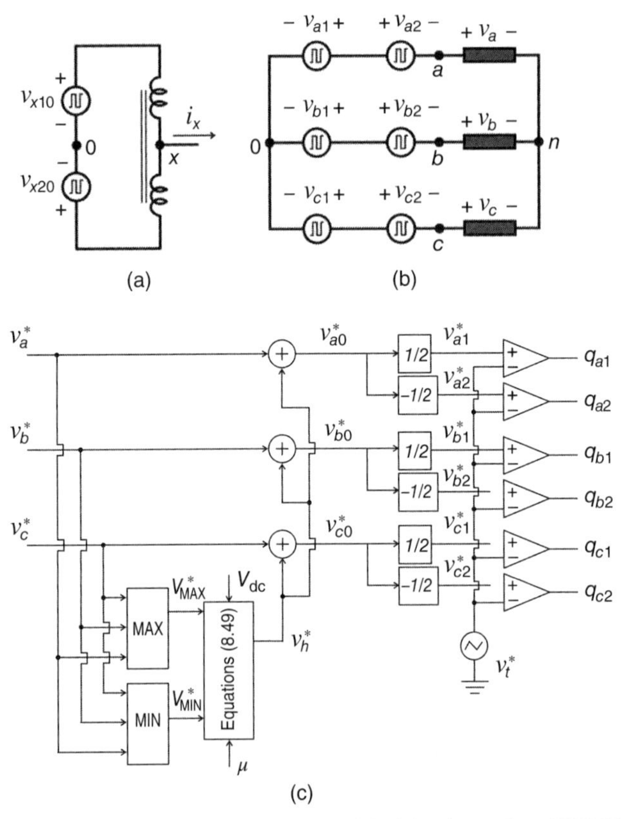

Figure 8.32 (a) Leg model. (b) Model of the three-phase PWM inverter with split-wound coupled inductors. (c) Analog PWM implementation.

8.6.2 Z-Source Converter

The PWM dc–ac power converters presented in this book are predominantly characterized as a voltage-source converter. However, in general terms, dc–ac power converters can be divided into two main groups: voltage-source and current-source converters, each one with its particular characteristics. The impedance-source converter is an alternative type of power converter, which provides the unique feature of buck and boost operation capability with the same number and same arrangement of the semiconductor devices. Such a converter employs an impedance circuit to connect the converter to the primary energy source, as seen in Fig. 8.33, and it is also known as Z-source inverter.

Unlike the conventional dc–ac converter in which there is a need for dead time to avoid short-circuit of the dc-link source, in the Z-source converter such short-circuit (named as shoot-through zero state) is required to guarantee the boost operation of the converter. In fact the Z-source converter is a buck-boost dc-ac converter with the operation mode (buck or boost) defined by the PWM strategy.

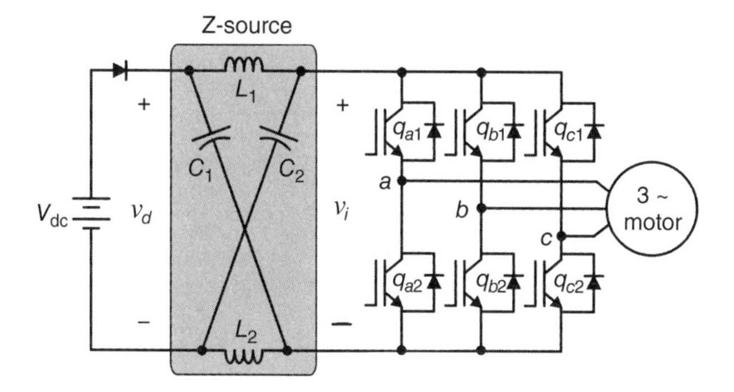

Figure 8.33 Z-source converter.

The circuit operates, boosting the input voltage (V_{dc}) if there is a short-circuit time between two switches of the same leg for one, two, or three legs. On the other hand, if there is no short-circuit time (as in a conventional converter) the circuit will operate as a buck converter. Figure 8.34(a) and 8.34(b) show the equivalent circuits considering the dc source (V_{dc}) point of view when the inverter is creating a short-circuit and without a short-circuit, respectively. Note that when there is no short-circuit, the inverter is represented by a current source i_i, since it is assumed that the load has inductive characteristics.

Considering an ideal and symmetrical circuit with $L_1 = L_2 = L$ and $C_1 = C_2 = C$, it leads to

$$V_{C1} = V_{C2} = V_C \tag{8.82}$$

$$v_{L1} = v_{L2} = v_L \tag{8.83}$$

From Fig. 8.34(a), that is, during the short-circuit time (τ_o), it is seen that

$$v_L = V_C \tag{8.84}$$

$$v_d = 2V_C \tag{8.85}$$

$$v_i = 0 \tag{8.86}$$

On the other hand, from Fig. 8.34(b), during the conventional operation (τ_N), it yields

$$v_L = V_{\mathrm{dc}} - V_C \tag{8.87}$$

$$v_d = V_{\mathrm{dc}} \tag{8.88}$$

$$v_i = 2V_C - V_{\mathrm{dc}} \tag{8.89}$$

From equations (8.84)–(8.86) and (8.87)–(8.89) and since the average value of the inductor voltage into one switching period (T_s) should be zero, it leads to

$$V_L = \frac{1}{T_s}\int_t^{t+T_s} v_L dt = \frac{\tau_o V_C + \tau_N(V_{\mathrm{dc}} - V_C)}{T_s} = 0 \tag{8.90}$$

where $T_s = \tau_N + \tau_o$.

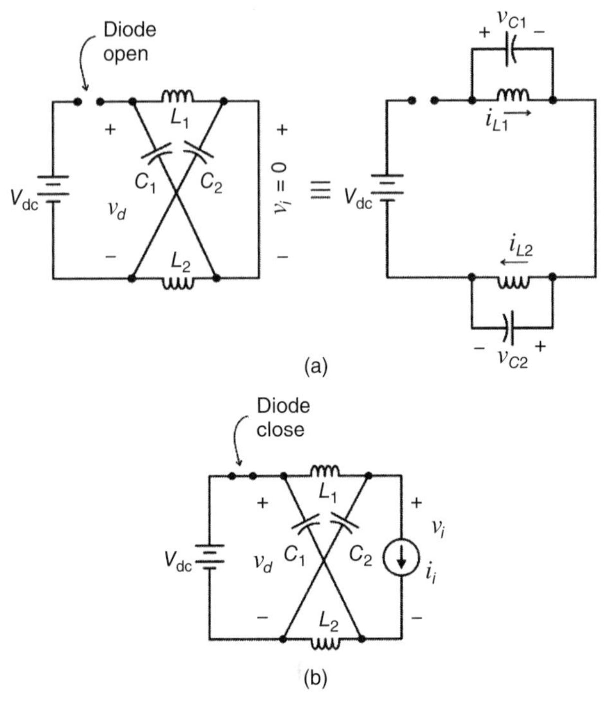

Figure 8.34 (a) Equivalent circuit during the short-circuit. (b) Equivalent circuit during the conventional operation.

Working on equation (8.90) it becomes

$$\frac{V_C}{V_{\text{dc}}} = \frac{\tau_N}{\tau_N - \tau_o} \tag{8.91}$$

The average voltage of the input converter side is given by

$$V_i = \frac{\tau_N}{\tau_N - \tau_o} V_{\text{dc}} \tag{8.92}$$

Finally, the peak dc-link voltage can be obtained as follows:

$$v_i = V_C - v_L \tag{8.93}$$

$$v_i = 2V_C - V_{\text{dc}} \tag{8.94}$$

$$v_i = \frac{1}{1 - 2\tau_{o/T_s}} V_{\text{dc}} = B \cdot V_{\text{dc}} \tag{8.95}$$

where B is the boost factor of the Z-source converter. Equation (8.95) means that for any τ_o different from 0 the voltage value of the input side of the converter will be higher than the voltage at the source side (V_{dc}), that is, boost operation. The capacitor voltages can also be obtained as a function of the shoot-through time (τ_o):

$$V_{C1} = V_{C2} = \frac{1 - \tau_{o/T_s}}{1 - 2\tau_{o/T_s}} V_{dc} \tag{8.96}$$

To generate the PWM signals for the Z-source converter allowing the boost operation it is necessary to include the short-circuit time (shoot-through). The CPWM technique presented throughout this chapter can be applied to this type of converter as well. In this case, in order to facilitate the demonstration of the PWM signal generation, the voltage of the input converter side (v_i) is divided into two parts, thus creating a virtual zero point ("0"), as observed in Fig. 8.35(a).

Then, the pole voltages are defined as the voltage from the central point of the leg to the point "0". The pole voltages are: v_{a0}, v_{b0}, and v_{c0}. Figure 8.35(b) shows the generation of pole voltages through the comparison between the triangular and sinusoidal waveforms for the conventional (left-side) and Z-source (right-side) inverters. The shoot-through zero states can be evenly distributed among the three-phase legs, while the equivalent active vectors are unchanged. It can be seen from Fig. 8.35(b) that τ_1 and τ_2 (time intervals of active vectors) have the same values for the conventional and Z-source converters. During the shoot-through zero state intervals of time the pole voltages are equal to zero, since $v_i = 0$.

Exercise 8.5

Assuming that the circuit in Fig. 8.33 (Z-source topology) should be designed to operate boosting this input voltage from 100 V to 150 V, what is the maximum modulation index (m_a) obtained in this scenario?

8.6.3 Open-End Winding Motor Drive System

As mentioned in previous chapters of this book, besides the Δ and Y connections there is another way to connect a three-phase machine in motor drive systems, which is called open-end winding connection. Such a connection consists of a series arrangement of two conventional two-level converters (inverters 1 and 2 in Fig. 8.36) through the open-end windings of a three-phase machine. Note that in this case it is necessary to have access to six terminals of the machine.

The converter pole voltages depend on the conduction states of the power switches, that is

$$v_{a_l 0_l} = (2q_{al} - 1)\frac{V_{dcl}}{2} \tag{8.97}$$

$$v_{b_l 0_l} = (2q_{bl} - 1)\frac{V_{dcl}}{2} \tag{8.98}$$

$$v_{c_l 0_l} = (2q_{cl} - 1)\frac{V_{dcl}}{2} \tag{8.99}$$

where $l = 1$ and 2.

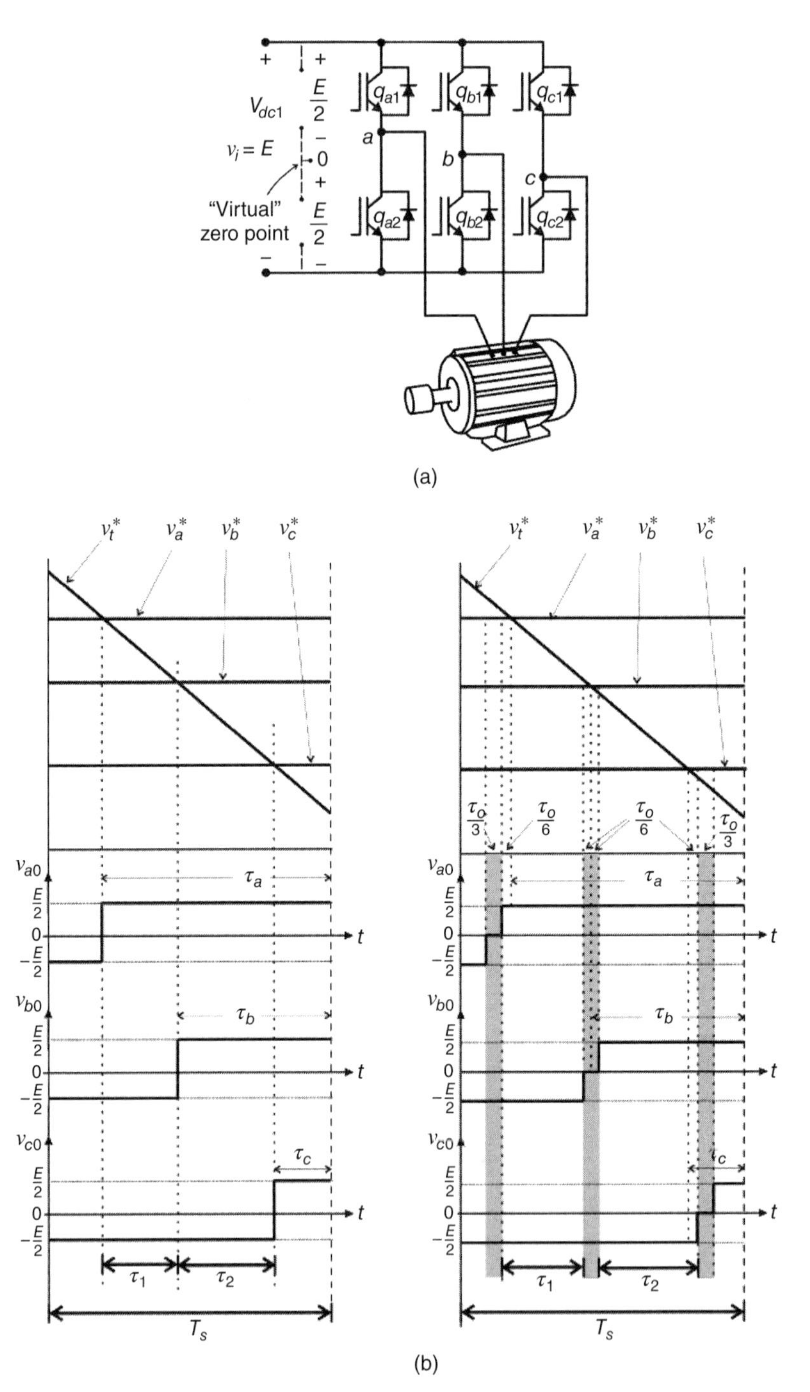

Figure 8.35 (a) Inverter highlighting the virtual zero point "0". (b) Generation of pole voltages through sine-triangle modulation for conventional (left-side) and Z-source (right-side) inverters.

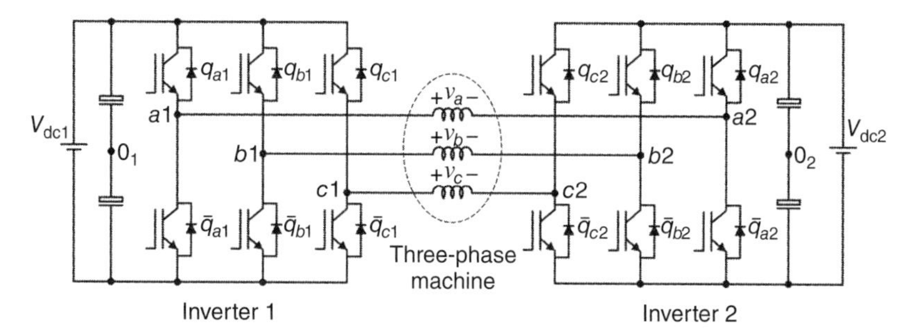

Figure 8.36 Open-end winding machine supplied by two inverter series connected.

The motor phase voltages can be written as a function of the pole voltages and as a function of the difference between the dc-link capacitor midpoints:

$$v_a = v_{a_1 0_1} - v_{a_2 0_2} + v_{0_1 0_2} \tag{8.100}$$

$$v_b = v_{b_1 0_1} - v_{b_2 0_2} + v_{0_1 0_2} \tag{8.101}$$

$$v_c = v_{c_1 0_1} - v_{c_2 0_2} + v_{0_1 0_2} \tag{8.102}$$

where, for a balanced case it becomes

$$v_{0_1 0_2} = -\frac{1}{3}\left(\sum_{j=a}^{c} v_{j_1 0_1} - \sum_{j=a}^{c} v_{j_2 0_2}\right) \tag{8.103}$$

Both power converters must generate appropriate reference voltages (v_a^*, v_b^*, and v_c^*) for the three-phase machine. Due to the inherent redundancy of the voltages demanded by a three-phase machine, the converters actually need to generate independently two line-to-line voltages (e.g., $v_{ab} = v_a - v_b$ and $v_{ca} = v_c - v_a$). For the system presented in Fig. 8.36 there are four degrees of freedom in terms of the PWM generation, since the converter has six legs and it needs to generate just two voltages. Considering that the voltages v_{ab}^* and v_{ca}^* should be imposed by the converter, four auxiliary variables are introduced as presented in equations (8.104)–(8.109)

$$v_{ab}^* = v_{a_1 0_1}^* - v_{b_1 0_1}^* - v_{a_2 0_2}^* + v_{b_2 0_2}^* \tag{8.104}$$

$$v_{ca}^* = -v_{a_1 0_1}^* + v_{c_1 0_1}^* + v_{a_2 0_2}^* - v_{c_2 0_2}^* \tag{8.105}$$

$$v_{h0}^* = v_{a_1 0_1}^* - v_{a_2 0_2}^* \tag{8.106}$$

$$v_{h1}^* = v_{a_2 0_2}^* \tag{8.107}$$

$$v_{h2}^* = v_{b_2 0_2}^* \tag{8.108}$$

$$v_{h3}^* = v_{c_2 0_2}^* \tag{8.109}$$

It can be seen from these equations that both desired and auxiliary voltages have been written as a function of the reference pole voltages. However, for the PWM generation purposes, it is necessary to define the reference pole voltages as a function of the desired and auxiliary voltages as follows:

$$v^*_{a_1 0_1} = v^*_{h0} + v^*_{h1} \tag{8.110}$$

$$v^*_{b_1 0_1} = -v^*_{ab} + v^*_{h0} + v^*_{h2} \tag{8.111}$$

$$v^*_{c_1 0_1} = v^*_{ca} + v^*_{h0} + v^*_{h3} \tag{8.112}$$

$$v^*_{a_2 0_2} = v^*_{h1} \tag{8.113}$$

$$v^*_{b202} = v^*_{h2} \tag{8.114}$$

$$v^*_{c_2 0_2} = v^*_{h3} \tag{8.115}$$

It is also possible to define effective reference pole voltages since

$$v^*_{a_{12}0} = v^*_{a_1 0_1} - v^*_{a_2 0_2} \tag{8.116}$$

$$v^*_{b_{12}0} = v^*_{b_1 0_1} - v^*_{b_2 0_2} \tag{8.117}$$

$$v^*_{c_{12}0} = v^*_{c_1 0_1} - v^*_{c_2 0_2} \tag{8.118}$$

Another convenient way to write the effective reference pole voltages is as follows:

$$v^*_{a_{12}0} = v^*_{h0} \tag{8.119}$$

$$v^*_{b_{12}0} = -v^*_{ab} + v^*_{h0} \tag{8.120}$$

$$v^*_{c_{12}0} = v^*_{ca} + v^*_{h0} \tag{8.121}$$

Equations (8.122)–(8.124) lead to a straightforward way to write the PWM equations:

$$v^*_{a_1 0_1} = v^*_{a_{12}0} + v^*_{h1} \tag{8.122}$$

$$v^*_{b_1 0_1} = v^*_{b_{12}0} + v^*_{h2} \tag{8.123}$$

$$v^*_{c_1 0_1} = v^*_{c_{12}0} + v^*_{h3} \tag{8.124}$$

$$v^*_{a_2 0_2} = v^*_{h1} \tag{8.125}$$

$$v^*_{b_2 0_2} = v^*_{h2} \tag{8.126}$$

$$v^*_{c_2 0_2} = v^*_{h3} \tag{8.127}$$

To finally define the reference pole voltages, it remains to determine the auxiliary voltages v^*_{h0}, v^*_{h1}, v^*_{h2}, and v^*_{h3}, as presented in the sequence:

The auxiliary voltage (v^*_{h0}) involves the desired line-to-line and effective pole voltages. It is possible to define v^*_{h0} respecting the following restriction

$$V_{0\mathrm{MIN}} \le v^*_{h0} \le V_{0\mathrm{MAX}} \tag{8.128}$$

with

$$V_{0\mathrm{MIN}} = \mathrm{MIN}\left\{-\frac{V_{\mathrm{dc1}} + V_{\mathrm{dc2}}}{2}, -\frac{V_{\mathrm{dc1}} + V_{\mathrm{dc2}}}{2} - v^*_{ab}, -\frac{V_{\mathrm{dc1}} + V_{\mathrm{dc2}}}{2} + v^*_{ca}\right\}$$

$$V_{0\mathrm{MAX}} = \mathrm{MAX}\left\{\frac{V_{\mathrm{dc1}} + V_{\mathrm{dc2}}}{2}, \frac{V_{\mathrm{dc1}} + V_{\mathrm{dc2}}}{2} - v^*_{ab}, \frac{V_{\mathrm{dc1}} + V_{\mathrm{dc2}}}{2} + v^*_{ca}\right\}$$

The voltages v_{h0}^* can be defined as a function of μ_0 as in equation (8.129), which is an adaption from equation (8.49):

$$v_{h0}^* = (1 - \mu_0)V_{0\mathrm{MIN}} + \mu_0 V_{0\mathrm{MAX}} \tag{8.129}$$

Using the auxiliary voltage v_{h0}^* and the desired line-to-line voltage it is possible to define the effective reference pole voltages. Finally to define the reference pole voltages (8.122)–(8.127) it is necessary to find v_{h1}^*, v_{h2}^*, and v_{h3}^* as in the sequence:

The auxiliary voltages (v_{h1}^*, v_{h2}^*, and v_{h3}^*) can be considered as three single-phase auxiliary voltages, which are also restricted by

$$V_{k\mathrm{MIN}} \leq v_{hk}^* \leq V_{k\mathrm{MAX}} \tag{8.130}$$

with $k = 1, 2, 3$; where the minimum and maximum values are given respectively by

$$v_{k\mathrm{MIN}}^* = \mathrm{MIN}\left\{ -\frac{V_{\mathrm{dc1}}}{2} - v_{l_{12}0}^*, -\frac{V_{\mathrm{dc2}}}{2} \right\}$$

$$v_{k\mathrm{MAX}}^* = \mathrm{MAX}\left\{ \frac{V_{\mathrm{dc1}}}{2} - v_{l_{12}0}^*, \frac{V_{\mathrm{dc2}}}{2} \right\}$$

when $k = 1 \rightarrow l = a$, $k = 2 \rightarrow l = b$, and $k = 3 \rightarrow l = c$.

The remaining auxiliary voltages (v_{hk}^*) can be calculated as presented below:

$$v_{hk}^* = (1 - \mu_k)V_{k\mathrm{MIN}} + \mu_k V_{k\mathrm{MAX}} \tag{8.131}$$

Equation (8.131) was obtained by following the same strategy as in equation (8.19).

8.7 SUMMARY

This chapter presented a systematic way to implement an optimized PWM applied to dc–ac converters [1–23]. It has been demonstrated that if the converter presents a number of pole voltages higher than the number of desired output voltages, a parameter v_h^* can be added to improve either THD or efficiency of the converter. Section 8.2 introduced such an optimization for the H-bridge converter along with its model. Sections 8.3, 8.5, and 8.6 showed the same PWM technique for other converters [24–32], that is: (i) three-leg converter feeding a two-phase load, (ii) four-leg converter supplying a three-phase four-wire load, (iii) split-wound coupled inductor converters, (iv) Z-source converter, and (v) open-end winding motor drive system. Section 8.4 described the space vector modulation applied to the conventional three-phase dc–ac converter.

REFERENCES

[1] Schönung A, Stemmler H. Static frequency changers with subharmonic control in conjunction with reversible variable speed a.c. drives. Brown Boweri Rev 1964;51:555–577.

[2] Jardan KR, Dewan SB, Slemon G. General analysis of three-phase inverters. IEEE Trans Ind Appl 1969;5(6):672–679.

[3] Jardan KR. Modes of operation of three-phase inverters. IEEE Trans Ind Appl 1969;5(6):680–685.

[4] Busse A, Holtz J. Multiloop control of a unity power factor fast-switching AC to DC Converter. Conference Record of IEEE Power Electronics Specialists Conference (PESC'82); 1982. p 171–179.

[5] Pffaf G, Weschta A, Wick A. Design and experimental results of a brushless AC servo drive. Conference Record of IEEE/IAS Annual Meeting; 1982. p 692–697.

[6] Zubek J, Abbondanti A, Nordby CJ. Pulsewidth modulated inverter motor drives with improved modulation. IEEE Trans Ind Appl 1975;IA-11(6):1224–1228.

[7] Bowes SR, Mount MJ. Microprocessor control of PWM inverters. IEE Proc B 1981:293–305.

[8] Holtz J. Pulsewidth modulation for electronic power conversion. Proc IEEE 1994;82:1194–1214.

[9] Buja G, Indri G. Improvement of pulse width modulation techniques. Arch Elektrotech 1977;57:281–289.

[10] Depenbrock M. Pulse width control of a 3-phase inverter with non-sinusoidal phase voltages. Proceedings of the IEEE International Semiconductor Power Converter Conference, ISPCC'07; 1977. p 399–403.

[11] Houldsworth JA, Grant DA. The use of harmonic distortion to increase the output voltage of a three-phase PWM inverter. IEEE Trans Ind Appl 1984;IA-20(5):1224–1228.

[12] Agelidis VG, Ziogas PD, Joos G. Dead-band PWM switching patterns. IEEE Trans Power Electron 1996;11:523–531.

[13] Boost MA, Ziogas PD. State-of-the-art carrier PWM techniques: a critical evaluation. IEEE Trans Ind Appl 1988;24:271–280.

[14] Ziogas PD, Moran L, Joos G, Vincenti D. A refined PWM scheme for voltage and current source converter. Proceedings IEEE PESC'90; 1990. p 977–983.

[15] Seixas P. Commande numérique d'une machine synchrone autopilotée [D.Sc. Thesis]. L'Intitut Nationale Polytechnique de Toulouse, INPT; 1988.

[16] Hava AM, Kerkman R, Lipo TA. A high-performance generalized discontinuous PWM algorithm. IEEE Trans Ind Appl 1998;34:1059–1071.

[17] Holmes DG. The significance of zero space vector placement for carrier-based PWM schemes. IEEE Trans Ind Appl 1996;32:1122–1129.

[18] Taniguchi K, Ogino Y, Irie H. PWM technique for power MOSFET inverter. IEEE Trans Power Elect 1988;3(3):328–334.

[19] Jacobina CB, Lima AMN, da Silva ERC, Alves RNC, Seixas PF. Digital scalar pulse-width modulation: a simple approach to introduce non-sinusoidal modulating waveforms. IEEE Trans Power Electron 2001;16:351–359.

[20] Blasko V. A hybrid PWM strategy combining modified space vector and triangle comparison methods. Proc Conf. Rec. PESC; 1996. p 1872–1878.

[21] Trzynadlowski A, Legowski S. Minimum-loss vector PWM strategy for three-phase inverter. IEEE Trans Power Electron 1994;9:26–34.

[22] van der Broeck HW, Skudelny HC, Stanke GV. Analysis and realization of a pulsewidth modulator based on voltage space vector. IEEE Trans Ind Appl 1988;24(1):142–150.

[23] Holmes DG. The general relationship between regular-sampled pulse-width-modulation and space vector modulation for hard switched converters. Conf. Rec. IEEE-IAS Annual Meeting; 1992. p 1002–1009.

[24] Matsui K, Murai Y, Watanabe M, Kaneko M, Ueda F. A pulsewidth modulated inverter with parallel-connected transistors by using current sharing reactors. IEEE Trans Ind Appl 1993;8(2):186–191.

[25] Salmon J, Ewanchuk J, Knight AM. PWM inverters using split-wound coupled inductors. IEEE Trans Ind Appl 2009;45(6):2001–2009.

[26] Teixeira, CA, McGrath BP, Holmes DG. Topologically reduced multilevel converters using complementary unidirectional phase-legs. Proc. of IEEE Int'l Symp on Industrial Electron; 2012. p 2007–2012.

[27] F. Z. Peng, Z-source inverter, IEEE Trans Ind Appl, vol. 39, pp. 504–510, 2003.

[28] dos Santos EC, Bradaschia F, Cavalcanti MC, da Silva ERC. Z-source converter applied for single-phase to three-phase conversion system. Proceedings of Applied Power Electron Conference and Exposition, APEC; 2011. p 216–223.

[29] Muniz JHG, da Silva ERC, dos Santos, EC. A hybrid PWM strategy for Z-source neutral-point-clamped inverter. Proceedings of Applied Power Electron Conference and Exposition, APEC; 2011. p 450–456.

[30] dos Santos EC, Pimentel Filho EPX, Oliveira AC, da Silva ERC. Hybrid pulse width modulation for Z-source inverters. Proceedings of Energy Conversion Congress and Exposition (ECCE); 2010. p 2888–2892.

[31] Shiny G, Baiju MR. Space vector PWM scheme without sector identification for an open-end winding induction motor based 3-level inverter. Proceedings of IEEE Industrial Electronic Conference, IECON; 2009. p 1310–1315.

[32] Park RM. Two-reaction theory of synchronous machines. AIEE Trans 1929;48:716–730.

CHAPTER 9

CONTROL STRATEGIES FOR POWER CONVERTERS

9.1 INTRODUCTION

After describing many topologies in Chapters 2–7, highlighting the circuits themselves, as well as optimized pulse width modulation (PWM) strategies in Chapter 8, this chapter deals with the control actions needed to keep a specific variable of the converter under control, for instance, in a motor drive system where the speed must be controlled electronically through a power converter, or in a circuit designed to correct the power factor actively. This chapter is strategically placed before the description of back-to-back converters, due to the need for regulation of some variables, such as grid current and dc-link capacitor voltage. It is worth mentioning that all power converters previously presented were without any feedback control.

The types of controllers employed in applications with power converters as controlled source can be basically classified into two main groups: linear and nonlinear controllers. Hysteresis and sliding-modes are examples of nonlinear controllers, while proportional (P), proportional–integral (PI), and proportional–integral–derivative (PID) regulators are examples of linear controllers. The principle and characteristics of both types of controllers are introduced here. Rather than providing deep analysis for each controller, this chapter aims to prepare the reader for the next chapters, where the converters demand feedback control.

A dynamic simulation is an important aspect to test and validate control strategies, since the behavior of the variables can be verified to predict the dynamics of the system. MATLAB codes are presented for the control strategies developed in this chapter as an effective way to verify the actions of the studied controllers. Also, Euler's integration method is employed to simulate the behavior of the state space variables, due to its simplicity.

Following this introduction, Section 9.2 presents some principles of the controllers for systems using power converters as actuators. Section 9.3 deals with a nonlinear control technique called hysteresis control, highlighting some advantages and disadvantages of this technique. Two applications are also considered in this section: dc motor drive system and regulation of an ac variable. Although this book deals with PWM converters processing ac voltage, the dc motor drive system plays an

Advanced Power Electronics Converters: PWM Converters Processing AC Voltages,
Forty Fifth Edition. Euzeli Cipriano dos Santos Jr. and Edison Roberto Cabral da Silva.

important role in understanding a critical feature of the hysteresis control, switching frequency changing with the load. The classic linear controllers (P, PI, and PID) are presented in Section 9.4 for dc variables, while Section 9.5 presents a linear type of controller for ac variables. A more complex control technique is presented in Section 9.6 with a cascade strategy for systems dealing with control of two variables with different dynamic responses. The cascade control strategy is applied to regulate both power factor and dc-link voltage in back-to-back converters for the next chapters. Finally, Section 9.7 summarizes this chapter.

9.2 BASIC CONTROL PRINCIPLES

The system presented in Fig. 9.1(a) shows a Controlled Voltage Source supplying an *RL* load. Let us assume that the control objective in this system is to guarantee the load current regulation, that is, the measured output current (i_o) must follow a specific reference value (i_o^*). Since the output voltage (v_o) has been employed to reach this goal, a simple control block diagram can be obtained, as in Fig. 9.1(b). In this figure, $C(s)$ and $H(s)$ are the controller and the *RL* load in the *s*-domain, respectively. The transfer function of the $H(s)$ is given by $\frac{1}{Ls+R}$.

A discretization method can be employed in order to simulate this system in a digital way (e.g., by using a MATLAB code). One of the simplest ways to discretize a specific process or system is to use the Euler discretization method. This is equivalent of approximating the integrals by using the left-point rule. Hence, the first integral is approximated as the product of the integrand at time t and the integration range dt, which is called step size (h) for the digital implementation.

In the system of Fig. 9.1(a), for a given v_o and considering a specific *RL* load, the following equations can be obtained:

$$v_o = Ri_o + L\frac{di_o}{dt} \tag{9.1}$$

$$\frac{di_o}{dt} = \frac{v_o - Ri_o}{L} \tag{9.2}$$

Considering the left-point rule, the discrete version of equation (9.2) can be written as

$$i_o(t+h) = i_o(t) + \frac{v_o - Ri_o}{L}h \tag{9.3}$$

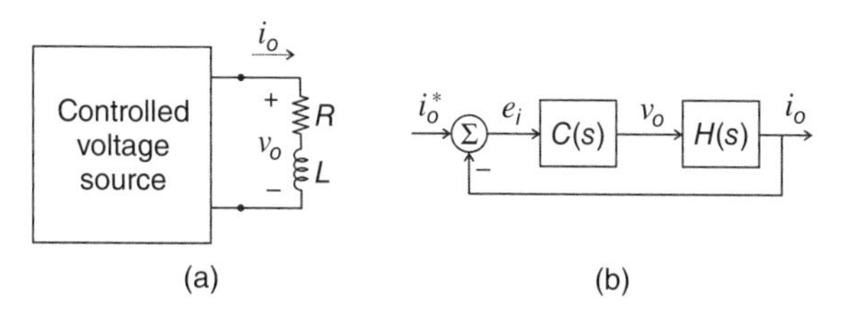

Figure 9.1 (a) Controlled voltage source feeding an *RL* load. (b) Control block diagram.

```
%Simulation of a RL load connected to a Controlled-
Voltage-Source
R=5;    %Resistance of the RL load
L=0.05; %Inductance of the RL load
io=0;      %state variable (load current)
vo=100;    %input variable (load voltage)
h=1e-4;    %step size
j=0;       %variable used to storage the outcomes
t=0;       %initial value of the digital time
tmax=1;    %maximum simulation time

while t<=tmax, %begining of the simulation loop
      t = t + h; %time incrementation
      io = io + (vo-R*io)*h/L;  %state space variable 1
      % storage of the variables
      j = j + 1;
      time(j)=t;
      current(j)=io;
      voltage(j)=vo;
end
   figure(1),plot(time,voltage)
   figure(2),plot(time,current)
```

Figure 9.2 MATLAB code for the system presented in Fig. 9.1(a) for open loop operation.

Figure 9.2 shows the MATLAB code for the system presented in Fig. 9.1(a), which simulates the behavior of the variables dynamically as well as plots their variables in the time domain. The output current for the open-loop operation has the expected exponential behavior for a step voltage response, as observed in Fig. 9.3. Note that the load current (i_o) is a function of both the load itself (which is normally not controllable) and the output voltage v_o, which is a variable used to control the load current.

To implement the control block diagram presented in Fig. 9.1(b), and considering the digital implementation as done in the MATLAB code of Fig. 9.2, it is necessary to add the code lines related to the controller $C(s)$. For the sake of illustration let us consider a simple proportional controller for $C(s)$, which means that the output of this controller will define the voltage v_o as follows:

$$v_o = Ke_i \tag{9.4}$$

where K is the gain of the controller and e_i is the current error.

Assuming a reference current as depicted in Fig. 9.4(a) with a desired step transient, the control system allows that the measured current i_o will follow its reference, as observed in Fig. 9.4(b). Those waveforms were obtained with $K = 200$.

Due to the inductive characteristics of the load, the value of voltage to allow this step transient of current is expected to be huge at the moment of the step transient (theoretically an infinite voltage is needed to allow such a variation of current). Figure 9.5(a) shows the output voltage of the controller, while Fig. 9.5(b) highlights the instant of the step transient [zoom of Fig. 9.5(a) between $t = 0.49$ s and $t = 0.51$ s]. Note that to guarantee a current step transient from 5 to −5 A as in Fig. 9.4, the value of voltage almost reaches −2000 V.

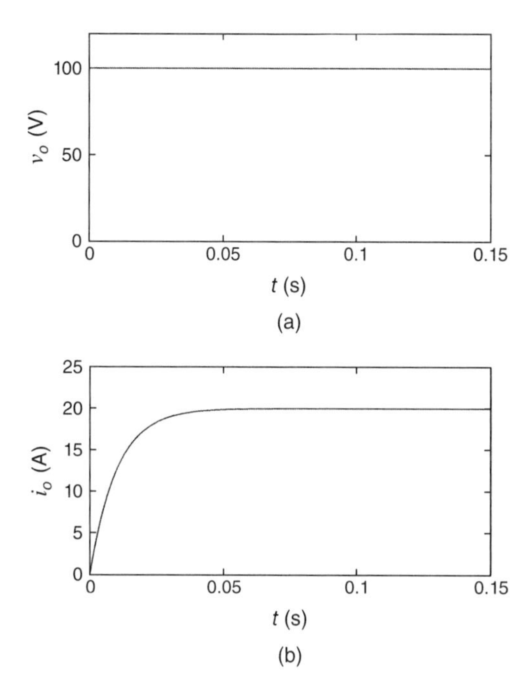

Figure 9.3 (a) Output voltage. (b) Output current.

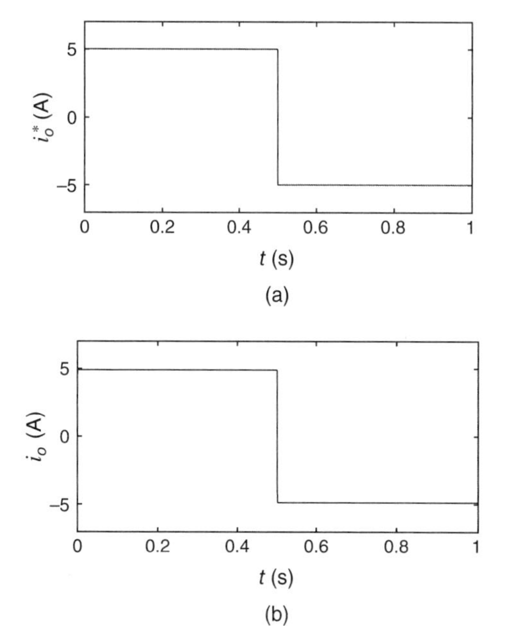

Figure 9.4 (a) Reference load current (b) Measured load current.

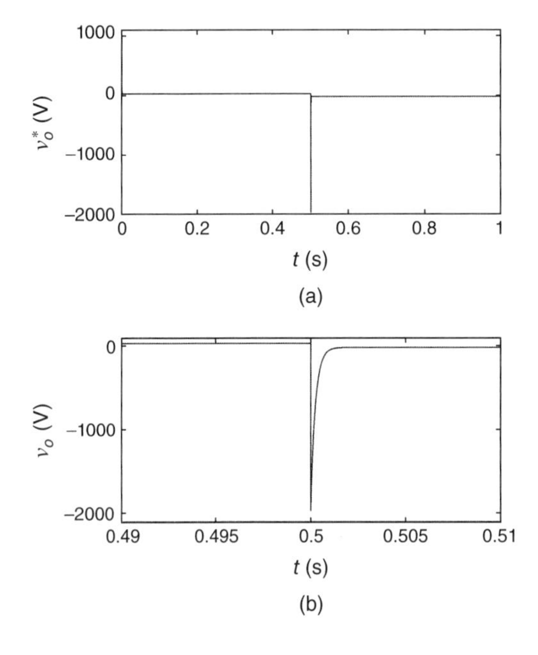

Figure 9.5 (a) Load voltage needed for a current step transient in Fig. 9.4. (b) Zoom of the voltage shown in Fig. 9.5(a).

Exercise 9.1

Write a MATLAB code to implement the control strategy shown in Fig. 9.1(b) for an *RL* load (the reader is encouraged to use the code furnished in Fig. 9.2). Plot the waveforms as in Figs 9.4 and 9.5. Additionally, explain the behavior of the output voltage shown in Fig. 9.5 through the equations of the inductor voltage.

In a real system, with the controlled voltage source having the maximum limit well defined in terms of voltage and power deliverable to the load, the situation observed in Fig. 9.5 is likely not possible. Even if a controller requires that amount of voltage to guarantee a small current error (ideally zero) during the step transient, there are practical limits for the voltage source, which will not allow such a condition.

Indeed, the control scheme presented in Fig. 9.1(b) does not show the controlled voltage source (actuator) needed to implement this control scheme in a real-world scenario. Note from Fig. 9.1(b) that the controller's output defines the load voltage directly. A more realistic system is considered in Fig. 9.6, in which the controlled voltage source is added for the system through block $G(s)$.

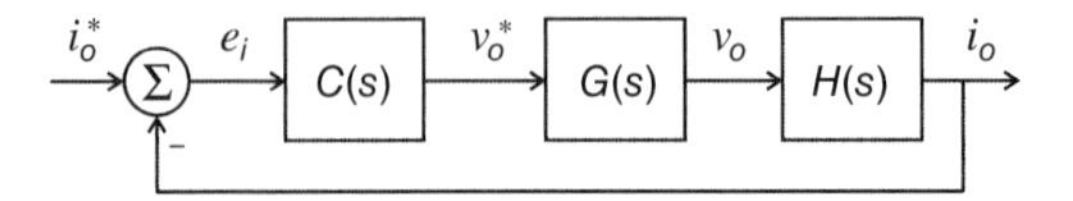

Figure 9.6 Control block diagram with inclusion of the controlled voltage source.

```
if t>tpwm, % begin of the pwm loop simulation
   tpwm=tpwm+hpwm; %definition of discrete time
   cont=0;         %auxiliar variable
   v10r = vor/2;    %pole voltage - leg 1
   v20r = -vor/2;   %pole voltage - leg 2
   tal1 = (v10r/Vdc + 1/2)*hpwm; %Eq. (24) - Chapter 8
   tal2 = (v20r/Vdc + 1/2)*hpwm; %Eq. (25) - Chapter 8
end% end of the pwm loop
   cont=cont+h; % updating auxiliary variable
   %Definition of the state of the switches
   if(cont<tal1)
     q1=1;
   else
     q1=0;
   end
   if(cont<tal2)
     q2=1;
   else
     q2=0;
   end
v10=(2*q1-1)*(Vdc/2); %Expression for the pole volt. 1
v20=(2*q2-1)*(Vdc/2); %Expression for the pole volt. 2
vo = v10 - v20; %Load voltage
```

Figure 9.7 MATLAB code for implementation of the digital PWM.

Block $G(s)$ could be implemented with one of the PWM converters described previously, but an H-bridge topology has been chosen and it will be used throughout this chapter. Applications in a dc motor drive system and in a controlled rectifier circuit will be considered later on in this chapter with the H-bridge converter implementing $G(s)$. Under the control point of view, the H-bridge converter must generate a voltage at the load side (v_o) that represents accurately the voltage required by the controller v_o^*. For an ideal controlled voltage source, it turns out that $v_o = v_o^*$ with $G(s) = 1$. However, in terms of average value, the H-bridge topology $G(s)$ can be represented more realistically by a first order transfer function, with its time constant equal to the switching period, $G(s) = \frac{1}{T_s s+1}$, where $T_s = 1/f_s$, and f_s being the switching frequency of the H-bridge topology.

A MATLAB code for an H-bridge converter can be written as in Fig. 9.7. The expressions (9.24), (9.25) deduced in Chapter 8 have been considered in this code for the digital implementation of the pulse widths (τ_1 and τ_2). Note from this code that there is a digital loop to emulate the interrupt in systems like Digital Signal Processors (DSP). In order to implement a control system as seen in Fig. 9.6, the code lines for the converter (Fig. 9.7) should be added in the program presented in Fig. 9.2.

Exercise 9.2

(a) Make a step transient for the reference voltage (v_o^*) applied to the H-bridge converter, that is, $v_o^* = -80$ V for $t < 0.1$ s and $v_o^* = 80$ V for $t \geq 0.1$ s.

Compare this reference with the average of the output voltage to demonstrate that this kind of converter can be represented approximately by a transfer function given by $G(s) = \frac{1}{T_s s+1}$. (b) Combine the code lines presented in Figs 9.2 and 9.7 to implement the closed-loop system as in Fig. 9.6, that is, with the controlled voltage source as actuator; assume $V_{dc} = 100$ V.

The results of the closed-loop transfer function considering the H-bridge topology as the controlled voltage source [$G(s)$] in Fig. 9.6 is presented in Fig. 9.8. When the controller requires a voltage higher than the maximum voltage available through the converter, the controlled voltage source (H-bridge topology) cannot follow this requirement due to physical limitations/restrictions. In this particular case, the maximum and minimum voltages available on the H-bridge topology are 100 and −100 V, respectively. As a consequence of using a non-ideal controlled voltage source, the

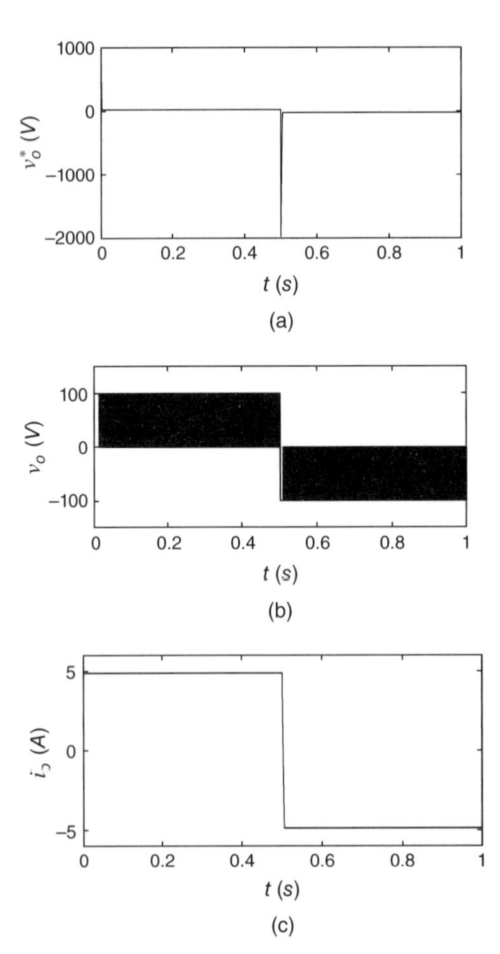

Figure 9.8 (a) Controller's output voltage (b) Converter's output voltage (c) Load current.

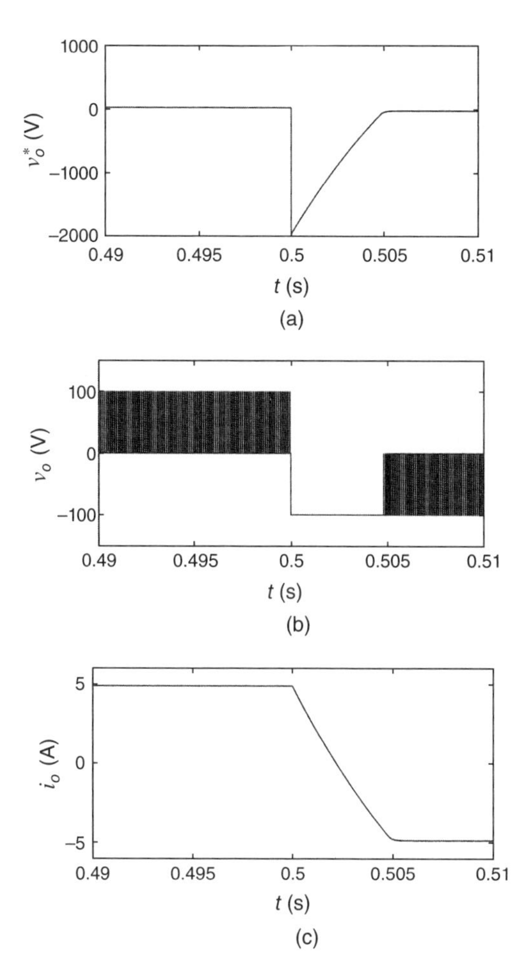

Figure 9.9 Zoom of the waveforms presented in Fig. 9.8 around the transient time ($t = 0.5$ s).

control actions will be delayed since instead of applying the necessary voltage to guarantee that transient for the current, that is, $v_o \cong -2000$ V, the converter will employ its minimum voltage of -100 V. The time transient will be affected, but the control action is still active, as observed in Fig. 9.9, which shows a zoom of the waveforms presented in Fig. 9.8 around the transient instant.

The proportional controller along with other linear regulators will be presented with more details later on in this chapter. However, prior to analyzing deeply the linear controllers, in the following section a hysteresis control strategy will be considered, due to its simplicity.

9.3 HYSTERESIS CONTROL

The hysteresis control is one of the simplest ways to guarantee a current control in systems as shown in Fig. 9.1(a). Unlike the linear controllers, such as the proportional one considered in Section 9.2 in which the gains should be designed to obtain specific characteristics, the hysteresis control will define directly the state of the switches of the H-bridge topology to make the current follow its reference with a fixed value for the maximum error given by Δi. Notice that the dynamics of the results (and consequently the error) obtained in Figs 9.8 and 9.9 can be changed if another gain for K is employed. Details of how to design the gains of linear controllers are furnished in Section 9.4.

Figure 9.10 shows the implementation of the hysteresis-type current controller, highlighting (i) the power and control parts of the circuit [Fig. 9.10(a)], (ii) the waveform of the variable under control (i_o) [Fig. 9.10(b) - top], and (iii) the switching states q_1 and q_2 [Fig. 9.10(b) - middle and bottom]. In this case, the states of the switches are defined as follows:

If $i_o > i_o^* + \Delta i$, the load current must be reduced, hence

$$q_2 = \text{on}$$

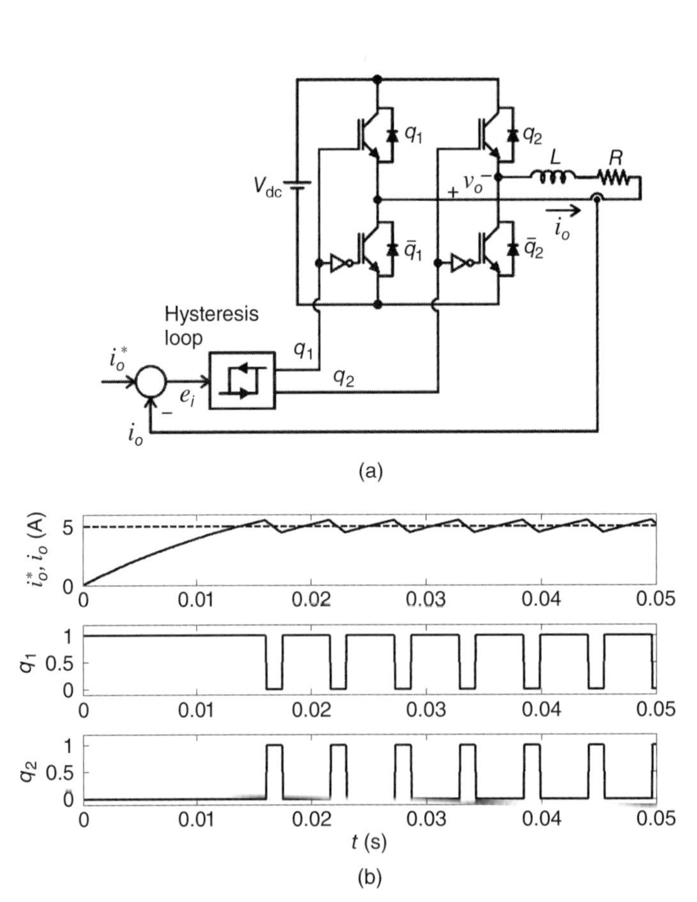

Figure 9.10 (a) Hysteresis controller defining the state of the switches. (b) Reference and measured currents (top), and the state of the switches q_1 and q_2 (middle and bottom).

and

$$q_1 = \text{off}$$

If $i_o < i_o^* - \Delta i$, the load current must be increase, hence

$$q_1 = \text{on}$$

and

$$q_2 = \text{off}$$

where Δi defines the hysteresis width below and above i_o^*. In steady state operation conditions, the load current ripple is constant. The results of the hysteresis control for the same current transient presented before are shown in Fig. 9.11.

The startup condition for the hysteresis control is highlighted in Fig. 9.12 where

t_r – interval of time in which i_o changes from 0 to i_o^*;

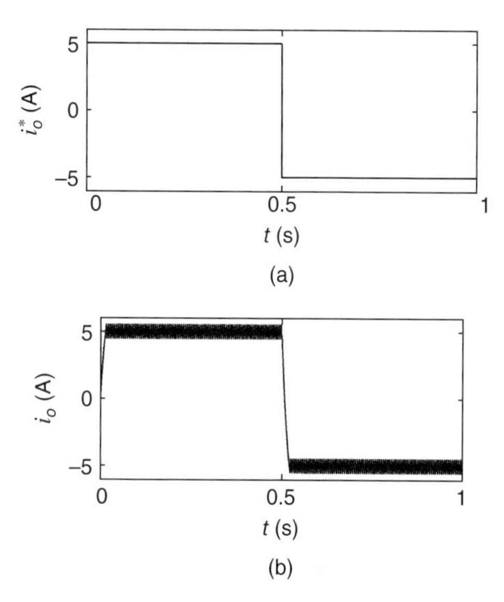

Figure 9.11 Hysteresis control: reference load current and (b) measured load current.

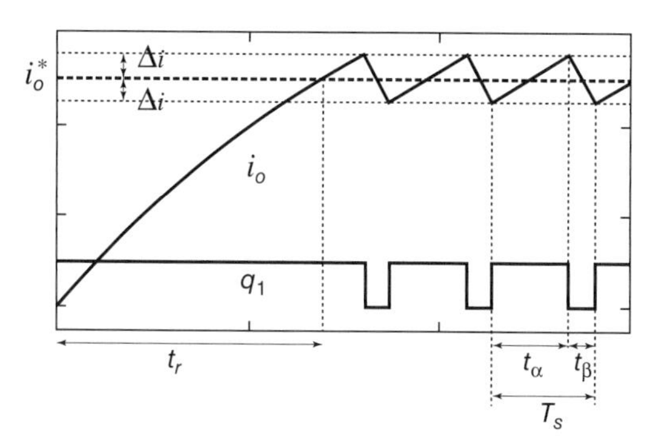

Figure 9.12 Startup transient for the hysteresis control.

t_α – time in steady state operation in which q_1 is *on;*

t_β – time in steady state operation in which q_2 is *on;*

T_s – switching period.

During t_r the behavior of the load current can be described as in equation (9.5)

$$i_o(t) = \frac{V_{dc}}{R}\left(1 - e^{\frac{-R}{L}t}\right) \tag{9.5}$$

when $t = t_r$ the load current will be $i_o = i_o^*$, then

$$i_o^* = \frac{V_{dc}}{R}\left(1 - e^{\frac{-R}{L}t_r}\right) \tag{9.6}$$

$$t_r = -\frac{L}{R}\ln\left(1 - \frac{Ri_o^*}{V_{dc}}\right) \tag{9.7}$$

The rising time (t_r) as well as the switching period ($T_s = t_\alpha + t_\beta$; in steady state operation) are functions of the load parameters (R and L). T_s is also a function of the hysteresis width (Δi).

Exercise 9.3

(a) Write a code in MATLAB to implement the hysteresis control as in Fig. 9.10(a). (b) Validate the expression given in equation (9.7) for two different *RL* loads. (c) By using the same simulation code show that the switching frequency of the hysteresis control is dependent of the *RL* parameters by filling the Table below.

R	5 Ω	10 Ω	50 Ω
L	5 mH	5 mH	5 mH
f_s			

SKSHS

Such a dependency of the switching period with the load can restrict the application of this type of controller. For instance, in dc motor drive applications, in addition to the static components as in the *RL* load presented in Fig. 9.1(a), the load (dc motor) has also a dynamic component, back electromotive force (*emf*), which is a function of the rotor speed. This type of load affects directly the performance of the hysteresis controller. This topic is addressed in Section 9.3.1.

Another characteristic of the hysteresis control is the constant ripple for the load current, which is defined by $2\Delta i$, that is, the instantaneous error is not zero in the steady state operation. Ideally, to reach null error, Δi should be zero, which implies a hypothetical scenario with infinite switching frequency.

Despite this disadvantage, it is possible to see some advantages of the hysteresis control, such as simple implementation and there is no need to define the gains of the controller as done for linear control approaches.

9.3.1 Application of the Hysteresis Control for dc Motor Drive

The model of the dc motor is presented in this section in order to understand the influence of a dynamic load in the switching frequency in a hysteresis control approach.

A dc motor is constituted by two main magnetic circuits: (i) field circuit located at the stator and (ii) armature circuit placed at the rotor. A sketch of this motor is presented in Fig. 9.13(a), while Fig. 9.13(b) shows the electrical model for the motor, which can be obtained as in equations (9.8) and (9.9):

$$v_f = R_f i_f + L_f \frac{di_f}{dt} \tag{9.8}$$

$$v_a = R_a i_a + L_a \frac{di_a}{dt} + e_a \tag{9.9}$$

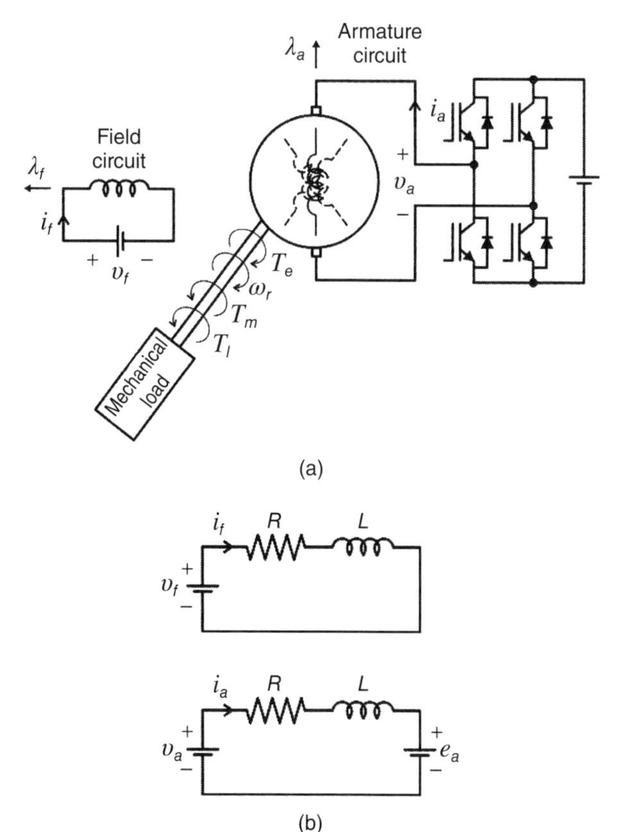

Figure 9.13 (a) Representation of a dc motor. (b) Equivalent circuit of the field (top) and armature (bottom) circuits.

where $e_a = k_c \lambda_f \omega_m$ is the back *emf* (electromotive force); k_c is the coupling constant, λ_f is the field flux and ω_m is the mechanical speed. The electromagnetic torque is produced due to the trend of the armature flux to align with the field flux, and then the magnitude of this torque is obtained as follows:

$$T_e = k_c' \lambda_a \lambda_f \tag{9.10}$$

or

$$T_e = k_c \lambda_f i_a \tag{9.11}$$

Applying Newton's second law for rotation, that is, the net external torque is equal to the moment of inertia times angular acceleration, then:

$$T_e - T_l = B_m \omega_m + J_m \frac{d\omega_m}{dt} \tag{9.12}$$

where T_l is the load torque, B_m is the damping coefficient and J_m is the moment of inertia.

From equations (9.8), (9.9), and (9.12) it is possible to write the state space representation for this type of machine, as in the following equations:

$$\frac{d}{dt}\begin{bmatrix} i_a \\ \omega_m \end{bmatrix} = \begin{bmatrix} -R_a/L_a & -k_c\lambda_f/L_a \\ k_c\lambda_f/J_m & -B_m/J_m \end{bmatrix}\begin{bmatrix} i_a \\ \omega_m \end{bmatrix} + \begin{bmatrix} 1/L_a & 0 \\ 0 & -1/J_m \end{bmatrix}\begin{bmatrix} v_a \\ T_l \end{bmatrix} \tag{9.13}$$

$$\omega_m = \begin{bmatrix} 0 & 1 \end{bmatrix}\begin{bmatrix} i_a \\ \omega_m \end{bmatrix} + [0]\begin{bmatrix} v_a \\ T_l \end{bmatrix} \tag{9.14}$$

where i_a and ω_m are the state space variables.

Figure 9.14 shows the MATLAB code for the motor equations presented in (9.13) and (9.14) with the Euler's integration method.

Figure 9.15, in turn, shows the simulation results highlighting the state space variables, that is, armature current and rotor angular speed. Besides the simplification brought up by this model, it describes the dynamics of the system reasonably well.

Considering the hysteresis control described in this chapter, let us observe the influence of the motor speed in the switching frequency for a constant hysteresis width ($2\Delta i$).

For low speed, when the amplitude of the back *emf* is small ($e_a = k_c \lambda_f \omega_m$), the switching frequency of the converter is high. On the other hand, for high speed, when the amplitude of the back *emf* is high, the switching frequency is low. To explain this behavior, a simulation has been performed with a dc motor connected to a mechanical load, which presents a variable characteristic, that is, the torque demand increases with speed squared. Examples of loads that exhibit this variable load torque characteristic are centrifugal fans, pumps, and blowers.

Also, to illustrate how the switching frequency of the H-bridge converter can be influenced by the motor speed with hysteresis control, a reference armature current is set equal to 10 A for operation in high speed, and set up equal to 2.5 A for low speed operation. The outcomes for the measured and reference currents are presented in

```
%constants
ra=0.06;
Laa=0.0018;
kf=0.1;
Bm=0.01;
Jm=0.001;
lambdaf=1;
%state variables
wm=0;
ia=0;
%input variables
va=1;  %Armature voltage
Tl=.1; %Load torque
h=1e-4;%step size
j=0;
t=0;
tmax=2;
while t<=tmax,
t = t + h; % simulation time
ia=ia+(va/Laa-ra*ia/Laa-kf*lambdaf*wm/Laa)*h;%state
space 1
wm = wm + (kf*lambdaf*ia/Jm-Tl/Jm-Bm*wm/Jm)*h;%state
space 2
% storage the variables
j = j + 1;
time(j)=t;
armature_current(j)=ia;
rotor_speed(j)=wm;
end
figure(1),subplot(2,1,1),plot(time,armature_current)
subplot(2,1,2),plot(time,rotor_speed)
```

Figure 9.14 MATLAB code for implementation of the dc motor.

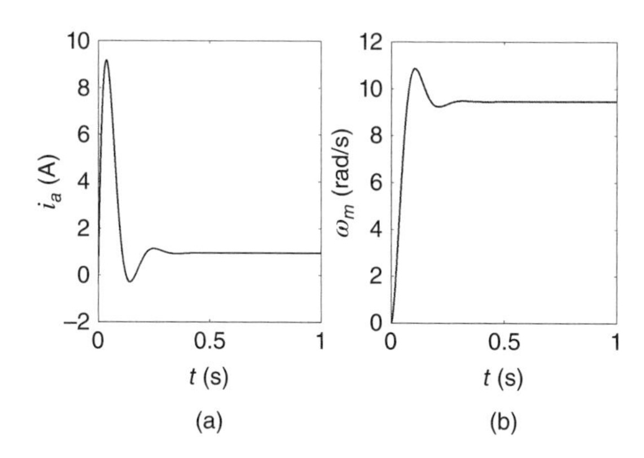

Figure 9.15 Dynamics of the dc motor showing (a) armature current and (b) rotor speed.

Fig. 9.16. As expected, the switching frequency is higher in low speed operation, since $T_s'' < T_s'$ then $f_s'' > f_s'$.

It is worth mentioning that the emphasis that has been given to show the variation of switching frequency in the hysteresis control approach is because it could

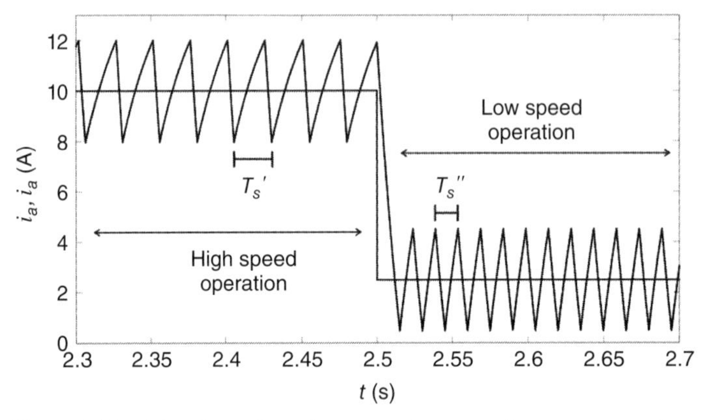

Figure 9.16 Waveforms for reference and measured motor current for operation in high and low speeds.

represent a problem in terms of the power electronics converter design. The specifications for the power switch normally consider current, blocking voltage, and switching frequency. Such a power switch may not operate adequately if it is outside the frequency range furnished by the manufacturer.

9.3.2 Hysteresis Control for Regulating an ac Variable

The previous discussions dealt with the hysteresis approach applied to control a dc variable, that is, a dc current for both *RL* load and dc machine. The hysteresis control is also useful for controlling ac variables with sinusoidal or non-sinusoidal waveforms. Examples of control systems requiring sinusoidal and non-sinusoidal waveforms are ac motor control systems (induction and synchronous motors) and active power filters, respectively.

Figure 9.17 shows the implementation of the hysteresis-type current controller, illustrating the circuit itself in Fig. 9.17(a) and the waveforms of the reference (i_o^*) and measured (i_o) variables in Fig. 9.17(b) and 9.17(c), respectively. Figure 9.17(a) is repeated here for convenience. Note in Fig. 9.17(a) that the definition of the states of the switches is exactly the same as seen previously for the dc control. In Fig. 9.17(b) there is an intentional transient in the current amplitude at $t = 0.5$ s to show the effectiveness of this control strategy.

Exercise 9.4

With the approach established earlier to control the dc load current (see the definition of the switching states in Section 9.3), the output voltage had only two levels. Determine and simulate a hysteresis method to guarantee the current control as well as to generate an output voltage with three levels.

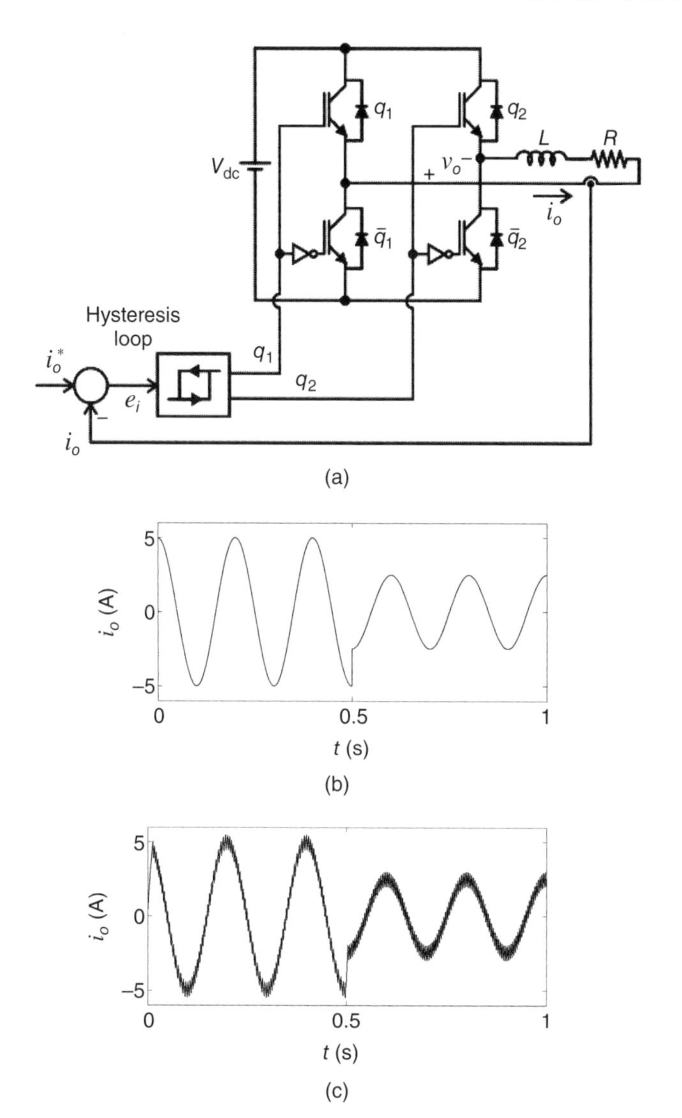

Figure 9.17 (a) Hysteresis controller defining the state of the switches (b) Reference current (c) Measured current.

9.4 LINEAR CONTROL – dc VARIABLE

9.4.1 Proportional Controller: RL Load

In Section 9.2 (Basic Control Principles) was presented a simple control approach with a proportional controller, which is the simplest example of linear control. As previously presented in Fig. 9.4, for an ideal, controlled voltage source, and in Fig. 9.8, for a nonideal controlled voltage source, it seems that the proportional controller is able to guarantee error equal to zero. However, this type of controller cannot reach null error for a system with a first order transfer function, like an *RL* load. Indeed, analyzing in detail the waveforms in Figs 9.4 and 9.8 it is possible to realize that those errors are not zero even for an ideal, controlled voltage source. The reader is encouraged to plot those results by using his/her own simulation codes and observe the error between the measured and reference currents.

Assuming an ideal controlled voltage source, as in Fig. 9.1(b), the open-loop transfer function $[F(s)]$ and closed-loop transfer function $[M(s)]$ are given respectively by

$$F(s) = \frac{k_p}{R + sL} \tag{9.15}$$

$$M(s) = \frac{k_p}{(R + k_p) + sL} \tag{9.16}$$

Equation (9.16) shows that in the steady state operation $M(s)$ is not equal to one. When $t \rightarrow \infty$ (steady state), $s = j\omega \rightarrow 0$ – meaning that $M(s)$ is equal to $k_p/(R + k_p)$. It is evident that null error will be obtained when $M(s) = 1$, which is only possible if $k_p \rightarrow \infty$. Also, for the case of using a non-ideal controlled voltage source (H-bridge topology) represented by $G(s) = \frac{1}{T_s s+1}$, null error will be obtained only when $k_p \rightarrow \infty$.

Unlike the hysteresis control, for which there is no need to specify gains, for the linear approach, the gains of the controller must be obtained to guarantee specific dynamic characteristics. A graphical analysis known as root locus can be used to determine stability of the system, as well as to visualize the poles and zeros behavior of the closed loop transfer function as a function of the proportional gain parameter, as in Fig. 9.18. These graphs are helpful to determine the design of the proportional controller (k_p). For this simple example presented in Fig. 9.18 (i.e., proportional controller plus a first order transfer function), the gain should be selected as large as possible to minimize the error. Higher values of proportional gains (k_p) guarantee smaller steady-state error and faster response as well, see Fig. 9.18 for k_p equal to 1, 5, and 20.

9.4.2 Proportional Controller: dc Motor Drive System

To analyze the proportional controller acting as a speed regulator in a dc motor drive system, it is useful to find the transfer function for the motor, as it was done for the *RL* load. Additionally, a detailed control block diagram will provide important information about how the motor variables and control actions interact among themselves.

From the state-space representation in equations (9.13) and (9.14), the transfer function can be obtained as follows:

$$\frac{(s)}{U(s)} = C(sI - A)^{-1}B + D \tag{9.17}$$

where (s) and $U(s)$ are the Laplace transformation for ω_m and input variables (v_a and T_l), respectively. The matrices for the state-space representation are given by:

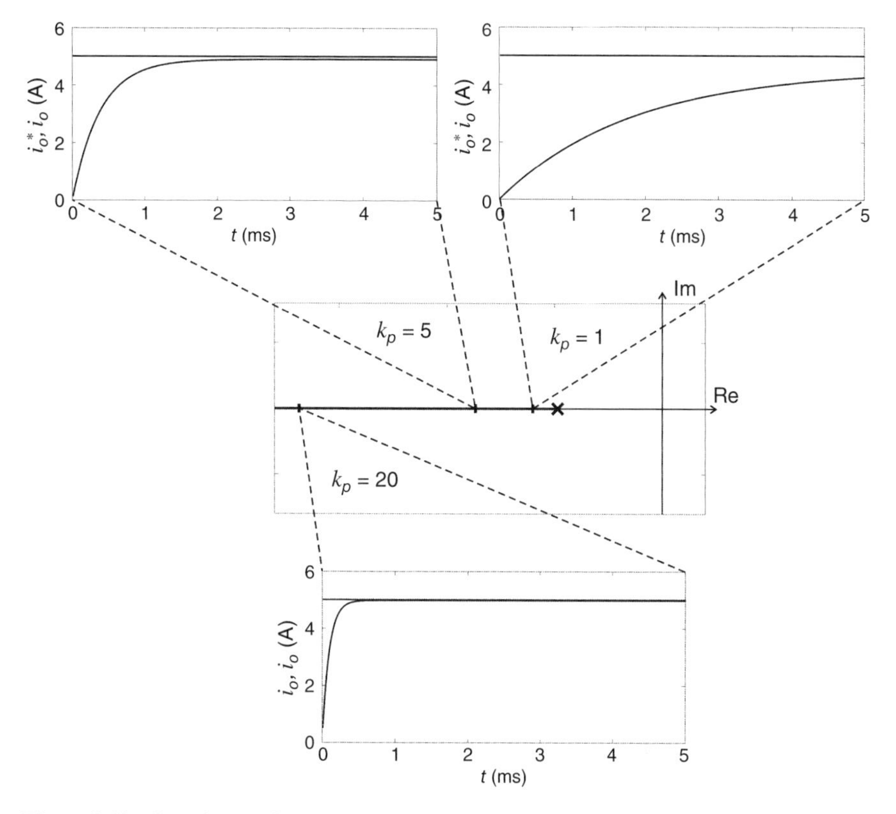

Figure 9.18 Root locus of the system in Fig. 9.1(b).

$A = \begin{bmatrix} -R_a/L_a & -k_c\lambda_f/L_a \\ k_c\lambda_f/J_m & -B_m/J_m \end{bmatrix}$, $B = \begin{bmatrix} 1/L_a & 0 \\ 0 & -1/J_m \end{bmatrix}$, $C = [0 \quad 1]$ and $D = 0$. The matrix I is a 2×2 identity matrix. Developing equation (9.17)

$$(s) = G_a(s)V_a(s) + G_m(s)T_l(s) \tag{9.18}$$

where $G_a(s) = \frac{k_a}{s^2 + (s_a + s_m)s + k} = \frac{k_a}{(s + s_1)(s + s_2)}$ with $k_a = \frac{k_c\lambda_f}{J_m L_a}$. The poles of the system can be written as follows:

$$s_1 = \left[-\left(s_a + s_m\right) + \sqrt{(s_a + s_m)^2 - 4k} \right] /2$$

$$s_2 = \left[-\left(s_a + s_m\right) - \sqrt{(s_a + s_m)^2 - 4k} \right] /2$$

with $s_a = r_a/L_a$, $s_m = B_m/J_m$, and $k = (r_a B_m + k_c^2 \lambda_f^2)/(J_m L_a)$.

Considering the model for a dc machine developed in equations (9.13) and (9.14) it is possible to determine its block diagram, as presented in Fig. 9.19, where k_c is the coupling constant. This block diagram helps to understand how the variables interact with each other. A simplified block diagram representation for this system

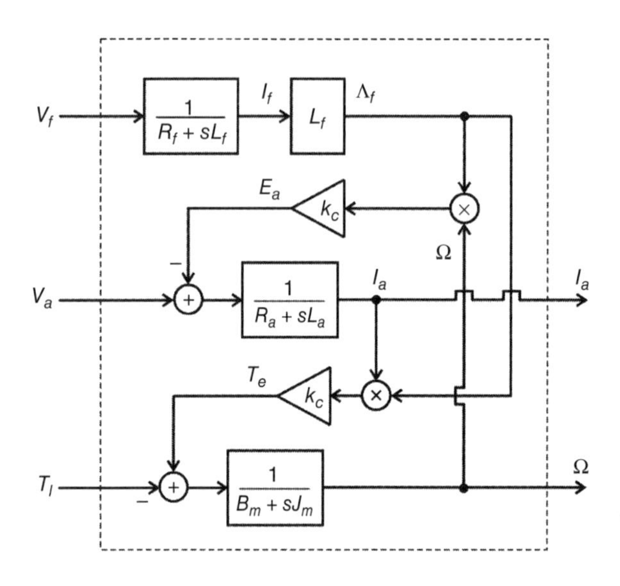

Figure 9.19 Detailed block diagram representation for a dc-motor.

with one output (Ω) and two inputs (V_a and T_l) is shown in Fig. 9.20(a). From the speed controller perspective, the term $G_m(s)T_l(s)$ can be considered as a disturbance which must be compensated by the feedback control system.

Figure 9.20(b) shows the speed closed-loop control block diagram, while Fig. 9.20(c) shows the same loop control without the disturbance term. The simplification brought up by Fig. 9.20(c) means that it has been assumed that the controller will be designed in such way that the disturbance term will be compensated by the controller itself, then for simplification purposes, the disturbance can be neglected.

For a proportional controller [$C(s) = k_p$] and assuming an ideal controlled voltage source, the open-loop transfer function [$F(s)$] and closed-loop transfer function [$M(s)$] for the system presented in Fig. 9.20(c) are given respectively by

$$F(s) = \frac{k_p k_a}{(T_1 s + 1)(T_2 s + 1)} \tag{9.19}$$

$$M(s) = \frac{k_p k_a}{(T_1 s + 1)(T_2 s + 1) + k_p k_a} \tag{9.20}$$

where $T_1 = 1/s_1$ and $T_2 = 1/s_2$.

Considering the steady state analysis ($t \to \infty \Rightarrow s = j\omega \to 0$) for equation (9.20), it turns out that $M(s) = k_p k_a/(1 + k_p k_a)$, which is different from a desirable $M(s) = 1$ (i.e., null error). The error is not null, unless when $k_p \to \infty$, which goes to an unrealistic scenario of infinite voltage needed to keep error equal to zero.

As observed in the root locus analysis presented in Fig 9.20(d), the consequence of increasing k_p (to reduce the steady state error) is that the imaginary parts of poles will be much higher than the real part of the poles, bringing oscillation issues for response.

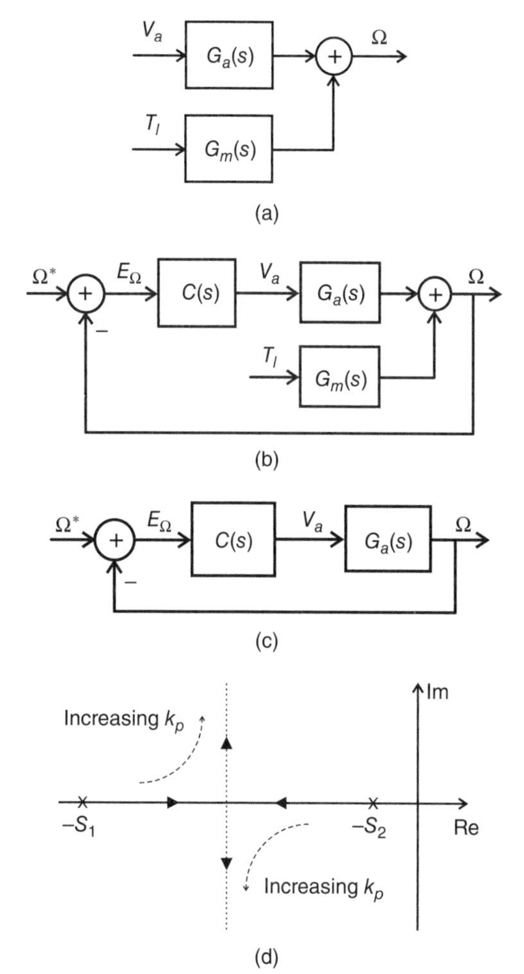

Figure 9.20 (a) Simple block diagram representation for a dc machine drive system (b) Speed closed-loop control (c) Speed closed-loop control without the disturbance term. (d) Root locus of the dc motor drive system with proportional controller.

9.4.3 Proportional-Integral Controller: RL Load

For a system with a first order transfer function (e.g., RL), the simplest linear controller to guarantee null error is the proportional-integral one. For this type of controller, the $F(s)$ is given by

$$F(s) = \left(k_p + \frac{k_i}{s}\right)\left(\frac{1}{R + sL}\right) \tag{9.21}$$

Equation (9.21) can be conveniently rewritten as below:

$$F(s) = \frac{k_i}{s}\left(\frac{k_p}{k_i}s + 1\right)\left(\frac{1/R}{\frac{L}{R}s + 1}\right) \tag{9.22}$$

From equation (9.22) it is evident that the open-loop transfer function has a pole placed at the origin of the complex plane, due to the k_i/s term, which means null error at steady state operation. It also suggests that it is possible to cancel the zero of the controller with the pole of the system (*RL* load), making

$$\frac{k_i}{k_p} = \frac{R}{L} \tag{9.23}$$

The zero-pole cancellation furnishes one possible design criterion for determining the gains of the proportional-integral controller. Then, considering equation (9.22) and the zero-pole cancellation, the $F(s)$ and $M(s)$ are given respectively by $\frac{k_i}{sR}$ and $\frac{k_i}{sR+k_i}$. The root locus of the system is given in Fig. 9.21, which highlights three-step responses for k_i equal to 0.5, 1.5, and 2.5. Once one of the characteristics is picked up from Fig. 9.21 for a specific value of k_i, the gain k_p will be defined from (9.23).

If the zero-pole cancellation is not employed to determine the gains of the controller, the root locus will be clearly different, which could bring up an undesired complex situation. Also, the determination of the gains of the controller is not straightforward. Without zero-pole cancellation $M(s)$ will be clearly a more complex

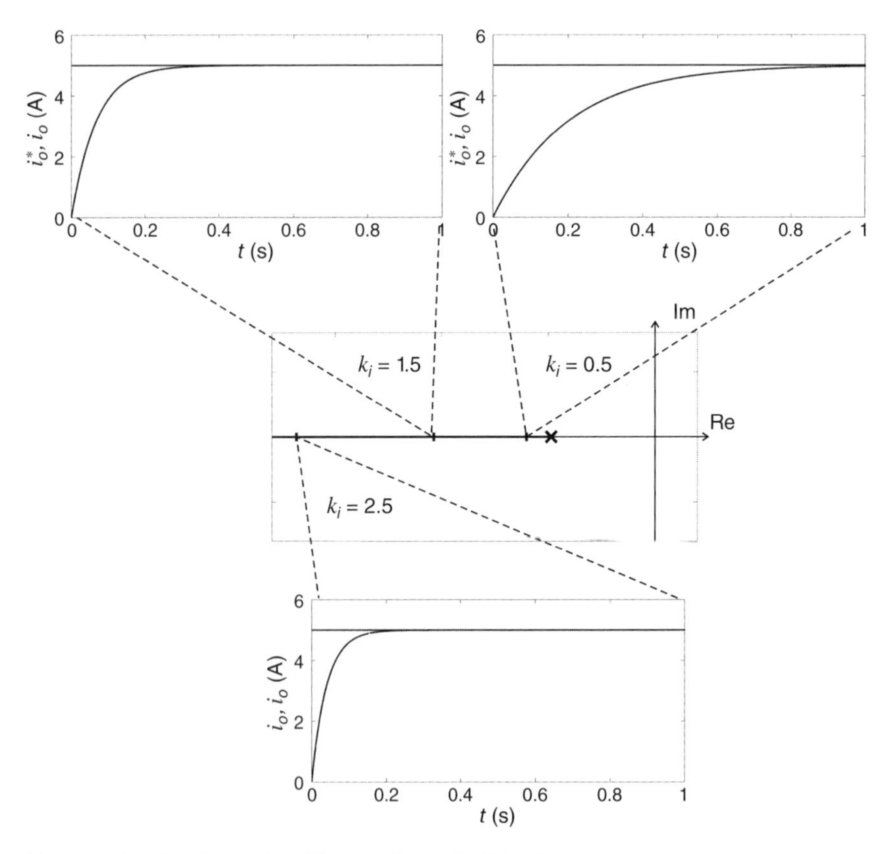

Figure 9.21 Root locus for a PI controller and *RL* load.

system, as below:

$$M(s) = \frac{sk_p + k_i}{s^2L + s(R + k_p) + k_i} \tag{9.24}$$

To test the proportional-integral controller by using the dynamic simulation (as done previously with the MATLAB code), it is necessary to discretize the controller itself. A discretization method using the Euler approach can also be used for simulation of the controller. The block diagram model for the PI controller is shown in Fig. 9.22, from which it is possible to determine the state space representation, as below:

$$\dot{x} = k_i e \tag{9.25}$$

$$y = x + k_p e \tag{9.26}$$

Exercise 9.5

Write a MATLAB code to implement a PI controller. Test this code by controlling an *RL* load current with both an ideal voltage source and with an H-bridge topology as actuator.

For the case where a non-ideal controlled voltage source is considered, the $F(s)$ obtained in (9.22) is now given by: $\frac{k_i}{s}\left(\frac{k_p}{k_i}s + 1\right)\left(\frac{1}{T_s s+1}\right)\left(\frac{1/R}{\frac{L}{R}s+1}\right)$. In this case, the zero of the controller can be used to cancel one of the two poles, that is, $s = -1/T_s$ and $s = -R/L$. Normally, due to the high switching frequency of the power converters, the dominant pole is $s = -R/L$, which must be cancelled if a faster response is desired.

9.4.4 Proportional-Integral Controller: dc Motor

As in the case of the *RL* type of load, a proportional-integral controller is employed to guarantee null error for a dc motor fed by an H-bridge converter. The control block diagram presented in Fig. 9.20(c) and the transfer function of the motor given by

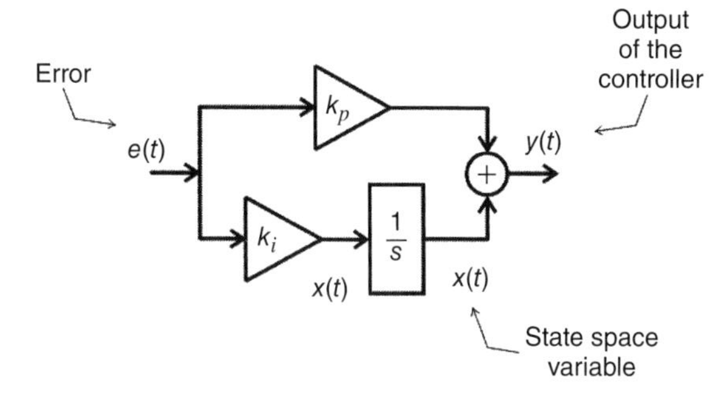

Figure 9.22 Block diagram representation for a PI controller.

$G_a(s) = \frac{k_a}{(T_1 s+1)(T_2 s+1)}$ are considered in this analysis. In this case, unlike the *RL* load, the system has two poles; one of these poles can be cancelled to obtain specific characteristics, such as guarantee of faster dynamic response in a closed loop operation. The open-loop transfer function $F(s)$ after the zero-pole cancelation $k_p/k_i = T_2$ is obtained as follows:

$$F(s) = \frac{k_i k_a}{s(T_1 s + 1)} \tag{9.27}$$

With the dominant pole cancelled ($s = -1/T_2$), the $M(s)$ is obtained as

$$M(s) = \frac{k_i k_a}{s(T_1 s + 1) + k_i k_a} \tag{9.28}$$

Figure 9.23(a) and 9.23(b) shows the speed response for two cases, when the slowest (dominant) pole is cancelled [Fig. 9.23(a)] and when fastest pole is cancelled [Fig. 9.23(b)]. These results were obtained considering the parameters of the dc machine in Fig. 9.14 assuming the same k_i gain for both results, that is, $k_i = 2$ and $k_p = T_1 k_i$ for Fig. 9.23(a); and with $k_i = 2$ and $k_p = T_2 k_i$ for Fig. 9.23(b).

Although the waveforms presented in Fig. 9.23(a) and 9.23(b) were obtained with the same number for the integral gain $k_i = 2$, it is also possible to determine the gains of the controllers analytically by using a specific design criterion. For example, Fig. 9.23(c) depicts the root locus of the dc motor with a PI controller after the cancellation of the dominant pole. Despite the fact that this controller guarantees null error, its limitation lies in the dynamics of the close-loop transfer function that is limited by $-1/2T_1$ regardless of the rise of the gain k_i. One possible analytical solution for designing the gains of the PI controller can be identical poles for close-loop transfer function, as in

$$k_i = \frac{1}{4 k_a T_1} \tag{9.29}$$

$$k_p = T_2 k_i \tag{9.30}$$

9.4.5 Proportional-Integral-Derivative Controller: dc Motor

As presented in the last section, the dc motor drive system with a PI controller has one of its poles placed at the origin of the complex plane (i.e., $s = 0$), which assures null error in steady state operation. From Fig. 9.23(c), however, it is evident that the fastest response obtained with this controller is reached when it leads to the real part of the dominant pole equal to $-1/(2T_1)$.

If a faster response is needed, another controller must be employed. Note that for the PI controller applied in this system, the dominant pole is always located on the right side of the pole $-1/T_1$. A PID controller is able to obtain faster response for a dc motor drive system than a PI type of controller, which has the following transfer function:

$$C(s) = k_p + \frac{k_i}{s} + k_d s \tag{9.31}$$

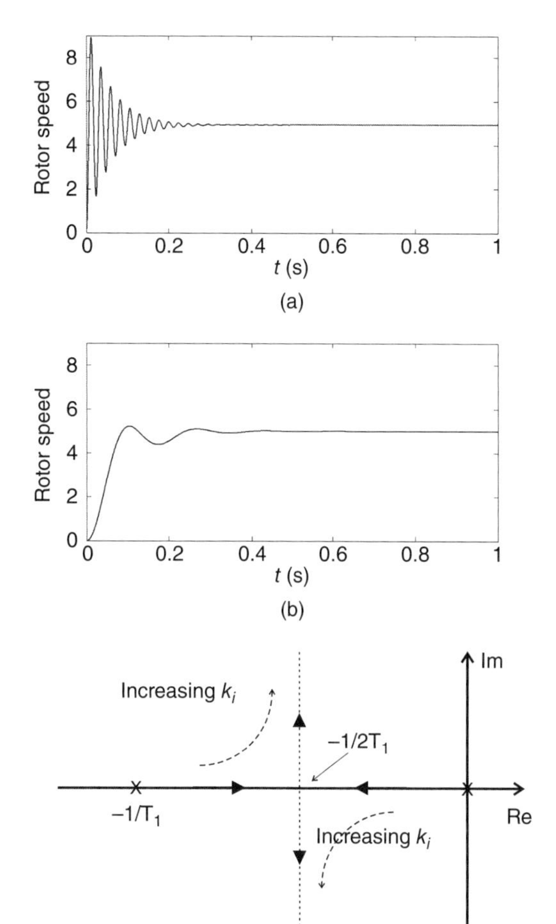

Figure 9.23 Speed dynamic response for cancellation of (a) slowest and (b) fastest pole (c) Root locus of the dc motor with a PI controller.

In fact, the derivative term of the controller cannot be implemented as it appears in (9.31), as the transfer function of the PID controller will be:

$$C(s) = k_p + \frac{k_i}{s} + \frac{k_d s}{sT_d + 1} \tag{9.32}$$

or

$$C(s) = \frac{(T_d k_p + k_d)s^2 + (T_d k_i + k_p)s + k_i}{s(T_d s + 1)} \tag{9.33}$$

where T_d indicates the quality of the derivative part. Ideally T_d must be equal to zero, but its real value depends on the physical limitation of both actuator and system as a whole.

The expression (9.33) shows that $C(s)$ presents two poles: $s = 0$ and $s = -1/T_d$, and two zeros. The zeros can be used to cancel both poles of the dc motor (as done with the PI for cancellation of one pole). Moreover, the parameters k_p and k_d can be employed to allocate the closed loop poles anywhere in the complex plane.

The following design of the controllers can be obtained for the condition of two real identical poles:

$$k_i = \frac{1}{T_i} = \frac{1/k_a}{4T_d} \tag{9.34}$$

$$k_p = \frac{T_1 + T_2 - T_d}{T_i} \tag{9.35}$$

$$k_d = \frac{T_1T_2 - (T_1 + T_2 - T_d)T_d}{T_i} \tag{9.36}$$

9.5 LINEAR CONTROL — ac VARIABLE

Note that the linear regulators considered so far in this chapter have been employed to control dc variables whether it is *RL* load or dc machine, current or speed. Two of these controllers were able to guarantee null steady state error (i.e., PI and PID), while the proportional one was not. Indeed, the integral part of the PI and PID controllers goes to infinite for dc variables; since the operation frequency is zero, $s = 0$ and consequently $k_i/s \to \infty$. Such a characteristic leads to a desirable zero error characteristic.

However, for ac variables, which means $s \neq 0$, there is no term in the controller that tends to infinite, which leads to an error different from zero. For the case of ac variables, it is necessary to use another PI type of controller, such as presented below:

$$C(s) = \frac{k_a s^2 + k_b s + k_c}{(s^2 + \omega^2)} \tag{9.37}$$

where k_a, k_b and k_c are the gains of the controller. In this case, the controller presents an infinite value for the rated frequency ω.

The state space representation of this controller can be obtained as in equations (9.38)–(9.40)

$$\dot{x}_1 = x_2 + k_b e \tag{9.38}$$

$$\dot{x}_2 = -\omega^2 x_1 + k_d e \tag{9.39}$$

$$y = x_1 + k_a e \tag{9.40}$$

where x_1 and x_2 are state variables and $k_d = k_c - \omega^2 k_a$.

After presenting controllers able to deal with dc and ac variables, it is important to present other control approaches for systems that have more than one variable to be regulated. For instance, a controlled rectifier circuit aims to guarantee power factor correction with sinusoidal current as well as keep the output voltage constant. The following section presents a control strategy regulating two variables known as cascade control.

9.6 CASCADE CONTROL STRATEGIES

The control strategies described in this chapter have considered a schematic as presented in Fig. 9.1(b). This leads to a controlled voltage source implemented by a dc–ac converter. For both types of loads (*RL* and dc motor) considered previously, there is just one variable to be regulated, either current or speed. However, in some applications there is a need to control two variables to reach specific requirements. In this section a control approach called cascade control is presented for two applications, that is, power factor rectifier and dc motor drive system.

9.6.1 Rectifier Circuit: Voltage-Current Control

Figure 9.24(a) shows the circuit for a single-phase controlled rectifier, which has two control objectives. They are: guarantee a regulated dc voltage (v_o) at the output and a sinusoidal grid current (i_g) with power factor close to one at input converter side. Both control objectives are obtained simultaneously through the H-bridge power converter.

Before showing details about the control operation of the rectifier circuit presented in Fig. 9.24(a), it is convenient and simple to show the operation of a similar

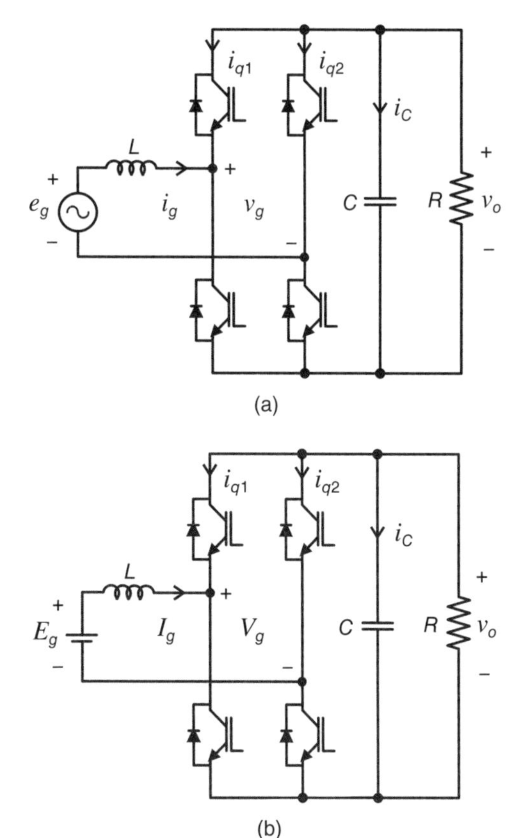

Figure 9.24 (a) Rectifier circuit. (b) Dc–dc converter.

circuit in terms of the control strategy, as in Fig. 9.24(b). This topology deals with a dc to dc conversion [instead of ac to dc, as in Fig. 9.24(a)] and the principal goal is to control the output voltage. Besides, there is no need to control the power factor; since the primary source of energy is dc, the control strategies for both circuits in Fig. 9.24 can be established following the same principle, that is, using control cascade scheme.

Equations (9.41) and (9.42) show that the capacitor current, and consequently the capacitor voltage, can be regulated through I_g.

$$i_C = -i_{q1} - i_{q2} - \frac{v_o}{R} \tag{9.41}$$

$$i_C = I_g(q_1 - q_2) - \frac{v_o}{R} \tag{9.42}$$

Figure 9.25(a) depicts a control block diagram for the converter in Fig. 9.24(b). The control structure for the controlled power factor rectifier in Fig. 9.24(a) is presented in Fig. 9.25(b). The difference between both control approaches for the converters in Fig. 9.24(a) and 9.24(b) is the synchronization method for the power factor correction.

9.6.2 Motor Drive: Speed-Current Control

The speed control of a dc motor, as presented in Fig. 9.20(c), was obtained directly by defining the armature voltage (V_a) as the output of the speed controller. Indeed,

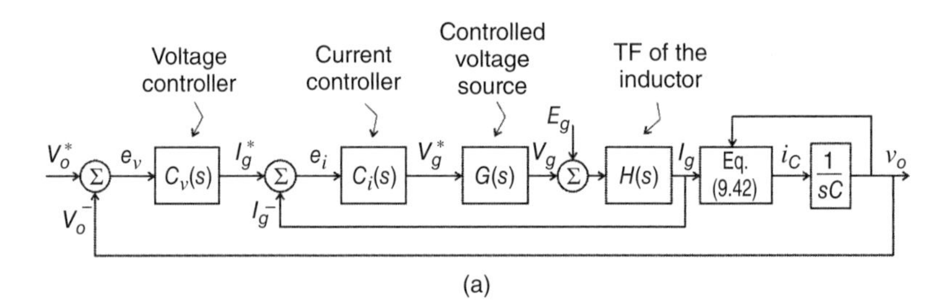

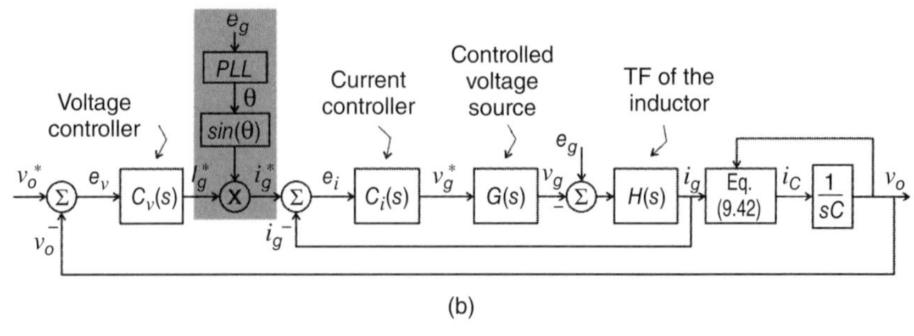

Figure 9.25 Control block diagrams: (a) for the converter in Fig. 9.24(b). (b) for the converter in Fig. 9.24(a).

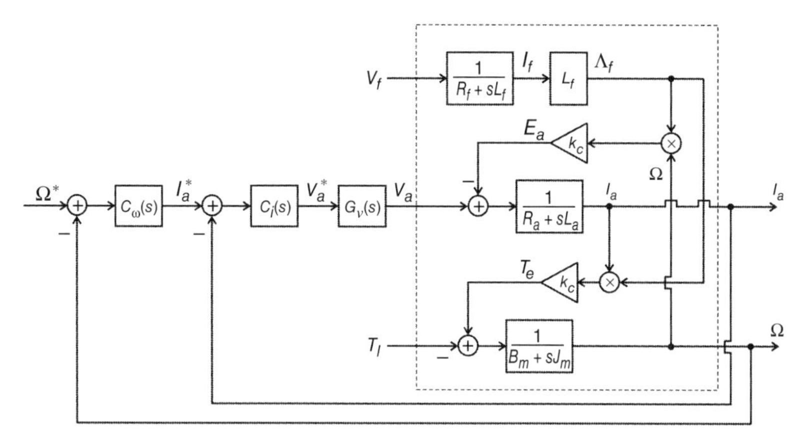

Figure 9.26 Cascade control for a dc machine.

this control strategy was possible because Ω is a function of V_a as shown in Fig. 9.19. However, there are two blocks between these two variables (Fig. 9.19), which represent the electrical and mechanical dynamics for the motor. Fig. 9.26 suggests another solution along with just one block in between the variable used to control (V_a) and the variable under control (I_a).

The control scheme using an outer controller for the speed (Ω) and an inner controller for the armature current (I_a) is also known as a cascade control, which is in fact the same principle as applied for the rectifier circuit presented in Section 9.6.1. The cascade control is possible in this case because the mechanical and electrical dynamics are completely different, with the internal loop much faster than the external one. Both controllers $C_\omega(s)$ and $C_i(s)$ are PI type of controllers, since it is enough to guarantee null error for the dc variables (I_a and Ω). Although using two controllers, instead of just one as in Fig. 9.20, the main advantage of the cascade control is the possibility to limit the armature current, which brings additional protection for the machine.

Assuming that the mechanical variables are constants for the electrical variables perspective, the internal control loop can be simplified as in Fig. 9.27. In this case, the open loop transfer function is given by

$$F_i(s) = \frac{k_i}{s}\left(s\frac{k_p}{k_I} + 1\right)\left(\frac{1}{T_s s + 1}\right)\left(\frac{1/r_a}{\frac{L_a}{r_a}s + 1}\right) \tag{9.43}$$

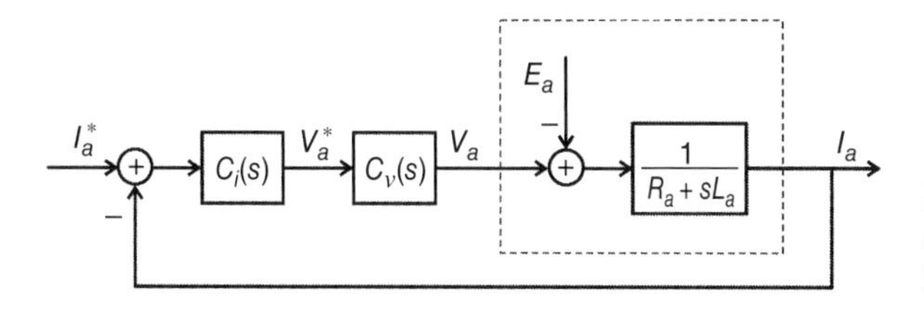

Figure 9.27 Internal control scheme for the block diagram of Fig. 9.26.

Canceling the zero of the controller with the pole of the motor, it yields

$$F_i(s) = \frac{k_i}{s}\left(\frac{1/r_a}{T_s s + 1}\right) \tag{9.44}$$

Consequently the close loop transfer function for the current control is given by

$$M_i(s) = \frac{k_i/r_a}{T_s s^2 + s + k_i/r_a} \tag{9.45}$$

As seen before, the gain k_i can be obtained by using either a root locus approach or making identical poles, which means $k_i = r_a/(4T_s)$; then

$$M_i(s) = \frac{1}{(2T_s s + 1)^2} \tag{9.46}$$

Since T_s is a small value, $M_i(s)$ becomes

$$M_i(s) = \frac{1}{4T_s s + 1} \tag{9.47}$$

The external loop of the block diagram shown in Fig. 9.26 is given in Fig. 9.28. Considering T_l as a perturbation to be compensated by the controller, the open loop transfer function for the speed controller is given by

$$F_\omega(s) = \frac{\frac{k_{i\omega}k_c\lambda_f}{B_m}\left(\frac{k_{p\omega}}{k_{i\omega}}s + 1\right)}{s\left(\frac{J_m}{B_m}s + 1\right)(4T_s s + 1)} \tag{9.48}$$

Cancelling the zero with the pole we have

$$F_\omega(s) = \frac{k_{i\omega}k_c\lambda_f}{B_m}\frac{1}{s(4T_s s + 1)} \tag{9.49}$$

Then the close loop transfer function is given by

$$M_\omega(s) = \frac{\frac{k_{i\omega}k_c\lambda_f}{B_m}}{4T_s s^2 + s + \frac{k_{i\omega}k_c\lambda_f}{B_m}} \tag{9.50}$$

Making identical poles we have

$$M_\omega(s) = \frac{1}{(8T_s s + 1)^2} \tag{9.51}$$

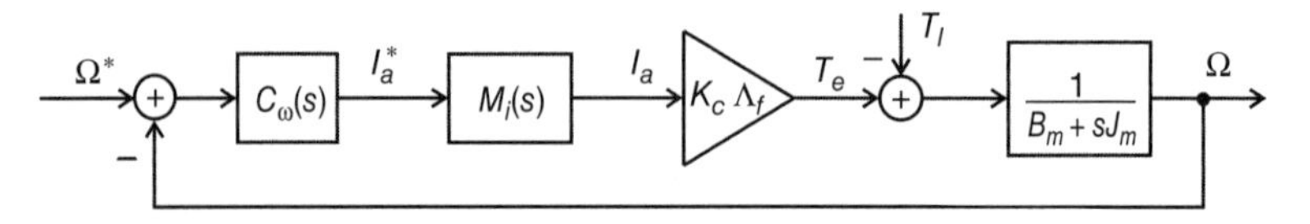

Figure 9.28 External loop of the controller in Fig. 9.26.

From this approach it is possible to design the gains of the controller as follows: $\frac{k_{p\omega}}{k_{i\omega}} = \frac{J_m}{B_m}$ with $k_{i\omega} = \frac{B_m}{16T_s k_c k_f}$.

9.7 SUMMARY

This chapter presented the basic principles of controllers for systems using power converters as actuators. This is indeed a preparation for the next chapters that will need feedback control actions to guarantee the correct operation of the back-to-back converters. This chapter is divided into two parts: nonlinear and linear control techniques. Section 9.3 dealt with a nonlinear control technique, that is, hysteresis control, highlighting some advantages and disadvantages. Two applications were also considered in this section: dc motor drive system and regulation of an ac variable. The classic linear controllers (P, PI, and PID) were presented in Section 9.4 for dc variables, while Section 9.5 presented a linear type of controller for ac variables. A more complex control technique was presented in Section 9.6 with a cascade strategy to introduce back-to-back converters, which are described in the following chapters. More details on the control techniques described can be found from 1 to 18.

REFERENCES

[1] L. Shang, D. Sun, J. Hu, Sliding-mode-based direct power control of grid-connected voltage-sourced inverters under unbalanced network conditions, Power Electron, IET, vol. 4, no. 5, pp. 570–579, May 2011.

[2] R.-J. Wai, C.-Y. Lin, Dual active low-frequency ripple control for clean-energy power-conditioning mechanism, IEEE Trans Ind Electron, vol. 58, no. 11, pp. 5172–5185, Nov. 2011.

[3] B. Labbe, B. Allard, X. Lin-Shi, D. Chesneau, An integrated sliding-mode buck converter with switching frequency control for battery-powered applications, IEEE Trans Power Electron, vol. 28, no. 9, pp. 4318–4326, Sept. 2013.

[4] J. Hu, L. Shang, Y. He, Z.Q. Zhu, Direct active and reactive power regulation of grid-connected DC/AC converters using sliding mode control approach, IEEE Trans Power Electron, vol. 26, no. 1, pp. 210–222, Jan. 2011.

[5] R. Haroun, A. Cid-Pastor, A.E. Aroudi, L. Martinez-Salamero, Synthesis of canonical elements for power processing in DC distribution systems using cascaded converters and sliding-mode control, IEEE Trans Power Electron, vol. 29, no. 3, pp. 1366–1381, March 2014.

[6] M. Khazraei, H. Sepahvand, M. Ferdowsi, K.A. Corzine, Hysteresis-based control of a single-phase multilevel flying capacitor active rectifier, IEEE Trans Power Electron, vol. 28, no. 1, pp. 154–164, Jan. 2013.

[7] D.G. Holmes, R. Davoodnezhad, B.P. McGrath, An improved three-phase variable-band hysteresis current regulator, IEEE Trans Power Electron, vol. 28, no. 1, pp. 441–450, Jan. 2013.

[8] T.-W. Chun, Q.-V. Tran, H.-H. Lee, H.-G. Kim, Sensorless control of BLDC motor drive for an automotive fuel pump using a hysteresis comparator, IEEE Trans Power Electron, vol. 29, no. 3, pp. 1382–1391, March 2014.

[9] S. Gautam, R. Gupta, Unified time-domain formulation of switching frequency for hysteresis current controlled AC/DC and DC/AC grid connected converters, Power Electron, IET, vol. 6, no. 4, pp. 683–692, April 2013.

[10] Dash AR, Babu BC, Mohanty KB, Dubey R. Analysis of PI and PR controllers for distributed power generation system under unbalanced grid faults. 2011 International Conference on Power and Energy Systems (ICPS); 2011 22–24 Dec. p. 1–6.

[11] Geethalakshmi B, Saraswathi A, Dananjayan P. Comparing and evaluating the performance of SSSC with Fuzzy Logic controller and PI controller for transient stability enhancement. IICPE 2006. India International Conference on Power Electronics; 2006 Dec 19–21. p. 140–143.

[12] Mikkili S, Panda AK, Patnaik SS, Suresh Y. Comparison of two compensation control strategies for SHAF in 3ph 4wire system by using PI controller. 2010 India International Conference on Power Electronics (IICPE); 2011 Jan 28–30. p. 1–7.

[13] Tsengenes G, Adamidis G. Comparative evaluation of Fuzzy-PI and PI control methods for a three phase grid connected inverter. Proceedings of the 2011-14th European Conference on Power Electronics and Applications (EPE 2011); Aug. 30 2011-Sept. 1 2011. p. 1–10.

[14] P. Kasinathan, R. Vairamani, S. Sundramoorthy, Dynamic performance investigation of d–q model with PID controller-based unified power-flow controller, Power Electron, IET, vol. 6, no. 5, pp. 843–850, May 2013.

[15] Z. Shen, N. Yan, H. Min, A multimode digitally controlled boost converter with PID autotuning and constant frequency/constant off-time hybrid PWM control, IEEE Trans Power Electronics, vol. 26, no. 9, pp. 2588–2598, Sept. 2011.

[16] M.Y.-K. Chui, W.-H. Ki, C.-Y. Tsui, A programmable integrated digital controller for switching converters with dual-band switching and complex pole-zero compensation, IEEE J Solid-State Circuits, vol. 40, no. 3, pp. 772–780, March 2005.

[17] M.M. Peretz, S. Ben-Yaakov, Time-domain design of digital compensators for PWM DC-DC converters, "IEEE Trans Power Electron," vol. 27, no. 1, pp. 284–293, Jan. 2012.

[18] Rasoanarivo I, Arab-Tehrani K, Sargos FM. Fractional Order PID and Modulated Hysteresis for high performance current control in multilevel inverters. 2011 IEEE Industry Applications Society Annual Meeting (IAS); 2011 Oct 9–13. p. 1–7.

CHAPTER 10

SINGLE-PHASE TO SINGLE-PHASE BACK-TO-BACK CONVERTER

10.1 INTRODUCTION

Single-phase two-stage energy conversion systems are considered in this chapter. Such conversion systems, also known as back-to-back converters, deal with rectification and inversion through a dc-link capacitor. The control goals are high power factor and sinusoidal current at the input side of the converter and regulated voltage at its output side. Several applications require this type of conversion system, such as line voltage regulators, universal active power filters, standby power supplies, and uninterruptible power supplies.

Power blocks are again employed in this chapter as an effective way to understand how the topologies presented in the technical literature have been conceived. Furthermore, many of the subjects considered in the previous chapters (e.g., optimized PWM and control strategies) are used to improve the converter's operation in terms of total harmonic distortion (THD) and efficiency, as well as to regulate the variables of the back-to-back converter, such as grid power factor and dc-link voltage.

This chapter describes the full-bridge topology for single-phase ac–dc–ac conversion in Section 10.2. In addition to the model, pulse width modulation (PWM), and control strategies, a power analysis and an expression for the dc-link capacitor oscillation are also presented in the section. Three sections of this chapter deal with nonconventional topologies, that is, topologies with reduced (Section 10.3) and increased (Sections 10.4 and 10.5) number of power semiconductor devices. Semiconductor component count reduction is obtained by either sharing a leg or by using the dc-link mid-point connection. On the other hand, topologies with increased number of components are addressed by connecting back-to-back converters both in parallel and series. This will allow THD improvements and reducing power processed by each converter. A summary of the chapter is presented in Section 10.6, while Section 10.7 shows the references.

Advanced Power Electronics Converters: PWM Converters Processing AC Voltages,
Forty Fifth Edition. Euzeli Cipriano dos Santos Jr. and Edison Roberto Cabral da Silva.

10.2 FULL-BRIDGE CONVERTER

10.2.1 Model

By connecting two PBs back-to-back, it is possible to build a single-phase to single-phase ac–dc–ac as presented in Fig. 10.1(a). The representation using switches is given in Fig. 10.1(b).

The conversion system consists of a single-phase utility grid, input inductor (with reactance X_g), capacitor bank at the dc-link with two capacitors (each one with capacitance equal to C), rectifier, inverter, and a single-phase load. Technically, the dc-link capacitor bank in Fig. 10.1 could be implemented with just one capacitance. However, for the sake of modeling, it is convenient to draw this circuit with two capacitors to have access to the point "0." The inductive element (L_g) between the utility grid and the input converter side is needed to avoid short-circuit between both voltage source type of element (grid and input converter side). Also, the dc-link bank is needed to decouple input and output converter sides. Moreover, it is expected that the input side of the converter in Fig. 10.1 operates as boost converter since the dc-link voltage should be bigger than the grid voltage to guarantee grid current control. The rectifier circuit constitutes switches q_{g1} and q_{g2}, and their complementary switches given respectively by $\overline{q}_{g1}$ and $\overline{q}_{g2}$. The inverter, in turn, constitutes switches q_{l1} and q_{l2} and their complementary switches $\overline{q}_{l1}$ and $\overline{q}_{l2}$. As considered in previous chapters, the conduction state of the switches is represented by a binary variable, where $q = 1$ indicates a closed switch while $q = 0$ is an open one with $= g1, g2, l1$, and $l2$. From Fig. 10.1(b), the following equations can be derived for both input and output converter sides as in equations (10.1) and (10.2), respectively:

$$v_g = v_{g10} - v_{g20} \tag{10.1}$$

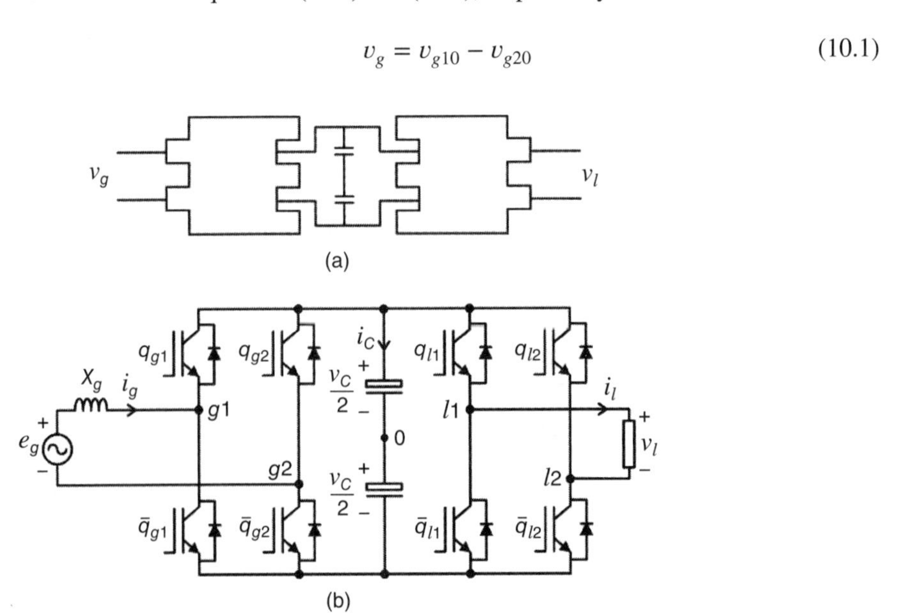

Figure 10.1 Two-leg converter supplying a single-phase load: (a) blocks and (b) switch representation.

$$v_l = v_{l10} - v_{l20} \tag{10.2}$$

where the pole voltages can be written as a function of the states of the switches, as follows:

$$v_{g10} = (2q_{g1} - 1)\frac{v_C}{2} \tag{10.3}$$

$$v_{g20} = (2q_{g2} - 1)\frac{v_C}{2} \tag{10.4}$$

$$v_{l10} = (2q_{l1} - 1)\frac{v_C}{2} \tag{10.5}$$

$$v_{l20} = (2q_{l2} - 1)\frac{v_C}{2} \tag{10.6}$$

Equation (10.7) furnishes the dynamic model for the variables of the grid converter side:

$$v_g = e_g - r_g i_g - l_g \frac{di_g}{dt} \tag{10.7}$$

where r_g and l_g represent the resistances and inductances of the input inductor.

The capacitor voltage can be obtained as follows:

$$v_c = \frac{1}{C}\int [(q_{g1} - q_{g2})i_g + (q_{l2} - q_{l1})i_l]dt \tag{10.8}$$

with the interval of integration equal to the switching frequency.

It is worth mentioning that the same approach used earlier in this book is considered in this chapter: (i) modeling the system by writing the voltage generated by the converter as a function of the switching states, (ii) defining the switching states by using a PWM technique, and (iii) obtaining the desired voltages for the PWM with closed-loop operation.

10.2.2 PWM Strategy

The optimized PWM strategy presented in Chapter 8 can be employed for the full-bridge configuration presented in Fig. 10.1 to improve either THD or efficiency at both sides (input and output) of the converter.

If the desired input and output voltages are given respectively by v_g^* and v_l^*, the reference pole voltages can be written as follows:

$$v_{g10}^* = v_g^* + v_{gh}^* \tag{10.9}$$

$$v_{g20}^* = v_{gh}^* \tag{10.10}$$

$$v_{l10}^* = v_l^* + v_{lh}^* \tag{10.11}$$

$$v_{l20}^* = v_{lh}^* \tag{10.12}$$

The auxiliary voltages v_{gh}^* and v_{lh}^* are given respectively by

$$v_{gh}^* = v_C\left(\mu_g - \frac{1}{2}\right) + (\mu_g - 1)V_{g\max}^* - \mu_g V_{g\min}^* \tag{10.13}$$

$$v_{lh}^* = v_C\left(\mu_l - \frac{1}{2}\right) + (\mu_l - 1)V_{l\max}^* - \mu_l V_{l\min}^* \tag{10.14}$$

where $V_{g\max} = \text{MAX}\{v_g^*, 0\}$, $V_{g\min} = \text{MIN}\{v_g^*, 0\}$ and $V_{l\max} = \text{MAX}\{v_l^*, 0\}$, $V_{l\min} = \text{MIN}\{v_l^*, 0\}$; μ_g and μ_l are the distribution factor and determine how the free-wheeling time can be distributed throughout the switching period. Note that v_{gh}^* and v_{lh}^* are independent of each other. Equations (10.13) and (10.14) were deduced in Section 8.2.2.

Once the reference pole voltages have been defined in equations (10.9)–(10.12), the pulse widths τ_{g1}, τ_{g2}, τ_{l1}, and τ_{l2} can be calculated by equations (10.15)–(10.18), as follows:

$$\tau_{g1} = \left(\frac{v_{g10}^*}{v_C} + \frac{1}{2}\right)T_s \tag{10.15}$$

$$\tau_{g2} = \left(\frac{v_{g20}^*}{v_C} + \frac{1}{2}\right)T_s \tag{10.16}$$

$$\tau_{l1} = \left(\frac{v_{l10}^*}{v_C} + \frac{1}{2}\right)T_s \tag{10.17}$$

$$\tau_{l2} = \left(\frac{v_{l20}^*}{v_C} + \frac{1}{2}\right)T_s \tag{10.18}$$

where T_s is the switching period.

10.2.3 Control Approach

Figure 10.2 presents the control block diagram for the system in Fig. 10.1. The input power factor and the dc-link capacitor voltage are both controlled by using this control approach. The output voltage is obtained by open-loop operation just setting up the PWM signals for the legs $l1$ (i.e., switches q_{l1}–$\overline{q}_{l1}$) and $l2$ (i.e., switches q_{l2}–$\overline{q}_{l2}$). Note that a cascade control approach is employed since the dynamics of the capacitor voltage is completely different from the inner loop that deals with grid current.

The dc-link voltage v_C is adjusted to its reference value v_C^* using the controller R_C, which is implemented with a standard PI controller. This controller provides the amplitude of the reference grid current I_g^*. To maintain the power factor close to one with a sinusoidal grid current, the instantaneous reference current i_g^* must be synchronized with the input voltage e_g. Such a synchronization scheme can be obtained via a PLL algorithm. The block G_e–i_g is responsible for generating the reference sinusoidal grid current in phase with the grid voltage $i_g^* = I_g^* \sin(\omega_g t + \Phi_g)$, which means that Φ_g is the phase angle of the grid voltage and ω_g is the grid frequency. The current controller loop must deal with sinusoidal waveforms, which means that the

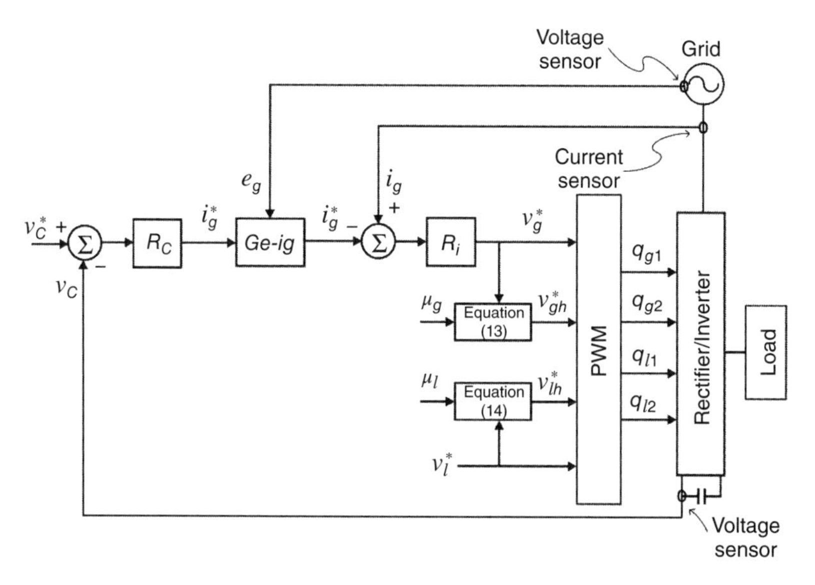

Figure 10.2 Control strategy for the single-phase to single-phase converter.

conventional PI controller will not guarantee null error in steady-state operation – a modified PI controller can be employed instead (see Section 9.5). As presented in Fig. 10.2 and mentioned earlier, the output voltage control is obtained by open-loop operation.

The PWM block presented in Fig. 10.2 can be implemented either from equations (10.15)–(10.18) or by comparing these reference pole voltages with high frequency carrier signals.

Exercise 10.1

Draw a block diagram system with a hysteresis current control instead of using the linear controller R_i in Fig. 10.2. Is there any change for the number of current and voltage sensors employed with the hysteresis control scheme?

10.2.4 Power Analysis

To find an expression of the capacitor voltage as a function of the system's variables including the power demanded by the load, let us assume the following general scenario:

Grid Voltage. $e_g(t) = E_g \cos(\omega_g t)$

Low Frequency Voltage of the Input Converter Side. $v_g(t) = V_g \cos(\omega_g t - \theta_g)$

Grid Current. $i_g(t) = I_g \cos(\omega_g t + \Phi_g)$

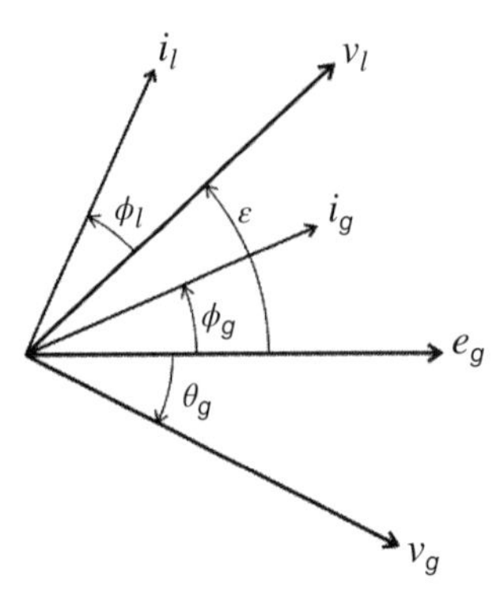

Figure 10.3 Phasor diagram for the low frequency converter variables of Fig. 10.1.

Low Frequency Voltage of the Output Converter Side. $v_l(t) = V_l \cos(\omega_l t + \varepsilon)$

Load Current. $i_l(t) = I_l \cos(\omega_l t + \varepsilon + \Phi_l)$

All these variables are depicted in the phasor diagram presented in Fig. 10.3. The effect of the high frequency voltages generated by the converters are neglected to simplify the analysis. Also, for the sake of simplification, the phasors of the input and output variables are considered in the same figure, which means that they have the same frequencies.

The input and output powers are given by $p_g(t) = v_g(t)i_g(t)$ and $p_l(t) = v_l(t)i_l(t)$, respectively. Neglecting the converter power losses, the dc-link capacitor power can be obtained as

$$p_c(t) = p_g(t) - p_l(t) \tag{10.19}$$

or

$$p_c(t) = \left[\frac{V_g I_g}{2}\cos\left(\theta_g + \Phi_g\right) - \frac{V_l I_l}{2}\cos(\Phi_l)\right] + \left[\frac{V_g I_g}{2}\cos\left(2\omega_g t - \theta_g + \Phi_g\right) - \frac{V_l I_l}{2}\cos(2\omega_l t + 2\epsilon + \Phi_l)\right] \tag{10.20}$$

As long as the converter losses are neglected, the active power demanded by the load is generated by the single-phase grid without losses through the conversion stage, that is,

$$V_g I_g \cos(\theta_g + \Phi_g) = V_l I_l \cos(\Phi_l) \tag{10.21}$$

Substituting (10.21) into (10.20), the capacitor power is given by

$$p_c(t) = \frac{V_g I_g}{2}\cos(2\omega_g t - \theta_g + \Phi_g) - \frac{V_l I_l}{2}\cos(2\omega_l t + 2\epsilon + \Phi_l) \tag{10.22}$$

or

$$p_c(t) = N_l \frac{\cos(\Phi_l)}{\cos(\theta_g + \Phi_g)}\cos(2\omega_g t - \theta_g + \Phi_g) - N_l \cos(2\omega_l t + 2\epsilon + \Phi_l) \tag{10.23}$$

where $N_l = V_l I_l / 2$.

Equation (10.23) suggests that the pulsating powers inherent from single-phase systems (at input and output sides of the converter) appear on the dc-link capacitor bank.

One of the desired operation conditions is input power factor close to one, which is obtained imposing $\Phi_g = 0$ (see Fig. 10.3). Although there is no control for some of the parameters in equation (10.23), such as N_l and Φ_l, the phase angle ϵ can be employed to reduce the pulsating power on the dc-link capacitors if both input and output converter sides have the same frequencies.

Exercise 10.2

Considering the input converter side modeled as a controlled voltage source (v_g), as depicted in the figure below, determine the value of X_g for an input power factor equal to one and assuming the following conditions: $e_g(t) = 170\cos(2\pi 60t + 12°)$, $v_g(t) = 204\ \cos(2\pi 60t)$ and with the amplitude of the grid current equal to 10 A. Assume an ideal inductance.

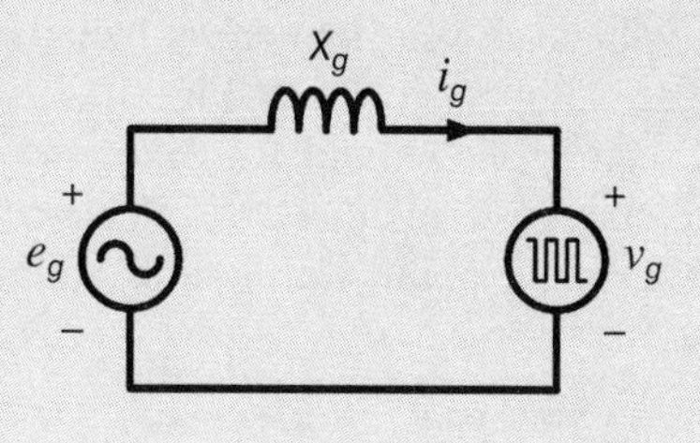

10.2.5 dc-link Capacitor Voltage

The capacitor voltage can be achieved from capacitor power. Notice that

$$p_c(t) = v_c i_c = C v_c \frac{dv_c}{dt} \tag{10.24}$$

this leads to

$$v_c \frac{dv_c}{dt} = \frac{N_l}{C}\left[\frac{\cos(\Phi_l)}{\cos(\theta_g + \Phi_g)}\cos(2\omega_g t - \theta_g + \Phi_g) - \cos(2\omega_l t + 2\epsilon + \Phi_l)\right] \tag{10.25}$$

Applying an integral at both sides of equation (10.25), it is possible to write

$$v_c^2 = N_l\left[\begin{array}{c}\dfrac{\cos(\Phi_l)}{\cos(\theta_g + \Phi_g)} X_{cg}\sin(2\omega_g t - \theta_g + \Phi_g) - \\ X_{cl}\sin(2\omega_l t + 2\epsilon + \Phi_l)\end{array}\right] + \overline{v}_c \tag{10.26}$$

where $X_{cg} = \frac{1}{\omega_g C}$, $X_{cl} = \frac{1}{\omega_l C}$ and $\overline{v}_c$ is a dc value of v_c.

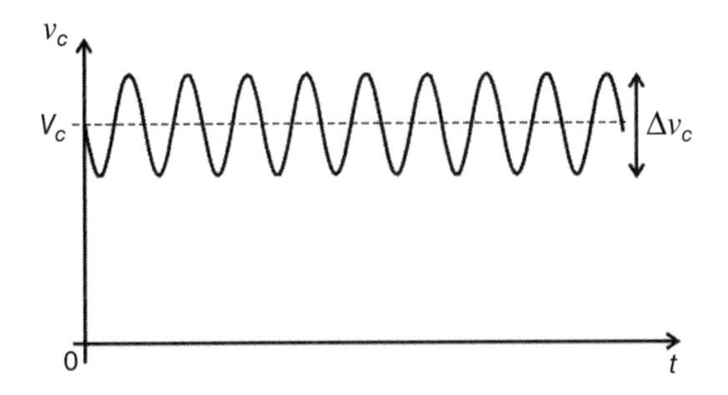

Figure 10.4 Typical dc-link voltage in single-phase back-to-back converters.

Therefore, the capacitor voltage can be written as follows:

$$v_c = \sqrt{\overline{v}_c + \widetilde{v}_c} \tag{10.27}$$

where $\widetilde{v}_c = N_l X_{cl}\left[\frac{\cos(\Phi_l)}{\cos(\theta_g+\Phi_g)}\frac{f_l}{f_g}\sin(2\omega_g t - \theta_g + \Phi_g) - \sin(2\omega_l t + 2\epsilon + \Phi_l)\right]$

Once the dc-link voltage is well known as being composed of a dc value overlapped by a small ac component, as in Fig. 10.4, it is possible to expand equation (10.27) in a Taylor series around V_c. Since the Taylor series represents a function as an infinite sum of terms that are calculated from the values of the functions derivatives at a single point, the dc-link voltage can be expressed as follows:

$$v_c = V_c + \frac{\widetilde{v}_c}{2V_c} - \frac{\widetilde{v}_c^2}{8V_c} + \sum_{n=3}^{\infty}\left[\frac{(-1)^{n+1}}{n!}\frac{\widetilde{v}_c^n}{2^n \widetilde{v}_c^{2n-1}}\prod_{j=2}^{n}(2j-3)\right] \tag{10.28}$$

The power series in (10.28) shows that the capacitor voltage consists of a dc component and an ac component with infinite number of elements. However, for practical values of the capacitor size and the dc-link reference voltage, the capacitor voltage is well represented by the first two terms in (10.28) (i.e., $V_c + (\widetilde{v}_c/2V_c)$). This approximation results in the following expression:

$$v_c \simeq V_c + \frac{N_l X_{cl}}{2V_c}\left[\frac{\cos(\Phi_l)}{\cos(\theta_g + \Phi_g)}\frac{f_l}{f_g}\sin(2\omega_g t - \theta_g + \Phi_g) - \sin(2\omega_l t + 2\epsilon + \Phi_l)\right] \tag{10.29}$$

In order to analyze the voltage fluctuation, it is convenient to define the maximum (peak-to-peak) capacitor voltage as $\Delta v_c = v_{cmax} - v_{cmin}$, where v_{cmax} and v_{cmin} are the maximum and minimum values of v_c in (10.29). The maximum value (v_{cmax}) occurs when $\sin(2\omega_g t - \theta_g + \Phi_g) = 1$ and $\sin(2\omega_l t + 2\epsilon + \Phi_l) = -1$ happen simultaneously, and the minimum value (v_{cmin}) occurs when $\sin(2\omega_g t - \theta_g + \Phi_g) = -1$ and $\sin(2\omega_l t + 2\epsilon + \Phi_l) = 1$ happen simultaneously. Hence,

$$v_{cmax} \simeq V_c + \frac{N_l X_{cl}}{2V_c}\left[\frac{\cos(\Phi_l)}{\cos(\theta_g + \Phi_g)}\frac{f_l}{f_g} + 1\right] \tag{10.30}$$

$$v_{c\min} \simeq V_c - \frac{N_l X_{cl}}{2V_c}\left[\frac{\cos(\Phi_l)}{\cos(\theta_g + \Phi_g)}\frac{f_l}{f_g} + 1\right] \tag{10.31}$$

The dc-link voltage fluctuation is given by

$$\Delta v_c \simeq \frac{N_l X_{cl}}{V_c}\left[\frac{\cos(\Phi_l)}{\cos(\theta_g + \Phi_g)}\frac{f_l}{f_g} + 1\right] \tag{10.32}$$

From (10.32), it can be seen that the increase of the capacitance size reduces the peak-to-peak voltage. For same grid and load frequencies ($\omega_g = \omega_l = \omega$) and unitary input power factor ($\cos\Phi_g = 1$), (10.29) can be written as follows:

$$v_c(t) \simeq V_c + \frac{N_l X_c}{2V_c} k_\epsilon \cos(2\omega + \beta) \tag{10.33}$$

where

$$k_\epsilon = \sqrt{k_\Phi^2 - 2k_\Phi \cos(\theta_g + 2\epsilon + \Phi_l) + 1}$$

$k_\Phi = \cos(\Phi_l)/\cos(\theta_g)$, $X_c = 1/\omega C$ and

$$\beta = \tan^{-1}\left[\frac{\cos\Phi_l - \cos(2\epsilon + \Phi_l)}{\cos\Phi_l \tan\theta_g + \sin(2\epsilon + \Phi_l)}\right]$$

The term k_ϵ plays an important role for reduction of the pulsating voltage (Δv_c) on the capacitor voltage. If both grid and load have the same frequencies (e.g., uninterrupted power supply (UPS)) it is possible to use synchronization of the input and output voltages to reduce k_ϵ and consequently reduce Δv_c.

The maximum, minimum, and fluctuation values of v_c are given respectively by

$$v_{c\max} \simeq V_c + \frac{N_l X_c}{2V_c} k_\epsilon \tag{10.34}$$

$$v_{c\min} \simeq V_c - \frac{N_l X_c}{2V_c} k_\epsilon \tag{10.35}$$

$$\Delta v_c \simeq \frac{N_l X_c}{V_c} k_\epsilon \tag{10.36}$$

The term k_ϵ reaches its maximum and minimum values for $\epsilon = -(\theta_g + \phi_l + \pi)/2$ and $\epsilon = -(\theta_g + \phi_l)/2$, respectively. Figure 10.5 shows k_ϵ as a function of the load power factor ϕ_l for the following conditions: $N_l = 1pu$ and $V_l = E_g = \sqrt{2}pu$. $\epsilon = 0$ also guarantees a considerable reduction for k_ϵ. In Fig. 10.5, $\phi_l < 0$ means a lagging power factor, while $\phi_l > 0$ means a leading power factor.

Exercise 10.3

UPS is an electrical apparatus that is designed to provide power to a load even when the input utility grid is not available. Such a device normally employs

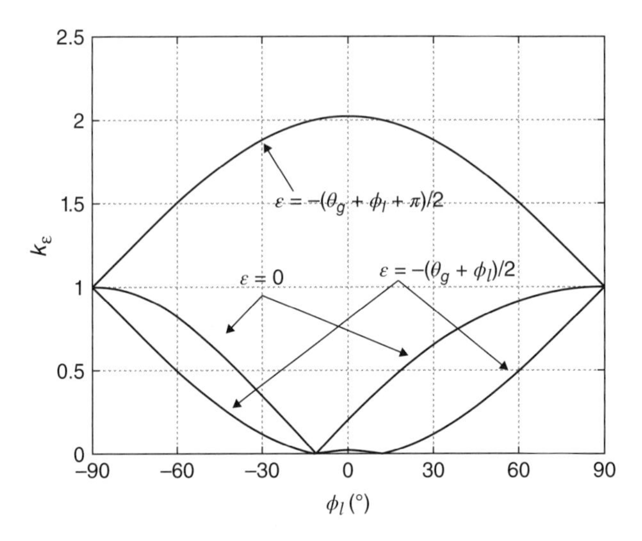

Figure 10.5 k_ε as a function of the load power factor ϕ_l.

energy storage elements like batteries to furnish energy to the load when the mains power fails. Assuming a single-phase to single-phase back-to-back converter (as in Fig. 10.1) employed as an UPS with a set of batteries connected to the dc-link voltage, and with a load frequency equal to the grid frequency, determine the phase angle of the load voltage (ε) to obtain the minimum value for the capacitor voltage oscillation. Consider that $\theta_g = 12°$, and that the input and output power factors are given respectively by 1 and 0.85.

10.2.6 Capacitor Bank Design

In order to apply the input voltage with amplitude V_g and the output voltage with amplitude V_l, the dc-link voltage v_c should be larger than both V_g and V_l at any instant of time. Figure 10.6(a) shows a case where an ideal scenario is considered with a dc-link voltage without any fluctuation. The grid reference voltage is also presented in this figure. On the other hand, Fig. 10.6(b) depicts the case in which the desired grid voltage cannot be generated appropriately because of the dc-link voltage fluctuation. Therefore, the minimum value of v_c given in (10.35) must satisfy the following equation:

$$v_{c\min} \geq \max\{V_g, V_l\} \tag{10.37}$$

By choosing equation (10.37) as a minimum value for v_c, the impedance X_{cl} can be written from (10.31) as

$$X_{cl} = \frac{V_c - \max\{V_g, V_l\}}{\frac{N_l}{2V_c}\left(\frac{\cos(\Phi_l)}{\cos(\theta_g + \Phi_g)}\frac{f_l}{f_g} + 1\right)} \tag{10.38}$$

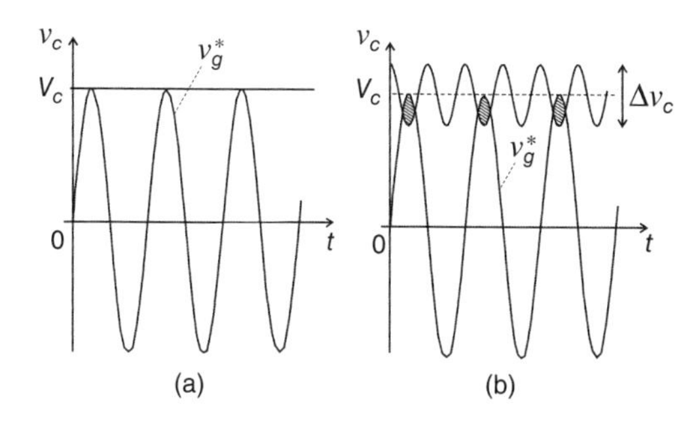

Figure 10.6 dc-Link voltage in single-phase back-to-back converters: (a) ideal scenario with no oscillation and (b) real scenario with a pulsating voltage on v_c.

Once load and grid voltages are specified, the average value of the capacitance must be chosen, based on the acceptable voltage fluctuation. Hence, the impedance X_{cl}, and consequently the capacitor C, can be calculated from (10.38).

Exercise 10.4

For a single-phase back-to-back converter supplying a load with apparent power equal to 2000 V-A, determine the capacitance value for the dc-link assuming: $\theta_g = 12°$ with input and output power factors given respectively by 1 and 0.85. It is also known that the input and output voltages are the same and are given by 120 V(RMS)−60Hz. Assume a maximum dc-link voltage oscillation equal to 10 V.

Application (Uninterrupted Power Supply – UPS)

As mentioned earlier in this chapter, the single-phase back-to-back converter can be employed as a UPS. The figure below (top) shows a simplified schematic of an online UPS highlighting the operation modes, while the bottom part of the figure shows the implementation of the online UPS with the configuration presented in Fig. 10.1. This system is characterized by its ability to supply conditioned and regulated power to a critical load during the normal and backup operations. Note that the inverter is between the primary source of energies (grid or battery) and the load – this system is also known as inverter-preferred or double-conversion UPS. In the case of malfunction of the inverter, the bypass switch is turned on.

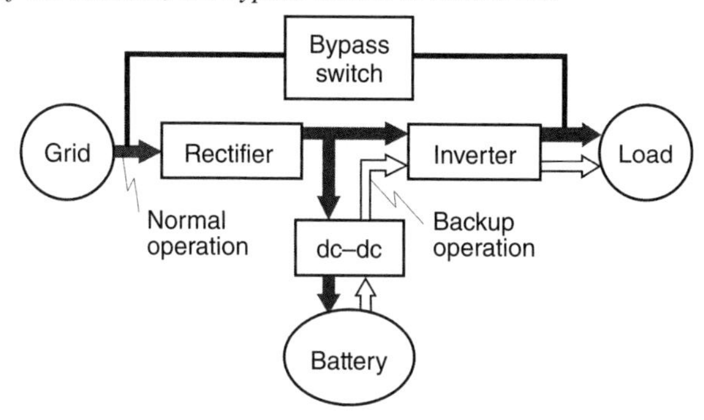

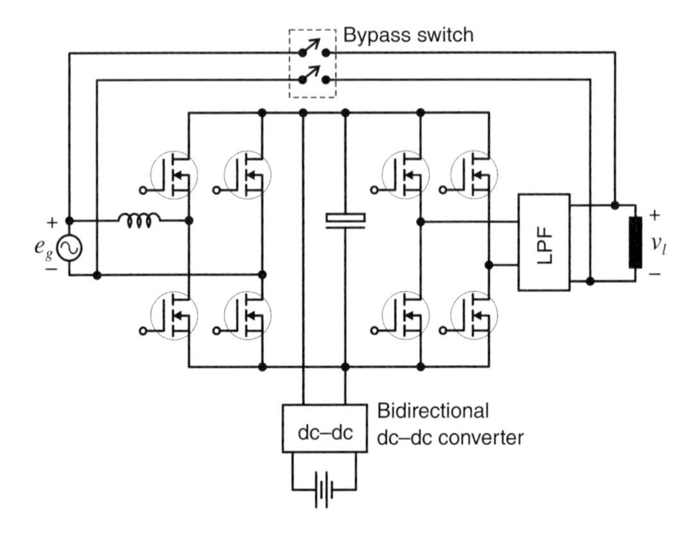

Another type of UPS is the line-interactive with the schematic presented in the figure below. During the normal operation, the grid is connected to the load directly through static switches. These switches are connected to multiple taps on the secondary transformer to compensate grid voltage variation. Also, during normal operation, the circuit should be able to charge the battery without any additional hardware – such a condition is obtained when the grid is feeding the load and consequently there is voltage at the primary transformer. If both bottom switches of the H-bridge converter are turned on simultaneously (i.e., short-circuit of the transformer at the primary side) the current i_{Bat} *will increase quickly. If the bottom switch is turned off* i_{Bat} *will flow through the antiparallel diode of the top switch to the battery and consequently charge it. The charging process is a high-frequency operation that allows the current to go from the grid to the battery without any additional component. During the backup operation, the load voltage is furnished with the following parts: battery, H-bridge configuration, transformer, and capacitor. In this case the static switches are all turned off.*

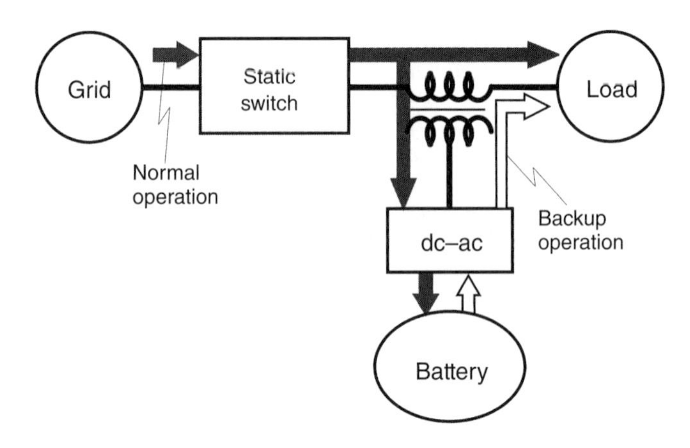

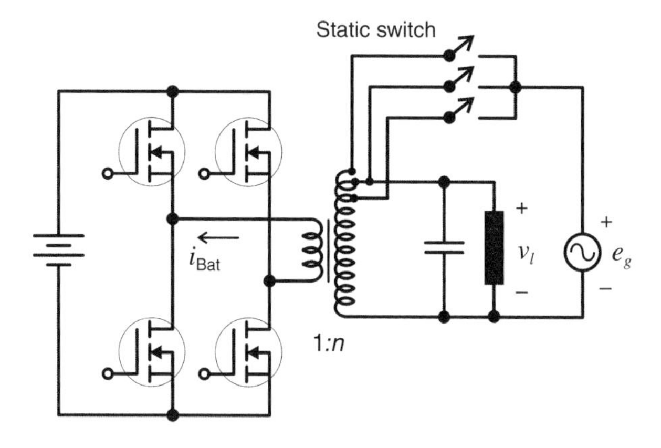

10.3 TOPOLOGY WITH COMPONENT COUNT REDUCTION

10.3.1 Model

Another single-phase back-to-back power converter is obtained by combining the first and third PB-ac as in Fig. 10.7(a). Figure 10.7(b) shows the same configuration with the switches represented. Note that the leg constituted by the switches q_s–$\overline{q}_s$ is shared between the input and output sides of the converter.

As compared to the topology in Fig. 10.1 there is a reduction of 25% of the number of power switches employed in the shared-leg converter. However, it is evident that the leg constituted by the switches q_s and $\overline{q}_s$ will handle a current that is dependent on the input and output currents $(i_g - i_l)$. Also, as discussed later in this section, the shared-leg plays an important role in the voltage capability of the converter. It is worth mentioning that the cost reduction will be effective if what is saved by reducing the number of semiconductor devices is not lost in higher device ratings and more electrolytic capacitor energy storage to keep the same conditions as in the full-bridge converter. The following equations can be derived from Fig. 10.7(b) for both input and output converter sides:

$$v_g = v_{g0} - v_{s0} \tag{10.39}$$

$$v_l = v_{l0} - v_{s0} \tag{10.40}$$

The pole voltages can be written as a function of the states of the switches:

$$v_{g0} = (2q_g - 1)\frac{v_C}{2} \tag{10.41}$$

$$v_{s0} = (2q_s - 1)\frac{v_C}{2} \tag{10.42}$$

$$v_{l0} = (2q_l - 1)\frac{v_C}{2} \tag{10.43}$$

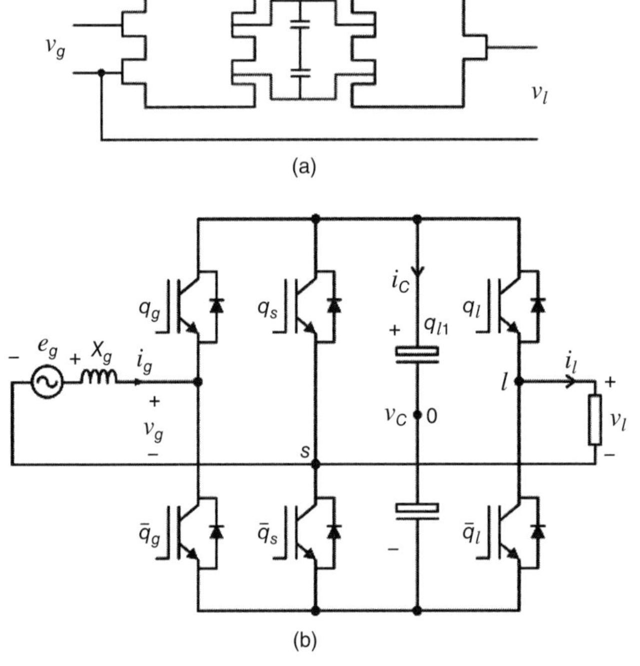

Figure 10.7 Shared-leg single-phase to single-phase back-to-back converter: (a) block and (b) switch representation.

Equation (10.7) is still valid for modeling the input converter side. However, the capacitor voltage is given by

$$v_c = \frac{1}{C}\int [i_g q_g + (i_l - i_g)q_s - i_l q_l]dt \tag{10.44}$$

with the interval of integration equal to the switching period.

10.3.2 PWM Strategy

If the desired input and output voltages are given respectively by v_g^* and v_l^*, one option for the reference pole voltages is as follows:

$$v_{g0}^* = v_g^* + v_h^* \tag{10.45}$$

$$v_{s0}^* = v_h^* \tag{10.46}$$

$$v_{l0}^* = v_l^* + v_h^* \tag{10.47}$$

Unlike the previous full-bridge converter that has employed two auxiliary voltages v_{gh}^* and v_{lh}^* to optimize both converter sides independently, the PWM strategy presented in (10.45)-(10.47) uses just one variable v_h^* due to the coupling between the input and output sides of the converter through the leg constituted by the switches

q_s–$\overline{q}_s$. v_h^* is given by

$$v_h^* = v_C\left(\mu - \frac{1}{2}\right) + (\mu - 1)V_{\max}^* - \mu V_{\min}^* \tag{10.48}$$

with $V_{\max} = \text{MAX}\{v_g^*, 0, v_l^*\}$ and $V_{\min} = \text{MIN}\{v_g^*, 0, v_l^*\}$.

Once the reference pole voltages have been defined, pulse widths τ_g, τ_s, and τ_l are given by

$$\tau_g = \left(\frac{v_{g0}^*}{v_C} + \frac{1}{2}\right)T_s \tag{10.49}$$

$$\tau_s = \left(\frac{v_{s0}^*}{v_C} + \frac{1}{2}\right)T_s \tag{10.50}$$

$$\tau_l = \left(\frac{v_{l0}^*}{v_C} + \frac{1}{2}\right)T_s \tag{10.51}$$

Exercise 10.5

Draw the control block diagram for the converter shown in Fig. 10.7 by adapting the cascade control strategy shown in Fig. 10.2.

10.3.3 dc-link Voltage Requirement

As aforementioned, the shared-leg presented in Fig. 10.7 will affect the dc-link voltage needed to guarantee the same input and output voltages as in the full-bridge circuit. To simplify the analysis let us assume that the dc-link capacitance is large enough to have $\Delta v_c \simeq 0$. The dc-link voltage for the converter shown in Fig. 10.1 must obey the requirement of $v_c \geq \max\{v_g, v_l\}$, where v_g and v_l are the voltage desirable at the input and output converter sides, respectively. Indeed, v_c is obtained from the statement that the maximum value of the difference between any combinations of pole voltages will define the dc-link voltage. For the topology with shared-leg in Fig. 10.7 it yields

$$v_c \geq \max\{v_{g0}^* - v_{s0}^*, v_{l0}^* - v_{s0}^*, v_{g0}^* - v_{l0}^*\} \tag{10.52}$$

or

$$v_c \geq \max\{v_g^*, v_l^*, v_g^* - v_l^*\} \tag{10.53}$$

If the load and grid frequencies are different, the term $v_g^* - v_l^*$ will be bigger than either v_g^* or v_l^* and consequently will define v_c. However, if both sides of the converter require the same frequency, the output voltage can be generated to minimize the effect of the term $v_g^* - v_l^*$. This will allow the shared-leg converter to operate with the same dc-link voltage as the full-bridge converter.

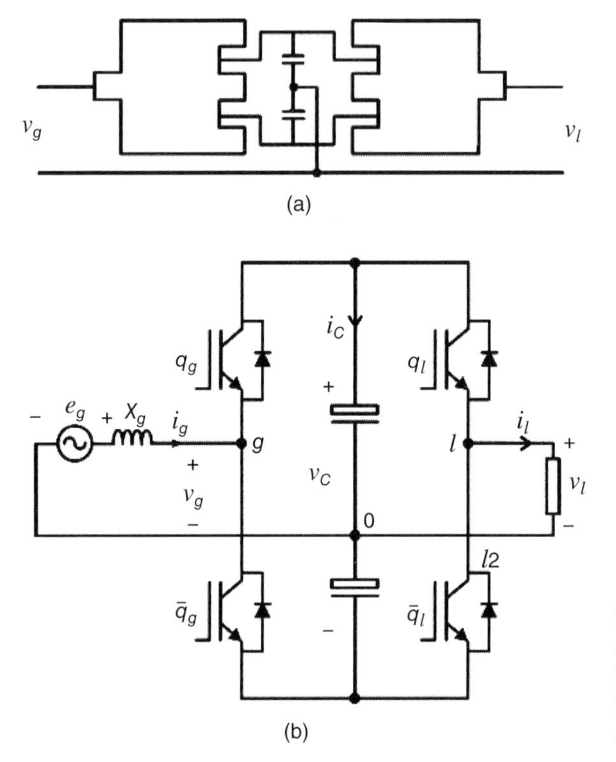

Figure 10.8 Half-bridge single-phase to single-phase back-to-back converter: (a) block and (b) switch representation.

10.3.4 Half-Bridge Converter

A back-to-back configuration with half-bridge at both converter sides (as in Fig. 10.8) guarantees another level of component count reduction as compared to the topology in Fig. 10.7. However, in this case there is a low frequency current flowing through the capacitors due to the dc-link midpoint connection to the load and grid, which will increase the voltage fluctuation on these elements and affect their losses.

The voltage at the input and output converter sides are determined directly from the pole voltages v_{g0} and v_{l0}, which are defined by equations (10.41) and (10.43), respectively. Also, the PWM equations are straightforward since there is no degree of freedom, that is, there are only two legs to process two voltages. In this case, it is not possible to use auxiliary voltage to optimize some parameters of the circuit.

10.4 TOPOLOGIES WITH INCREASED NUMBER OF SWITCHES (CONVERTERS IN PARALLEL)

The search for topologies with a reduced number of components continued over a considerable period of time. This can be, in part, explained by the high cost of the power switch when compared to other elements of the converter. However, as the price of the semiconductor was going down, this tendency changed, and recently

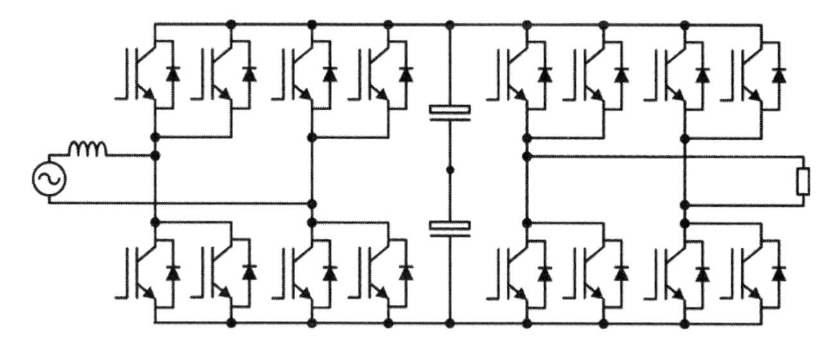

Figure 10.9 Single-phase to single-phase back-to-back converter with switches in parallel.

the configurations with an increased number of components appear as an interesting option, especially in terms of reliability, efficiency, and distortions improvement.

Particularly, the connection of converters in parallel can be justified if the ratings of the power switches available in the market cannot handle the level of current demanded for a specific application. Indeed, connecting switches in parallel is the simplest solution when the currents needed are higher than the rated currents of the switches, as shown in Fig. 10.9 for the single-phase back-to-back converter.

Although the number of power switches is twice as compared to that in Fig. 10.1, the characteristics of the converter in Fig. 10.9 in terms of quality of the waveform generated are the same as in the conventional solution. As described in the following sections, connecting back-to-back converters in parallel instead of switches in parallel have the following benefits: (i) increase the current capability of the converter and (ii) improve the quality of the waveforms generated by the converter at the input and output sides.

10.4.1 Model

Connecting converters in parallel allows reducing the level of current and consequently power processed by each converter. Another advantage is reducing outer current ripple with interleaved technique. Figure 10.10 shows a single-phase to single-phase energy conversion unit with two back-to-back converters in parallel. Such a conversion unit consists of four PBs-ac [see Fig. 10.10(a)], or in other words, the converters 1, 2, 3, and 4 [see Fig. 10.10(b)], single-phase load, four inductive filters (L_{1a}, L_{1b}, L_{3a}, and L_{3b}) at the grid side, four more inductive filters (L_{2a}, L_{2b}, L_{4a}, and L_{4b}) at the load side, and two dc-link capacitor banks.

Considering converters 1 and 3 (converters at the grid side), it is possible to write the following equations:

$$e_g = z_{1a}i_{1a} - z_{1b}i_{1b} + v_1 \tag{10.54}$$

$$e_g = z_{3a}i_{3a} - z_{3b}i_{3b} + v_3 \tag{10.55}$$

$$i_g = i_{1a} + i_{3a} \tag{10.56}$$

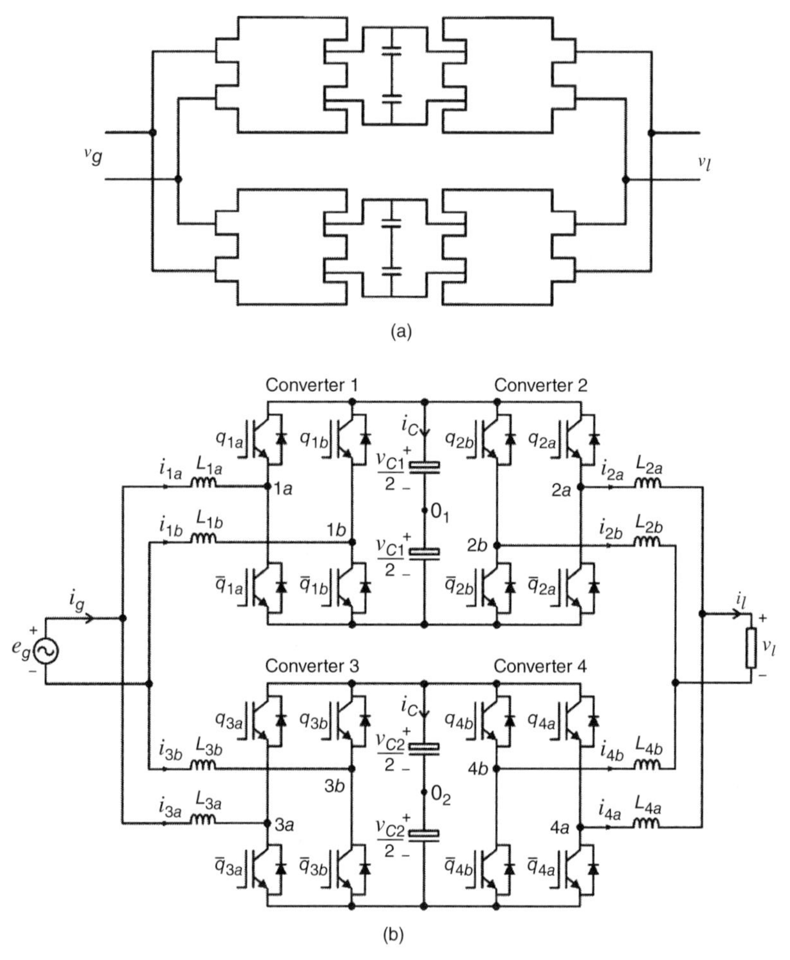

Figure 10.10 Single-phase to single-phase back-to-back converter connected in parallel: (a) block and (b) switch representation.

where $z_{1a} = r_{1a} + l_{1a}p$, $z_{1b} = r_{1b} + l_{1b}p$, $z_{3a} = r_{3a} + l_{3a}p$, and $z_{3b} = r_{3b} + l_{3b}p$ are the impedances of the inductors L_{1a}, L_{1b}, L_{3a}, and L_{3b}, respectively; with $p = d/dt$. The voltages v_1 and v_3 are given by $v_{1a0_1} - v_{1b0_1}$ and $v_{3a0_2} - v_{3b0_2}$, respectively.

For the converters at the load side, it yields:

$$v_l = -z_{2a}i_{2a} + z_{2b}i_{2b} + v_2 \tag{10.57}$$

$$v_l = -z_{4a}i_{4a} + z_{4b}i_{4b} + v_4 \tag{10.58}$$

$$i_l = i_{2a} + i_{4a} \tag{10.59}$$

where $z_{2a} = r_{2a} + l_{2a}p$, $z_{2b} = r_{2b} + l_{2b}p$, $z_{4a} = r_{4a} + l_{4a}p$, and $z_{4b} = r_{4b} + l_{4b}p$ are the impedances of the inductors L_{2a}, L_{2b}, L_{4a}, and L_{4b}, respectively. The voltages v_2 and v_4 are given by $v_{2a0_1} - v_{1b0_1}$ and $v_{4a0_2} - v_{4b0_2}$, respectively.

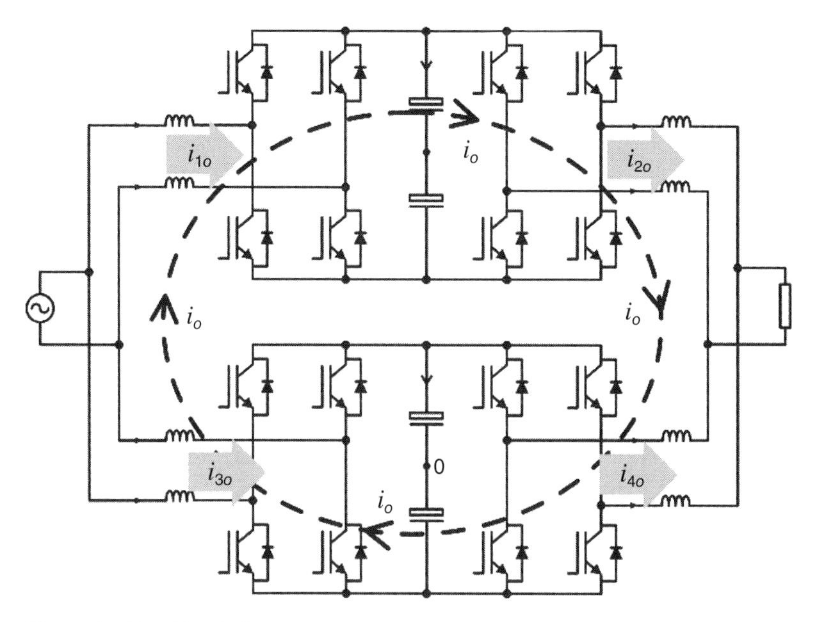

Figure 10.11 Circulating current between both back-to-back converters in parallel.

Since there is no transformer for isolation purpose, there exists a circulation current between both back-to-back converters as highlighted in Fig. 10.11. The following equations can be written from the circuit in Fig. 10.10:

$$z_{1a}i_{1a} - z_{3a}i_{3a} + z_{2a}i_{2a} - z_{4a}i_{4a} + v_{1a0_1} - v_{3a0_2} - v_{2a0_1} + v_{4a0_2} = 0 \tag{10.60}$$

$$z_{1a}i_{1a} - z_{3a}i_{3a} + z_{2b}i_{2b} - z_{4b}i_{4b} + v_{1a0_1} - v_{3a0_2} - v_{2b0_1} + v_{4b0_2} = 0 \tag{10.61}$$

$$z_{1b}i_{b} - z_{3b}i_{3b} + z_{2a}i_{2a} - z_{4a}i_{4a} + v_{1b0_1} - v_{3b0_2} - v_{2a0_1} + v_{4a0_2} = 0 \tag{10.62}$$

$$z_{1b}i_{1b} - z_{3b}i_{3b} + z_{2b}i_{2b} - z_{4b}i_{4b} + v_{1b0_1} - v_{3b0_2} - v_{2b0_1} + v_{4b0_2} = 0 \tag{10.63}$$

Equations (10.60)–(10.63) were obtained with the following meshes (see Fig. 10.10) $1a$–$2a$–$4a$–$3a$, $1a$–$2b$–$4b$–$3a$, $1b$–$2a$–$3b$–$4a$, and $1b$–$2b$–$3b$–$4b$, respectively.

An expression for the circulating voltage can be obtained by adding equations (10.60)–(10.63):

$$\begin{aligned} v_o = {} & z_{1a}i_{1a} + z_{1b}i_{1b} - z_{3a}i_{3a} - z_{3b}i_{3b} + z_{2a}i_{2a} \\ & + z_{2b}i_{2b} - z_{4a}i_{4a} - z_{4b}i_{4b} \end{aligned} \tag{10.64}$$

with

$$v_o = -v_{1a0_1} - v_{1b0_1} + v_{3a0_2} + v_{3b0_2} + v_{2a0_1} + v_{2b0_1} - v_{4a0_2} - v_{4b0_2} \tag{10.65}$$

The circulating currents in Fig. 10.11 are given by

$$i_{1o} = i_{1a} + i_{1b} \tag{10.66}$$
$$i_{3o} = i_{3a} + i_{3b} \tag{10.67}$$
$$i_{2o} = i_{2a} + i_{2b} \tag{10.68}$$
$$i_{4o} = i_{4a} + i_{4b} \tag{10.69}$$

As seen in Fig. 10.11, the circulating current is in fact a single current that circulates among the converters. Such a current can be written as

$$i_o = i_{1o} = -i_{3o} = i_{2o} = -i_{4o} \tag{10.70}$$

Then substituting equations (10.66)–(10.69) into (10.54)–(10.59), it is possible to model the parallel converter as follows:

$$e_g = (z_{1a} + z_{1b})i_{1a} - z_{1b}i_o + v_1 \tag{10.71}$$
$$e_g = (z_{3a} + z_{3b})i_{3a} + z_{3b}i_o + v_3 \tag{10.72}$$
$$e_l = -(z_{2a} + z_{2b})i_{2a} + z_{2b}i_o + v_2 \tag{10.73}$$
$$e_l = -(z_{4a} + z_{4b})i_{4a} + z_{4b}i_o + v_4 \tag{10.74}$$
$$\begin{aligned} v_o = {} & (z_{1a} - z_{1b})i_{1a} - (z_{3a} - z_{3b})i_{3a} - (z_{2a} - z_{2b})i_{2a} + (z_{4a} - z_{4b})i_{4a} \\ & + (z_{1b} + z_{3b} + z_{2b} + z_{4b})i_o \end{aligned} \tag{10.75}$$

There are two components for the circulating current, that is, high and low frequencies components. The high-frequency component is due to the difference of the deadtime for each converter and also due to asymmetry of the power switches employed in both stages. Such high-frequency components are not modeled in equation (10.75). However, the model furnished in (10.75) considers the low frequency component for the circulating current. For instance, any level of imbalance between L_{xa} and L_{xb} (with $x = 1, 2, 3,$ and 4) can be identified from this model. The current i_o can be compensated (tracking a zero reference) with an additional controller, as described in Section 10.4.3. Also, the dynamic model obtained from equations (10.71)–(10.75) suggests that the voltage v_1 is employed to control i_{1a}, and the voltage v_3 is employed to control i_{3a}. With these two currents (i_{1a} and i_{3a}) under control, the grid current is also controlled [see equation (10.56)]. The voltage v_o is employed to control i_o.

Considering a balanced case where $z_{1a} = z_{1b} = z_{3a} = z_{3b} = z_1$ and $z_{2a} = z_{2b} = z_{4a} = z_{4b} = z_2$, the dynamic model is simplified by the following equations:

$$e_g = 2z_1i_{1a} - z_1i_o + v_1 \tag{10.76}$$
$$e_g = 2z_1i_{3a} + z_1i_o + v_3 \tag{10.77}$$
$$e_l = -2z_2i_{2a} + z_2i_o + v_2 \tag{10.78}$$
$$e_l = -2z_2i_{4a} - z_2i_o + v_4 \tag{10.79}$$
$$v_o = 2(z_1 + z_2)i_o \tag{10.80}$$

10.4.2 PWM Strategy

From the model presented earlier, it is possible to write the reference voltages as a function of the reference pole voltages:

$$v_1^* = v_{1a0_1}^* - v_{1b0_1}^* \quad (10.81)$$

$$v_2^* = v_{2a0_1}^* - v_{2b0_1}^* \quad (10.82)$$

$$v_3^* = v_{3a0_2}^* - v_{3b0_2}^* \quad (10.83)$$

$$v_4^* = v_{4a0_2}^* - v_{4b0_2}^* \quad (10.84)$$

$$v_o^* = -v_{1a0_1}^* - v_{1b0_1}^* + v_{3a0_2}^* + v_{3b0_2}^* + v_{2a0_1}^* + v_{2b0_1}^* - v_{4a0_2}^* - v_{4b0_2}^* \quad (10.85)$$

There are eight pole voltages and just five variables to control, two variables associated to the grid current (v_1^* and v_3^*), one variable to deal with the circulating current (v_o^*), and finally two more variables to handle load voltage control (v_2^* and v_4^*). There exist, therefore, three degrees of freedom, that is, three auxiliary variables. Those auxiliary variables can be defined as follows:

$$v_x^* = \frac{v_{1a0_1}^* + v_{1b0_1}^*}{2} \quad (10.86)$$

$$v_y^* = \frac{v_{2a0_1}^* + v_{2b0_1}^*}{2} \quad (10.87)$$

$$v_z^* = \frac{v_{4a0_2}^* + v_{4b0_2}^*}{2} \quad (10.88)$$

From equations (10.86)–(10.88) it is possible to define the reference pole voltages as follows:

$$v_{1a0_1}^* = \frac{v_1^*}{2} + v_x^* \quad (10.89)$$

$$v_{1b0_1}^* = -\frac{v_1^*}{2} + v_x^* \quad (10.90)$$

$$v_{3a0_2}^* = \frac{v_3^*}{2} + \frac{v_o^*}{2} - v_y^* + v_z^* + v_x^* \quad (10.91)$$

$$v_{3b0_2}^* = -\frac{v_3^*}{2} + \frac{v_o^*}{2} - v_y^* + v_z^* + v_x^* \quad (10.92)$$

$$v_{2a0_1}^* = \frac{v_2^*}{2} + v_y^* \quad (10.93)$$

$$v_{2b0_1}^* = -\frac{v_2^*}{2} + v_y^* \quad (10.94)$$

$$v_{4a0_2}^* = \frac{v_4^*}{2} + v_z^* \quad (10.95)$$

$$v_{4b0_2}^* = -\frac{v_4^*}{2} + v_z^* \quad (10.96)$$

The reference pole voltages are defined as a function of the desirable voltages (defined by the controllers) v_1^*, v_2^*, v_3^*, and v_4^* as well as the auxiliary voltages v_x^*, v_y^*, and v_z^*. Note that the auxiliary voltages can be defined independently as follows, if equations (10.100)–(10.105) are satisfied:

$$v_x^* = \mu_x v_{x\mathrm{MAX}}^* + (1-\mu_x)v_{x\mathrm{MIN}}^* \tag{10.97}$$

$$v_y^* = \mu_y v_{y\mathrm{MAX}}^* + (1-\mu_y)v_{y\mathrm{MIN}}^* \tag{10.98}$$

$$v_z^* = \mu_z v_{z\mathrm{MAX}}^* + (1-\mu_z)v_{z\mathrm{MIN}}^* \tag{10.99}$$

where

$$v_{x\mathrm{MAX}}^* = \frac{v_{c1}^*}{2} - \max|V_x^*| \tag{10.100}$$

$$v_{x\mathrm{MIN}}^* = -\frac{v_{c1}^*}{2} - \min|V_x^*| \tag{10.101}$$

$$v_{y\mathrm{MAX}}^* = \frac{v_{c1}^*}{2} - \max|V_y^*| \tag{10.102}$$

$$v_{y\mathrm{MIN}}^* = -\frac{v_{c1}^*}{2} - \min|V_y^*| \tag{10.103}$$

$$v_{z\mathrm{MAX}}^* = \frac{v_{c2}^*}{2} - \max|V_z^*| \tag{10.104}$$

$$v_{z\mathrm{MIN}}^* = -\frac{v_{c2}^*}{2} - \min|V_z^*| \tag{10.105}$$

with $V_x^* = \left\{\frac{v_1^*}{2}, \frac{-v_1^*}{2}, \frac{v_3^*}{2} + \frac{v_o^*}{2} - v_y^* + v_z^*, -\frac{v_3^*}{2} + \frac{v_o^*}{2} - v_y^* + v_z^*\right\}$, $V_y^* = \left\{\frac{v_2^*}{2}, -\frac{v_2^*}{2}\right\}$ and $V_z^* = \left\{\frac{v_4^*}{2}, -\frac{v_4^*}{2}\right\}$; and $0 \le \mu_x \le 1$, $0 \le \mu_y \le 1$, and $0 \le \mu_z \le 1$.

The gating signals of the power switches can be obtained by either comparing the reference pole voltages (10.89)–(10.96) with carrier waveforms or in a digital way by programming timers. Since there are two converters connected in parallel, an interleaved technique can be employed to reduce the current ripple. Figure 10.12 shows the carrier signals (v_{t1}^* and v_{t2}^*) for the interleaved technique. In this case, v_{t1}^* is employed for converters 1 and 2, while v_{t2}^* is employed for converters 3 and 4. Figure 10.13 shows the output of the parallel topology highlighting the influence of the interleaved technique. Notice that since the ripple of i_{2a} is shifted by 180 electrical degrees (considering the switching frequency) from the ripple of i_{4a}, the output current will have a smoother ripple since $i_l = i_{2a} + i_{4a}$.

10.4.3 Control Strategy

The control block diagram for the parallel back-to-back converter is given in Fig. 10.14. The dc-link voltages are regulated by the controllers R_{c1} and R_{c3}. The output of these controllers defines the amplitude of the currents in converters 1 and 3, respectively. The block G_i creates sinusoidal currents (i_{1a}^* and i_{3a}^*) in phase with the grid voltage; this is a *PLL*-based block. The controllers R_{i1} and R_{i3} regulate the currents i_{1a} and i_{3a}, and the output of these controllers define the voltages (v_1^* and v_3^*) employed in the PWM technique described previously. The circulating current is

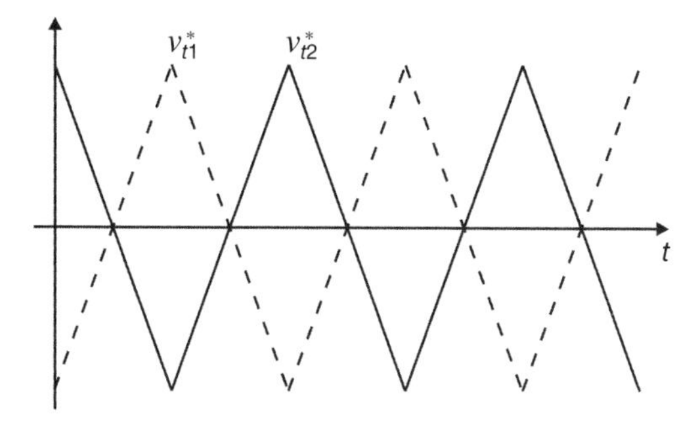

Figure 10.12 Carrier signals for the interleaved technique applied to the parallel converters.

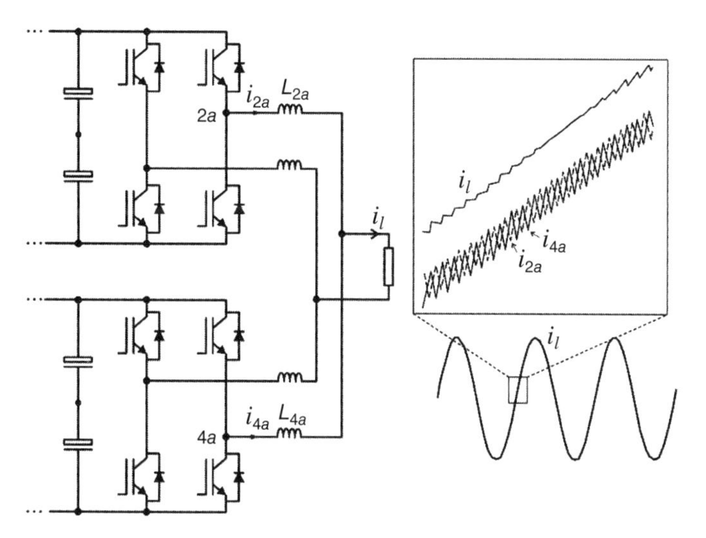

Figure 10.13 Output of the parallel back-to-back converter with interleaved technique.

regulated by using the controller R_o. The block k_v can be employed to compensate imbalance among the inductors; for a balanced case $k_v = 0.5$.

Exercise 10.6

Assuming well-designed controllers for the parallel topology presented in Fig. 10.10, that is, power factor equal to one, circulating current equal to zero, and dc-link capacitor voltages under control, specify the power switches employed in this configuration in terms of current and voltage with: (i) the active power of the load is $P_l = 1700\,W$ with $\cos(\Phi_l) = 0.85$, load voltage with $120V_{\text{rms}}$ – 60 Hz, (ii) dc-link voltage fluctuation equal to zero, and (iii) a drop voltage due to the inductors L_{xa} and L_{xb} $(x = 1, 2, 3, 4)$ equal to $5V_{\text{rms}}$ on each inductor.

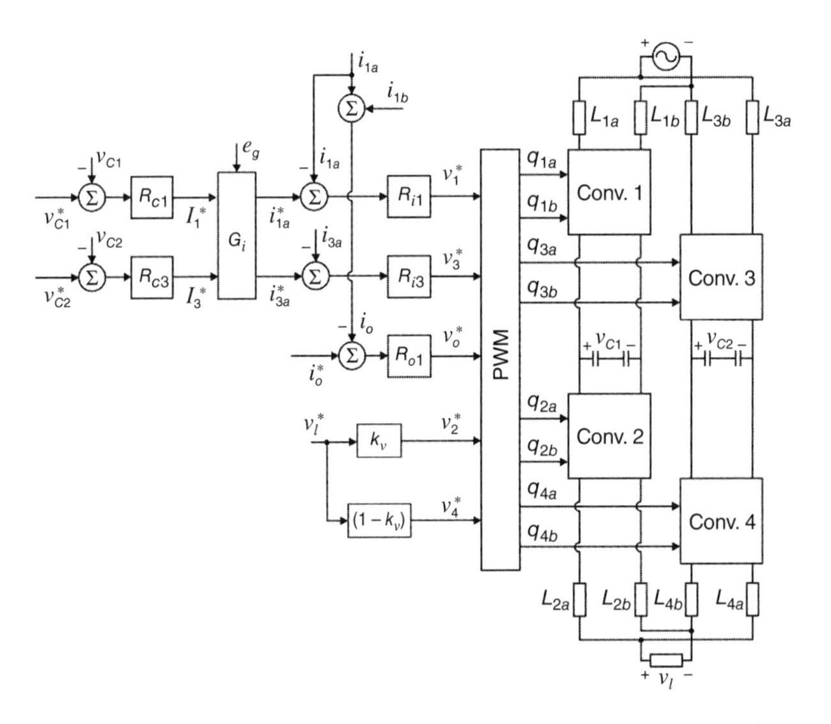

Figure 10.14 Control block diagram of the parallel single-phase back-to-back converter.

10.5 TOPOLOGIES WITH INCREASED NUMBER OF SWITCHES (CONVERTERS IN SERIES)

Connecting single-phase to single-phase converters in parallel brings up interesting characteristics such as reduction of the current processed by each power switch and interleaved PWM, which allows current ripple reduction. Following a similar approach, it is also possible to connect single-phase to single-phase back-to-back converters in series as seen in Fig. 10.15. Figure 10.15(a) shows the power block representation while Fig. 10.15(b) depicts the representation with switches. In this case, it is necessary for half of the dc-link voltage to have the same voltage capability of the conventional single back-to-back converter. Also the load and grid will give the back-to-back converters an H-bridge cascade configuration, that is, as a multilevel converter (see Chapter 5). The only difference between the input/output of the converter in Fig. 10.15 and the cascade topology in Chapter 5 is a circulating current between both back-to-back units since there is no isolation. Such a circulating current can be determined from the model, as described in the later. It is worth mentioning that the majority of the topologies presented in Chapter 5 have isolated dc sources.

The following equations can be derived from the converter in Fig. 10.15(b).

$$e_g = z_1 i_g + v_{1ab} + z_2 i'_g + v_{2ab} \tag{10.106}$$

$$v_l = v_{1cd} + v_{2cd} \tag{10.107}$$

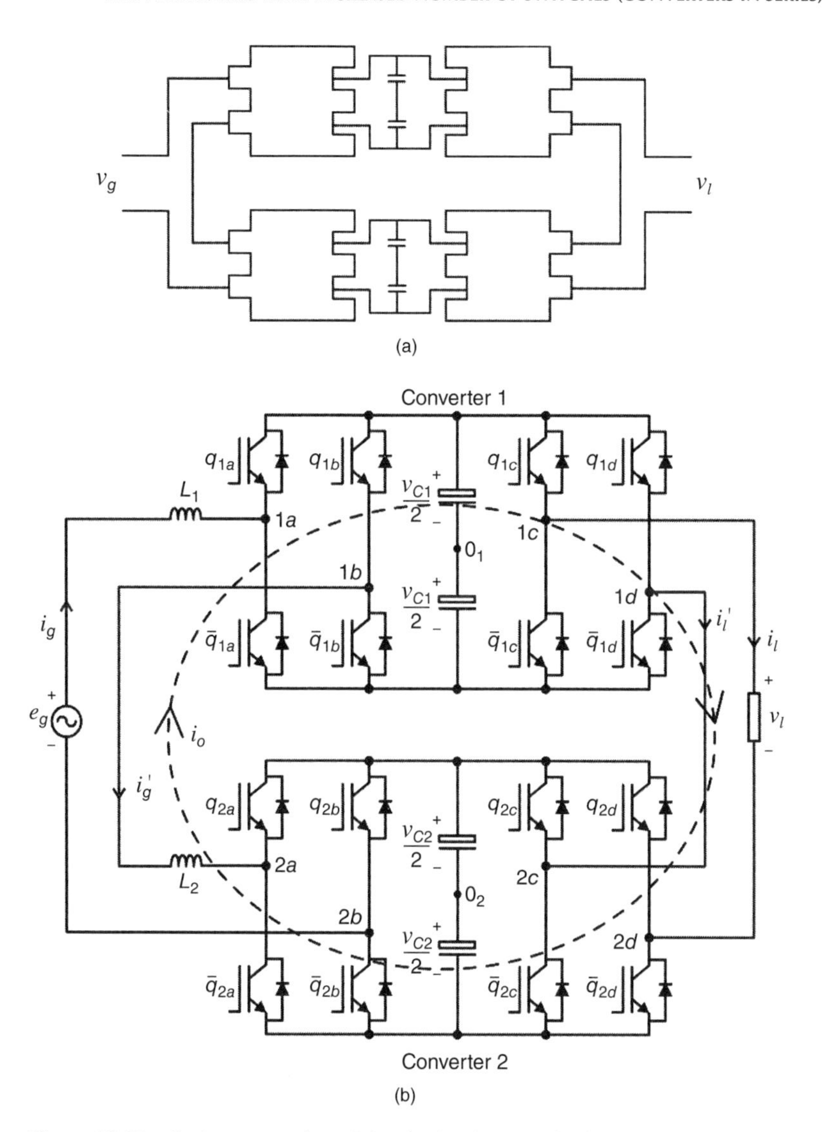

Figure 10.15 Series connection of the single-phase to single-phase back-to-back converters.

where $z_1 = r_1 + pl_1$, $z_2 = r_2 + pl_2$ with $p = d/dt$; $v_{1ab} = v_{1a0_1} - v_{1b0_1}$, $v_{2ab} = v_{2a0_2} - v_{2b0_2}$, $v_{1cd} = v_{1c0_1} - v_{1d0_1}$, and $v_{2cd} = v_{2c0_2} - v_{2d0_2}$.

Since this is a transformerless configuration, there are circulating meshes, which can be modeled by equations (10.108) and (10.109):

$$e_g = z_1 i_g + v_{1ac} + v_l - v_{2bd} \tag{10.108}$$

$$0 = v_{1bd} - v_{2ac} - z_2 i_g' \tag{10.109}$$

Adding (10.108) and (10.109), it is possible to define the circulating voltage as follows:

$$v_o = e_g - z_1 i_g + z_2 i'_g = v_{1ac} + v_{1bd} - v_{2ac} - v_{2bd} + v_l \tag{10.110}$$

Ideally $i_g = i'_g$ and $i_l = -i'_l$. However, due to the circulating current we have

$$i'_g = i_g - i_o \tag{10.111}$$

$$i'_l = -i_l + i_o \tag{10.112}$$

The dynamic model of the converter can be rewritten from (10.106) as follows:

$$e_g = z_1 i_g + z_2 i_g - z_2 i_o + v_g \tag{10.113}$$

where $v_g = v_{1ab} + v_{2ab}$. From equation (10.110), the model for the circulating voltage can be written in terms of i_o as

$$v_o = e_g + (z_2 - z_1) i_g - z_2 i_o \tag{10.114}$$

For a balanced case with $z_1 = z_2$ we have

$$e_g = 2z_1 i_g - z_1 i_o + v_g \tag{10.115}$$

$$v_o = e_g - z_2 i_o \tag{10.116}$$

The PWM approach described for the series back-to-back converter can be adapted from the solution employed in the parallel converter presented in the previous section.

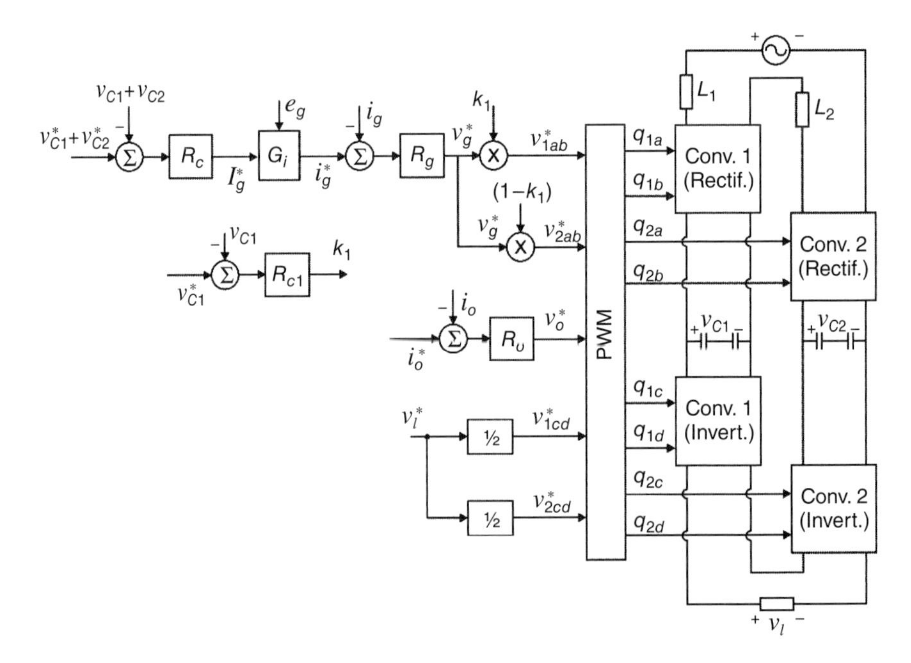

Figure 10.16 Control block diagram of the single back-to-back converter series connected.

The control strategy is shown in Fig. 10.16. Since the current i_g affects both dc-link voltages v_{C1} and v_{C2}, the controller R_c has been employed to regulate ($v_{C1} + v_{C2}$). The output of this controller defines I_g^*, the instantaneous sinusoidal current i_g^* synchronized with the grid voltage (in order to obtain high power factor) is determined by the PLL-based block G_i. The inner control loop should be designed to guarantee grid current control; the output of the block R_g defines v_g^*. It is important to highlight that v_g is the voltage generated by both rectifiers in converters 1 and 2 to guarantee grid current control [see equation (10.115)]. In fact, although v_g is viewed from the grid as only one-component voltage, it is constituted by two terms, that is, $v_g = v_{1ab} + v_{2ab}$. The term k_1 determines the level of voltage that should be processed by the rectifiers of converters 1 and 2 to keep the voltage v_{C1} under control. Hence, the controllers R_c and R_{c1} are employed to control the dc-link voltages (v_{C1} and v_{C2}), while the controller R_g along with the block G_i are used to control grid power factor. The circulating current must track a zero reference ($i_g^* = 0$) with the controller R_o, and the output voltage v_l^* defines the voltages v_{1cd}^* and v_{2cd}^*. Although the load voltage control in Fig. 10.16 is done in open-loop operation, a closed-loop can be easily obtained from this control block diagram.

Exercise 10.7

Repeat Exercise 10.6 with the same conditions for the series topology in Fig. 10.15.

10.6 SUMMARY

This chapter presented several circuits able to implement a single-phase back-to-back converter. A comprehensive analysis for the full-bridge topology was presented in Section 10.2. The following aspects were considered for this topology: (i) model, (ii) PWM, (iii) control, (iv) power analysis, and (v) dc-link capacitor voltage, and (vi) capacitor bank design. Section 10.3 dealt with topologies with reduced number of power switches. Two strategies were considered: utilization of shared-leg and dc-link capacitor midpoint connections as valid ways to reduce the number of switches. On the other hand, the topologies with increased number semiconductor devices were presented in Sections 10.4 and 10.5, with configurations in parallel and series, respectively. The model, PWM and control approaches highlighting the presence of the circulating current were also presented for both topologies. More information can be found in references from 1 to 35.

REFERENCES

[1] Asiminoaei L, Aeloiza E, Enjeti PN, Blaabjerg F, Danfoss G. Shunt active-power-filter topology based on parallel interleaved inverters. IEEE Trans Ind Electron 2008;55(3):1175–1189.

[2] Ramos R, Biel D, Fossas E, Guinjoan F. Interleaving quasisliding-mode control of parallel-connected buck-based inverters. IEEE Trans Ind Electron 2008;55(11):3865–3873.
[3] Chen Y-T, Jiang R-S, Liang R-H. Analysis and design of zero-voltage-switching parallel interleaved current-doubler converters with coupled output inductors. IET Power Electron 2012;5(4):467–476.
[4] Chunkag V, Kamnarn U. Parallelling three-phase ac to dc converter using cuk rectifier modules based on power balance control technique. IET Power Electron 2010;3(4):511–524. DOI: 10.1049/iet-pel.2008.0209.
[5] Ye Z, Boroyevich D, Choi J-Y, Lee FC. Control of circulating current in two parallel three-phase boost rectifiers. IEEE Trans Power Electron 2002;17(5):609–615.
[6] Vilathgamuwa DM, Gajanayake CJ, Loh PC. Modulation and control of three-phase paralleled z-source inverters for distributed generation applications. IEEE Trans Energy Convers 2009;24(1):173–183.
[7] Pan C-T, Liao Y-H. Modeling and control of circulating currents for parallel three-phase boost rectifiers with different load sharing. IEEE Trans Ind Electron 2008;55(7):2776–2785.
[8] Wang J. Design a parallel buck derived converter system using the primary current droop sharing control. IET Power Electron 2011;4(5):491–502. DOI: 10.1049/iet-pel.2010.0042.
[9] Singh RP, Khambadkone AM. Current sharing and sensing in n-paralleled converters using single current sensor. IEEE Trans Ind Appl 2010;46(3):1212–1219.
[10] Cuzner RM, Nowak DJ, Bendre A, Oriti G, Julian AL. Mitigating circulating common-mode currents between parallel soft-switched drive systems. IEEE Trans Ind Appl 2007;43(5):1284–1294.
[11] Li R, Xu D. Parallel operation of full power converters in permanent-magnet direct-drive wind power generation system. IEEE Trans Ind Electron 2013;60(4):1619–1629.
[12] Ko Y-J, Lee K-B, Lee D-C, Kim J-M. Fault diagnosis of three-parallel voltage-source converter for a high-power wind turbine. IET Power Electron 2012;5(7):1058–1067.
[13] Hou C-C. A multicarrier pwm for parallel three-phase active front-end converters. IEEE Trans Power Electron 2013;28(6):2753–2759.
[14] Xu Z, Li R, Zhu H, Xu D, Zhang C. Control of parallel multiple converters for direct-drive permanent-magnet wind power generation systems. IEEE Trans on Power Electron 2012;27(3):1259–1270. DOI: 10.1109/TPEL.2011.2165224.
[15] Zhang D, Wang F, Burgos R, Lai R, Boroyevich D. Interleaving impact on ac passive components of paralleled three-phase voltage-source converters. IEEE Trans Ind Appl 2010;46(3):1042–1054.
[16] Itkonen T, Luukko J, Sankala A, Laakkonen T, Pollanen R. Modeling and analysis of the dead-time effects in parallel pwm two-level three-phase voltage-source inverters. IEEE Trans Power Electron 2009;24(11):2446–2455.
[17] Genc N, Iskender I. DSP-based current sharing of average current controlled two-cell interleaved boost power factor correction converter. IET Power Electronics 2011;4(9):1015–1022. DOI: 10.1049/iet-pel.2010.0349.
[18] Chen Y-T, Jiang R-S, Liang R-H. Analysis and design of zero-voltage-switching parallel interleaved current-doubler converters with coupled output inductors. IET Power Electronics 2012;5(4):467–476. DOI: 10.1049/iet-pel.2010.0163.
[19] Park HW, Park SJ, Park JG, Kim CU. A novel high-performance voltage regulator for single-phase AC sources. IEEE Trans Ind Electron 2001;48(3):554–562.
[20] Choi J-H, Kwon J-MB, Jung J-H, Kwon B-H. Highperformance online UPS using three-leg-type converter. IEEE Trans Ind Electron 2005;52(3):889–897.
[21] Rocha N, Jacobina CB, dos Santos EJ, de Cavalcanti RMB. Parallel connection of two single-phase ac-dc-ac three-leg converter with interleaved technique. Proc. of IEEE Ind Electron, IECON 2012; p 639–644.
[22] Holtz J. Pulsewidth modulation for electronic power conversion. Proc IEEE 1994;82(8):1194–1214.
[23] Ojo O, Kshirsagar PM. Concise modulation strategies for four-leg voltage source inverters. IEEE Trans Power Electron 2004;19(1):46–53.
[24] Jacobina CB, Lima AMN, da Silva ERC, Alves RNC, Seixas PF. Digital scalar pulse-width modulation: a simple approach to introduce non-sinusoidal modulating waveforms. IEEE Trans Power Electron 2001;16(3):351–359.
[25] Panda AK, Suresh Y. Research on cascade multilevel inverter with single dc source by using three-phase transformers. Int J Electr Power Energy Syst 2012;40(1):9–20.

[26] Malinowski M, Kazmierkowski MP, Trzynadlowski AM. A comparative study of control techniques for PWM rectifiers in AC adjustable speed drives. IEEE Trans Power Electron 2003;18(6):1390–1396.

[27] Filho RMS, Seixas PF, Cortizo PC, Torres LAB, Souza AF. Comparison of three single-phase pll algorithms for ups applications. IEEE Trans Ind Electron 2008;55(8):2923–2932.

[28] Verdelho P, Marques GD. Four-wire current-regulated PWM voltage converter. IEEE Trans Ind Electron 1998;45(5):761–770.

[29] Abu-Rub H, Guzinski J, Krzeminski Z, Toliyat H. Predictive current control of voltage-source inverters. IEEE Trans Ind Electron 2004;51(3):585–593.

[30] Dong G, Ojo O. Current regulation in four-leg voltage-source converters. IEEE Trans Ind Electron 2007;54(4):2095–2105.

[31] Jacobina CB, de Correa MB, Oliveira TM, Lima AMN, da Silva ERC. Current control of unbalanced electrical systems. IEEE Trans Ind Electron 2001;48(3):517–525.

[32] Babaeia E, Hosseinia SH, Gharehpetianb GB. Reduction of the and low order harmonics with symmetrical output current for single-phase ac/ac matrix converters. Int J Electr Power Energy Syst 2010;32(3):225–235.

[33] Kieferndorf F, Forster M, Lipo T. Reduction of dc-bus capacitor ripple current with PAM/PWM converter. IEEE Trans Ind Applic 2004;40(2):607–614.

[34] Kolar J, Round S. Analytical calculation of the rms current stress on the dc-link capacitor of voltage-PWM converter systems. Electric Power Applications, IEE Proceedings 2006;153(4):535–543.

[35] Cavalcanti M, da Silva E, Boroyevich D, Dong W, Jacobina C. A feasible loss model for igbt in soft-switching inverters. Proc IEEE PESC 2003;3:1845–1850.

CHAPTER 11

THREE-PHASE TO THREE-PHASE AND OTHER BACK-TO-BACK CONVERTERS

11.1 INTRODUCTION

A three-phase back-to-back converter is an important electrical apparatus with a wide range of applications on systems demanding a processed three-phase voltage. Although the operation principle is similar to the single-phase to single-phase back-to-back converter presented in the previous chapter, the characteristics of the three-phase version are quite different, especially due to the nonexistent pulsating power on the dc-link capacitor voltages. Note that the power at the input (grid-side) and output (load-side) of the converter are constant for a balanced case. The concept of power blocks is considered to provide a broader perception of how the topologies presented in the technical literature have been conceived.

Following the introduction, this chapter is organized in six sections. Section11.2 deals with the full-bridge ac–dc–ac converter presenting its model, pulse width modulation (PWM), and control strategies. Topologies with component count reduction are presented in Section 11.3, while Sections 11.4 and 11.5 describe the model, PWM, and control for configurations with an increased number of semiconductor devices. A similar approach as in the previous chapter is employed for the three-phase back-to-back converters, that is, shared-leg and dc-link capacitor midpoint connections for reducing the number of semiconductor devices, while connecting back-to-back converters in parallel and series are considered in Sections 11.4 and 11.5, respectively. Section 11.6 presents other converters, especially for single-phase to three-phase conversion and for a six-phase motor drive system. Finally, Section 11.7 summarizes the chapter.

Advanced Power Electronics Converters: PWM Converters Processing AC Voltages,
Forty Fifth Edition. Euzeli Cipriano dos Santos Jr. and Edison Roberto Cabral da Silva.

11.2 FULL-BRIDGE CONVERTER

11.2.1 Model

Since the control goals are active power factor and regulation of the dc-link capacitor voltage, the equations derived in this section highlight both grid current and capacitor voltage variables. Although an open loop operation is assumed for the output of the converter, a closed loop control can be easily implemented for the output converter side.

The three-phase back-to-back converter can be obtained directly by using two PBs back-to-back connected per phase as presented in Fig. 11.1(a). The representation of the three-phase back-to-back converter using switches is furnished in Fig. 11.1(b). The conversion system is composed of a three-phase utility grid, input inductors (with reactance equal to X_{g1}, X_{g2}, and X_{g3}), capacitor bank at the dc-link with two capacitors (each one with capacitance equal to C), controlled rectifier, inverter, and a three-phase load. The rectifier circuit comprises the switches q_{g1}, q_{g2}, and q_{g3} and their complementary switches: $\overline{q}_{g1}$, $\overline{q}_{g2}$, and $\overline{q}_{g3}$. Similarly, the inverter comprises the switches q_{l1}, q_{l2}, and q_{l3} and their complementary switches $\overline{q}_{l1}$, $\overline{q}_{l2}$, and $\overline{q}_{l3}$. Binary variables are employed to define the conduction states of the switches.

The model of the back-to-back converter can be obtained from Fig. 11.1(b) as follows:

$$v_{g1} = v_{g10} - v_{n_g0} \tag{11.1}$$

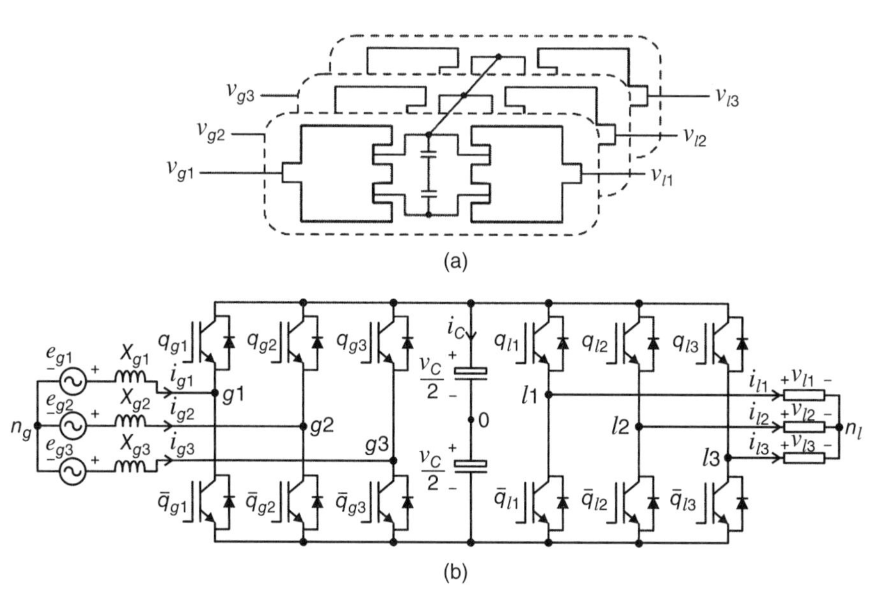

Figure 11.1 Three-phase to three-phase back-to-back converter: (a) block and (b) switch representation.

$$v_{g2} = v_{g20} - v_{n_g0} \quad (11.2)$$

$$v_{g3} = v_{g30} - v_{n_g0} \quad (11.3)$$

$$v_{l1} = v_{l10} - v_{n_l0} \quad (11.4)$$

$$v_{l2} = v_{l20} - v_{n_l0} \quad (11.5)$$

$$v_{l3} = v_{l30} - v_{n_l0} \quad (11.6)$$

where the pole voltages can be written as a function of the states of the switches, as below:

$$v_{g10} = (2q_{g1} - 1)\frac{v_C}{2} \quad (11.7)$$

$$v_{g20} = (2q_{g2} - 1)\frac{v_C}{2} \quad (11.8)$$

$$v_{g30} = (2q_{g3} - 1)\frac{v_C}{2} \quad (11.9)$$

$$v_{l10} = (2q_{l1} - 1)\frac{v_C}{2} \quad (11.10)$$

$$v_{l20} = (2q_{l2} - 1)\frac{v_C}{2} \quad (11.11)$$

$$v_{l30} = (2q_{l3} - 1)\frac{v_C}{2} \quad (11.12)$$

with $v_{ng0} = \frac{1}{3}(v_{g10} + v_{g20} + v_{g30})$ and $v_{nl0} = \frac{1}{3}(v_{l10} + v_{l20} + v_{l30})$ for a balanced case.

Equations (11.13)–(11.15) furnish the dynamic model for the variables at the grid side:

$$v_{g1} = e_{g1} - r_g i_{g1} - l_g \frac{di_{g1}}{dt} \quad (11.13)$$

$$v_{g2} = e_{g2} - r_g i_{g2} - l_g \frac{di_{g2}}{dt} \quad (11.14)$$

$$v_{g3} = e_{g3} - r_g i_{g3} - l_g \frac{di_{g3}}{dt} \quad (11.15)$$

where r_g and l_g represent the resistances and inductances of the input inductor.

The capacitor voltage can be obtained as

$$v_c = \frac{1}{C}\int (q_{g1}i_{g1} + q_{g2}i_{g2} + q_{g3}i_{g3} - q_{l1}i_{l1} - q_{l2}i_{l2} - q_{l3}i_{l3})dt \quad (11.16)$$

with the interval of integration equal to the switching period.

11.2.2 PWM Strategy

The optimized PWM strategy presented in Chapter 8 can be employed for the full-bridge configuration presented in Fig. 11.1 to improve either total harmonic distortion (THD) or efficiency at both sides.

If the desired input and output voltages are given respectively by v_{g1}^*, v_{g2}^*, v_{g3}^* and v_{l1}^*, v_{l2}^*, v_{l3}^*, the reference pole voltages can be written as follows:

$$v_{g10}^* = v_{g1}^* + v_{gh}^* \tag{11.17}$$

$$v_{g20}^* = v_{g2}^* + v_{gh}^* \tag{11.18}$$

$$v_{g30}^* = v_{g3}^* + v_{gh}^* \tag{11.19}$$

$$v_{l10}^* = v_{l1}^* + v_{lh}^* \tag{11.20}$$

$$v_{l20}^* = v_{l2}^* + v_{lh}^* \tag{11.21}$$

$$v_{l30}^* = v_{l3}^* + v_{lh}^* \tag{11.22}$$

The auxiliary voltages v_{gh}^* and v_{lh}^* are given respectively by

$$v_{gh}^* = v_C\left(\mu_g - \frac{1}{2}\right) + (\mu_g - 1)V_{g\max}^* - \mu_g V_{g\min}^* \tag{11.23}$$

$$v_{lh}^* = v_C\left(\mu_l - \frac{1}{2}\right) + (\mu_l - 1)V_{l\max}^* - \mu_l V_{l\min}^* \tag{11.24}$$

where $V_{g\max} = \mathrm{MAX}\{v_{g1}^*, v_{g2}^*, v_{g3}^*\}$, $V_{g\min} = \mathrm{MIN}\{v_{g1}^*, v_{g2}^*, v_{g3}^*\}$ and $V_{l\max} = \mathrm{MAX}\{v_{l1}^*, v_{l2}^*, v_{l3}^*\}$, $V_{l\min} = \mathrm{MIN}\{v_{l1}^*, v_{l2}^*, v_{l3}^*\}$; μ_g and μ_l are the distribution factor and determine how the free-wheeling time can be distributed throughout the switching period. Note that, since the grid and load sides of the converter are decoupled by the dc-link capacitor bank, v_{gh}^* and v_{lh}^* are independent of each other.

Once the reference pole voltages have been defined, pulse widths τ_{g1}, τ_{g2}, τ_{g3}, τ_{l1}, τ_{l2}, and τ_{l3} can be calculated from equations (11.25)–(11.30), as below:

$$\tau_{g1} = \left(\frac{v_{g10}^*}{v_C} + \frac{1}{2}\right)T_s \tag{11.25}$$

$$\tau_{g2} = \left(\frac{v_{g20}^*}{v_C} + \frac{1}{2}\right)T_s \tag{11.26}$$

$$\tau_{g3} = \left(\frac{v_{g30}^*}{v_C} + \frac{1}{2}\right)T_s \tag{11.27}$$

$$\tau_{l1} = \left(\frac{v_{l10}^*}{v_C} + \frac{1}{2}\right)T_s \tag{11.28}$$

$$\tau_{l2} = \left(\frac{v_{l20}^*}{v_C} + \frac{1}{2}\right)T_s \tag{11.29}$$

$$\tau_{l3} = \left(\frac{v^*_{l30}}{v_C} + \frac{1}{2} \right) T_s \tag{11.30}$$

where T_s is the switching period. The state of the switches q_{g1}, q_{g2}, q_{g3}, q_{l1}, q_{l2}, and q_{l3} can be defined directly from the previous equations by programming timers in a digital implementation of the PWM.

Note that the reference voltages v^*_{g1}, v^*_{g2}, and v^*_{g3} employed in (11.25)–(11.27) are obtained as the output of controllers (as presented in Section 11.2.3.) Although the voltages v^*_{l1}, v^*_{l2}, and v^*_{l3} can be defined with either open-loop or closed-loop, the open loop operation is considered in this section.

11.2.3 Control Approach

Figure 11.2 presents the control block diagram for the three-phase back-to-back converter. While the output voltage is controlled by open-loop operation, the dc-link capacitor voltage and the grid power factor are regulated by the controllers R_C and R_{i1}, R_{i2}, R_{i3}, respectively. Since the dynamic of the capacitor voltage is slow as compared to the grid current, a cascade control approach has been employed to guarantee regulation of both variables.

The dc-link voltage v_C is adjusted to its reference value v^*_C by using the controller R_C, which could be implemented with a standard PI controller. This controller

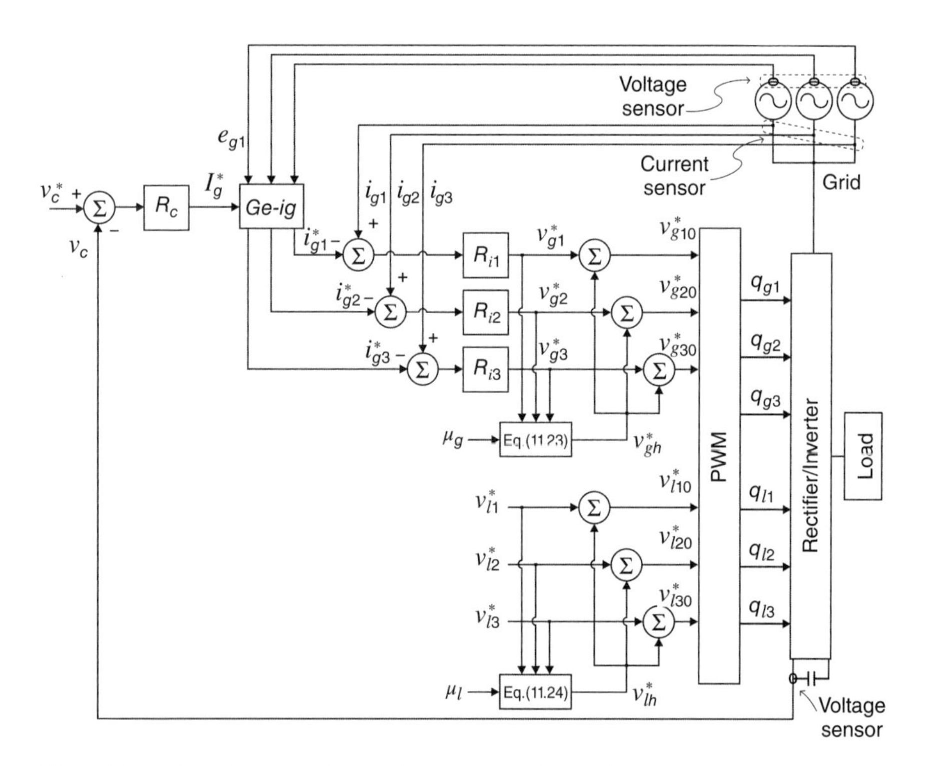

Figure 11.2 Control strategy for the three-phase back-to-back converter.

provides the amplitude of the reference grid current I_g^*. To maintain the power factor close to one with a sinusoidal grid current, the instantaneous reference currents i_{g1}^*, i_{g2}^* and i_{g3}^* must be synchronized with the input voltage e_{g1}, e_{g2}, and e_{g3}, respectively. Such a synchronization scheme can be obtained through a phase locked loop (PLL)-based control. The block $G_e - i_g$ is responsible for generating the reference sinusoidal grid current in phase with the grid voltage, that is, $i_{g1}^* = I_g^* \sin(\omega_g t + \Phi_g)$, $i_{g2}^* = I_g^* \sin(\omega_g t + \Phi_g + 120°)$, $i_{g3}^* = I_g^* \sin(\omega_g t + \Phi_g + 240°)$. The current controller loop must deal with sinusoidal waveforms, which means that a conventional PI controller will not guarantee null error in steady-state operation – a modified PI controller can be employed instead.

The PWM block shown in Fig. 11.2 can be implemented either from equations (11.17)–(11.22) or by comparing these reference pole voltages with high-frequency carrier signals.

Application (Universal Active Power Filter)

Also known as Universal Power Quality Conditioner (UPQC), the Universal Active power filter uses the back-to-back schematic circuit to improve the quality of both voltage and current in a three-phase electrical system. The electrical energy generators in power plants furnish voltages nearly sinusoidal. However, the quality of this voltage can be deteriorated due to steady-state and/or transient phenomena at a given point of the electrical system. Examples of these phenomena are nonlinear loads and faults in part of the power circuit resulting in current distortion and voltage sags/swells. The universal active power filter can be employed to compensate dynamically such disturbances for both steady-state and transient problems, with the ability to deal with the correction of grid voltage disturbance and fluctuation. As presented in the figure below, one side (grid side) of the back-to-back converter is employed to compensate the voltage disturbance by adding voltage through a three-phase transformer, while the other side is used to inject the current at the load side to guarantee sinusoidal grid current with power factor close to one. The universal active power filter is especially recommended to protect critical loads at the point of installation as well as compensate the grid from current harmonics generated at the load side. The voltage v_g *is used to compensate any disturbance on* e_g*, since* $v_l = e_g + v_{n_g n_l} - v_g$*. The current* i_f *has been employed to guarantee a sinusoidal grid current in phase with the grid voltage, since* $i_g = i_l + i_f$*.*

Although several techniques are presented in the literature for active power filter control, the main structure is based on two control loops: a dc-link voltage outer loop and a current inner loop. The outer loop regulates the dc-link voltage while the inner loop tracks the reference current necessary to compensate harmonics and/or fundamental reactive power. Two basic methods are considered in the literature to generate current reference signal: direct method and indirect method. In the direct method the disturbance to be compensated is sensed directly from the load and employed as the filter reference current signal. In the indirect method the disturbance to be compensated is sensed on the grid and controlled to a minimum value through a closed-loop. The last method has reduced cost since it uses less number of sensors, while the first one is more accurate.

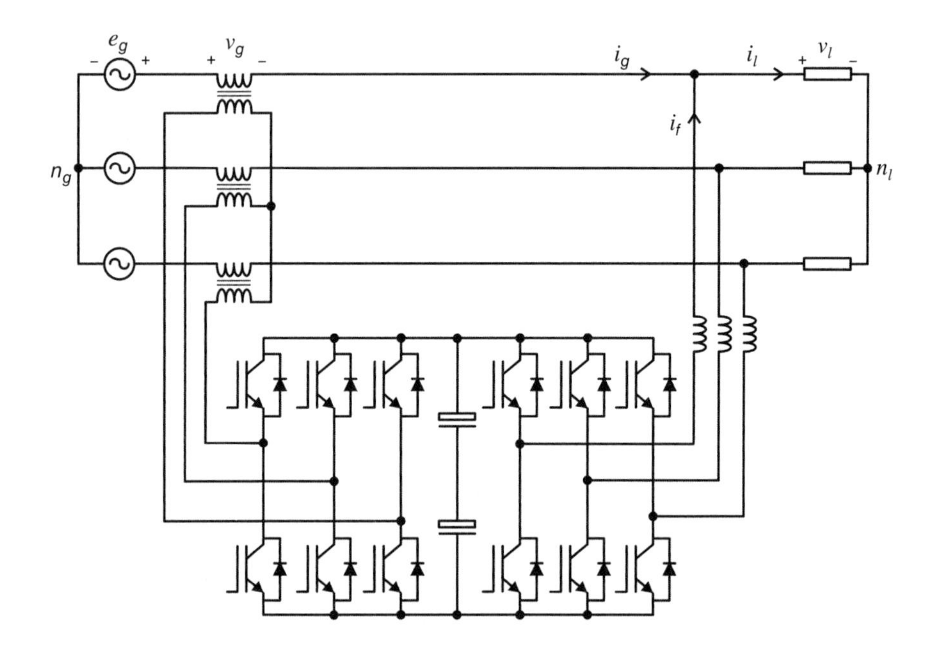

11.3 TOPOLOGY WITH COMPONENT COUNT REDUCTION

11.3.1 Model

Another three-phase to three-phase ac–dc–ac power converter is obtained by removing one of the PBs-ac as shown in Fig. 11.3(a). Figure 11.3(b) shows the same configuration with switch representation. Note that the leg constituted by the switches $q_s - \overline{q}_s$ is shared between the input and output sides of the converter, and consequently such a leg will handle a current that is a sum of input and output currents $(i_{l3} - i_{g3})$. Also, the shared-leg will play an important role for the voltage capability of the converter. Unlike the previous back-to-back converter where both input and output converter sides are decoupled, the shared-leg topology in Fig. 11.3 presents coupling due to the leg $q_s - \overline{q}_s$.

The following equations can be derived from Fig. 11.3(b) for both input and output converter sides:

$$v_{g1} = v_{g10} - v_{ng0} \tag{11.31}$$

$$v_{g2} = v_{g20} - v_{ng0} \tag{11.32}$$

$$v_{g3} = v_{s0} - v_{ng0} \tag{11.33}$$

$$v_{l1} = v_{l10} - v_{nl0} \tag{11.34}$$

$$v_{l2} = v_{l20} - v_{nl0} \tag{11.35}$$

$$v_{l3} = v_{s0} - v_{nl0} \tag{11.36}$$

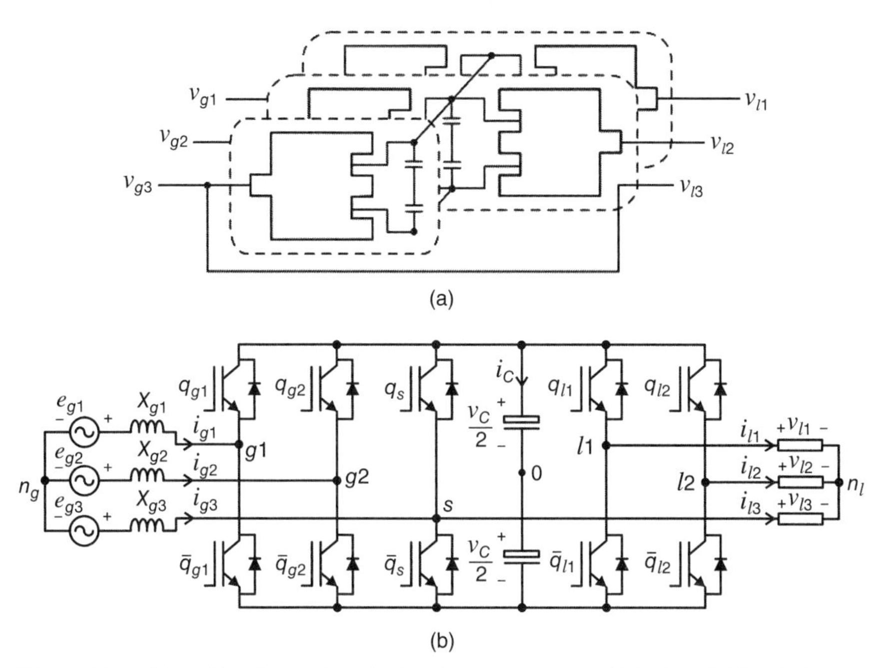

Figure 11.3 Shared-leg three-phase back-to-back converter: (a) block and (b) switch representation.

The pole voltages can be written as in equations (11.7)–(11.12) with $v_{ng0} = \frac{1}{3}(v_{g10} + v_{g20} + v_{s0})$ and $v_{nl0} = \frac{1}{3}(v_{l10} + v_{l20} + v_{s0})$. Also, equations (11.13)–(11.15) are still valid for modeling the input converter side. However, the capacitor voltage for the shared-leg topology in Fig. 11.3 is given by

$$v_c = \frac{1}{C}\int [q_{g1}i_{g1} + q_{g2}i_{g2} + q_s(i_{g3} - i_{l3}) - q_{l1}i_{l1} - q_{l2}i_{l2}]dt \tag{37}$$

with the interval of integration equal to the switching period.

11.3.2 PWM Strategies

If the desired input and output voltages are given respectively by v_{g1}^*, v_{g2}^*, v_{g3}^* and v_{l1}^*, v_{l2}^*, v_{l3}^*, one option for the reference pole voltage is given below:

$$v_{g10}^* = v_{g1}^* - v_{g3}^* + v_h^* \tag{11.38}$$

$$v_{g20}^* = v_{g2}^* - v_{g3}^* + v_h^* \tag{11.39}$$

$$v_{s0}^* = v_h^* \tag{11.40}$$

$$v_{l10}^* = v_{l1}^* - v_{l3}^* + v_h^* \tag{11.41}$$

$$v_{l20}^* = v_{l2}^* - v_{l3}^* + v_h^* \tag{11.42}$$

Due to the shared-leg, which connects both input and output sides of the converter, there is only one voltage v_h^* that optimizes both converter sides. The v_h^* can be defined as follows:

$$v_h^* = v_C\left(\mu - \frac{1}{2}\right) + (\mu - 1)V_{\max}^* - \mu V_{\min}^* \tag{11.43}$$

where $V_{\max} = \mathrm{MAX}\{v_{g1}^* - v_{g3}^*, v_{g2}^* - v_{g3}^*, 0, v_{l1}^* - v_{l3}^*, v_{l2}^* - v_{l3}^*\}$ and $V_{\min} = \mathrm{MIN}\{v_{g1}^* - v_{g3}^*, v_{g2}^* - v_{g3}^*, 0, v_{l1}^* - v_{l3}^*, v_{l2}^* - v_{l3}^*\}$.

Once the reference pole voltages have been defined, pulse widths can be obtained as in equations (11.25)–(11.30). In this case, instead of defining τ_{g3} and τ_{l3}, the pulse width τ_s is defined by $(v_s^*/v_C + 1/2)T_s$.

11.3.3 dc-link Voltage Requirement

To guarantee the same input and output voltages as in the full-bridge converter, the minimum dc-link voltage for the shared leg topology must obey the following requirement:

$$v_c \geq \max\{v_{gx0}^* - v_{gy0}^*, v_{gx0}^* - v_{s0}^*, v_{lx0}^* - v_{ly0}^*, v_{lx0}^* - v_{s0}^*, v_{gx0}^* - v_{ly0}^*\} \tag{44}$$

with $x = 1, 2$; $y = 1, 2$; and $x \neq y$.

If the load frequency is different from the grid frequency, the term $v_{gx0}^* - v_{ly0}^*$ is dominant and defines v_c. However, if both sides of the converter require the same frequency, the output voltage can be generated, synchronized to the input voltage to minimize the effect of the term $v_{gx0}^* - v_{ly0}^*$. This will allow the shared-leg converter to operate with the same dc-link voltage as the full-bridge converter.

11.3.4 Half-Bridge Converter

The half-bridge configuration at both converter sides is presented in Fig. 11.4). In this case there is a low frequency current flowing through the capacitors due to the dc-link midpoint connection, which increases the voltage fluctuation on these elements and consequently affects their losses.

The voltage at the input and output converter sides is defined directly from the poles voltages v_{g10}, v_{g20} and v_{l10}, v_{l20}, which are defined by equations (11.7), (11.8), (11.10), and (11.11), respectively. Also, the PWM equations are defined straightforward since there is no degree of freedom to be explored, that is, there are only two legs to process two voltages.

11.4 TOPOLOGIES WITH INCREASED NUMBER OF SWITCHES (CONVERTERS IN PARALLEL)

The connection of converters in parallel can be employed when the power ratings of the switches available do not handle the level of current needed by the circuit. The simplest solution when the currents demanded by the load are higher than the rated currents of the switches is connecting switches in parallel, as shown in Fig. 11.5 for the three-phase back-to-back converter. Although the number of power switches is twice as compared to those in Fig. 11.1, the characteristics of this converter in terms

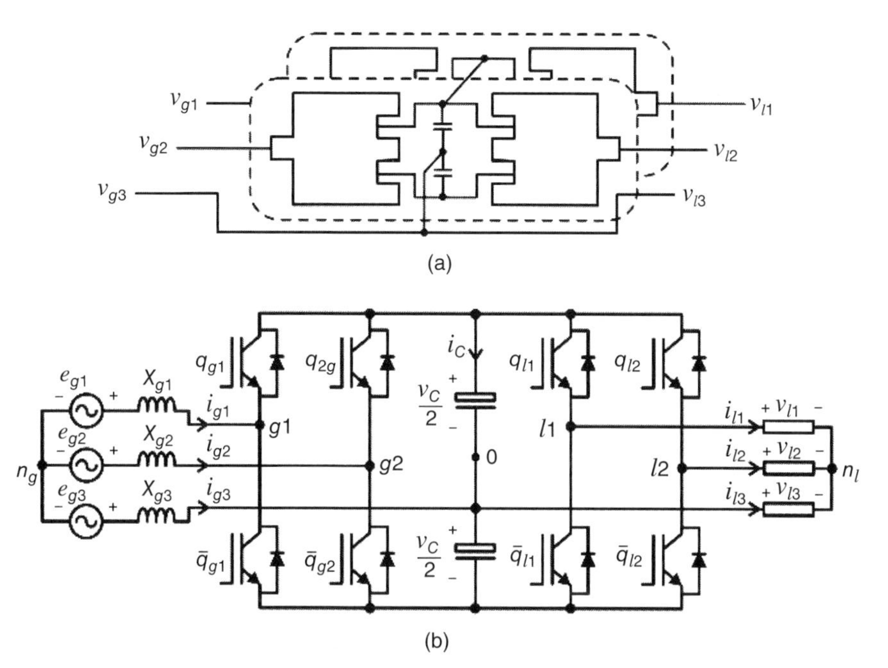

Figure 11.4 Half-bridge three-phase back-to-back converter: (a) block and (b) switch representation.

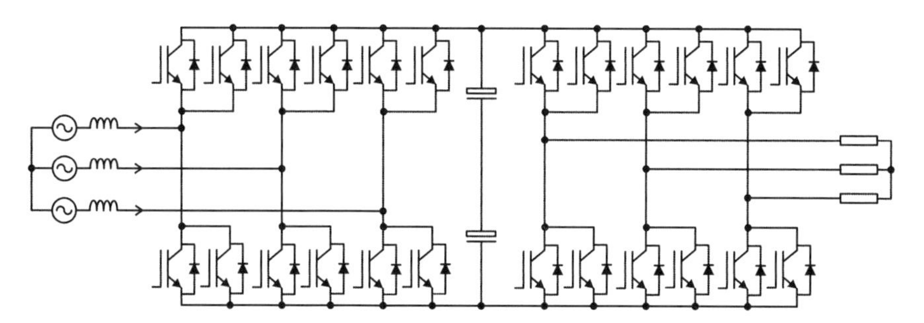

Figure 11.5 Three-phase back-to-back converter with switches in parallel.

of quality of the waveform generated are the same as in the conventional solution (see Fig. 11.1).

As considered in Chapter 10 for the single-phase back-to-back converters, connecting converters in parallel instead of switches in parallel will bring the following benefits: (i) increase the current capability of the conversion system, (ii) improve the quality of the waveforms generated at the input and output sides of the converter, and (iii) reduce the stress on the dc-link capacitors.

11.4.1 Model

Although connecting converters in parallel allows reducing the level of power processed by each converter, as well as current ripple reduction, a circulating current

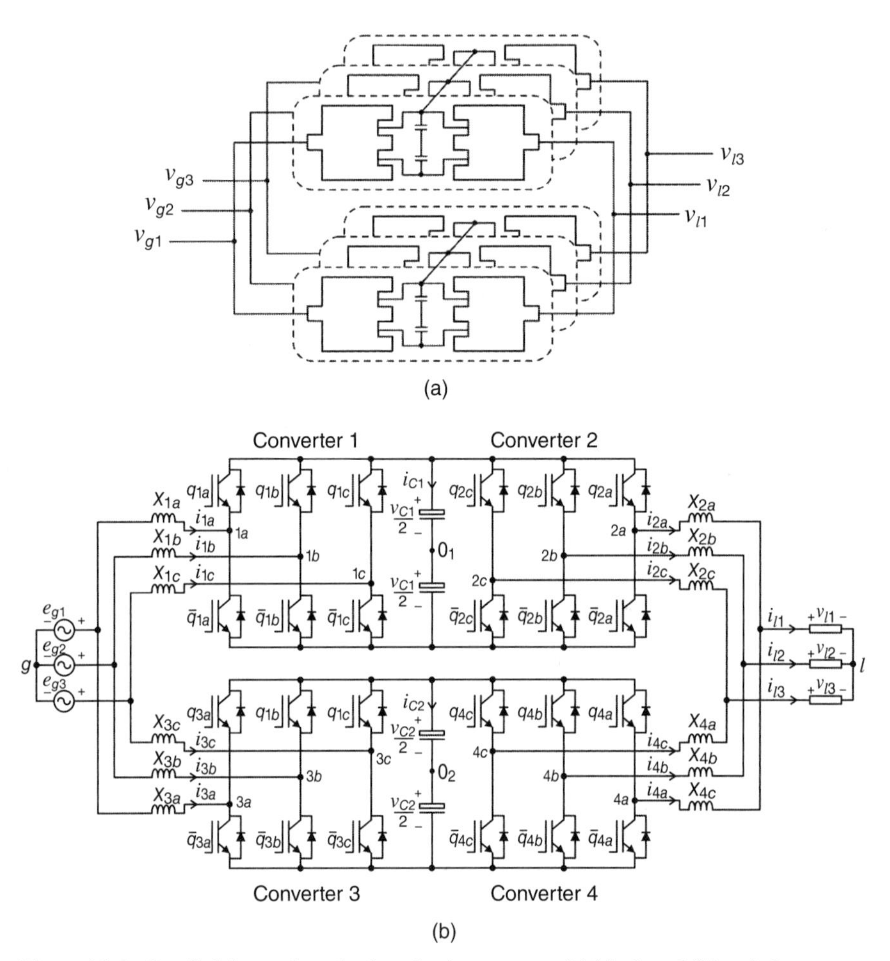

Figure 11.6 Parallel three-phase back-to-back converter: (a) block and (b) switch representation.

exists between both back-to-back converters. To model such a circulating current it is necessary to write down all equations of the system. Figure 11.6 shows a three-phase to three-phase energy conversion unit with two back-to-back converters connected in parallel. Note that, although increasing the complexity of the system, other back-to-back converters can be added in parallel, which reduces the power processed by each unit. Such a conversion unit can be represented by either power blocks [Fig. 11.6(a)] or power switches [Fig. 11.6(b)]. This conversion system consists of a three-phase load, six inductance filters (X_{ja}, X_{jb}, and X_{jc} with $j = 1, 3$) at the grid side, six inductance filters (X_{ma}, X_{mb}, and X_{mc} with $m = 2, 4$) at the load side, and two dc-link capacitor banks.

Considering converters 1 and 3 (converters connected at the grid side), it is possible to write the following equations:

$$e_{g1} = z_{1a} i_{1a} + v_{1a0_1} - v_{g0_1} \tag{11.45}$$

$$e_{g2} = z_{1b}i_{1b} + v_{1b0_1} - v_{g0_1} \quad (11.46)$$

$$e_{g3} = z_{1c}i_{1c} + v_{1c0_1} - v_{g0_1} \quad (11.47)$$

$$e_{g1} = z_{3a}i_{3a} + v_{3a0_2} - v_{g0_2} \quad (11.48)$$

$$e_{g2} = z_{3b}i_{3b} + v_{3b0_2} - v_{g0_2} \quad (11.49)$$

$$e_{g3} = z_{3c}i_{3c} + v_{3c0_2} - v_{g0_2} \quad (11.50)$$

for a balanced case v_{g0_1} and v_{g0_2} are given respectively by:

$$v_{g0_1} = \frac{1}{3}(v_{1a0_1} + v_{1b0_1} + v_{1c0_1}) \quad (11.51)$$

$$v_{g0_2} = \frac{1}{3}(v_{3a0_2} + v_{3b0_2} + v_{3c0_2}) \quad (11.52)$$

where $z_{1a} = r_{1a} + l_{1a}p$, $z_{1b} = r_{1b} + l_{1b}p$, $z_{1c} = r_{1c} + l_{1c}p$, $z_{3a} = r_{3a} + l_{3a}p$, $z_{3b} = r_{3b} + l_{3b}p$, and $z_{3c} = r_{3c} + l_{3c}p$ are the impedances of the elements X_{1a}, X_{1b}, X_{1c}, X_{3a}, X_{3b}, and X_{3c}, respectively; with $p = d/dt$. v_{g0_1} and v_{g0_2} are the voltages between the neutral of the grid and the dc-link midpoints 0_1 and 0_2, respectively. Similar equations are obtained for converters 2 and 4 connected to the load, as follows:

$$v_{l1} = -z_{2a}i_{2a} + v_{2a0_1} - v_{l0_1} \quad (11.53)$$

$$v_{l2} = -z_{2b}i_{2b} + v_{2b0_1} - v_{l0_1} \quad (11.54)$$

$$v_{l3} = -z_{2c}i_{2c} + v_{2c0_1} - v_{l0_1} \quad (11.55)$$

$$v_{l1} = -z_{4a}i_{4a} + v_{4a0_2} - v_{l0_2} \quad (11.56)$$

$$v_{l2} = -z_{4b}i_{4b} + v_{4b0_2} - v_{l0_2} \quad (11.57)$$

$$v_{l3} = -z_{4c}i_{4c} + v_{4c0_2} - v_{l0_2} \quad (11.58)$$

for a balanced case v_{l0_1} and v_{l0_2} are given respectively by:

$$v_{l0_1} = \frac{1}{3}(v_{2a0_1} + v_{2b0_1} + v_{2c0_1}) \quad (11.59)$$

$$v_{l0_2} = \frac{1}{3}(v_{4a0_2} + v_{4b0_2} + v_{4c0_2}) \quad (11.60)$$

As expected, due to the connection of both converters in parallel, there is a circulating mesh that originates a circulating current between both ac–dc–ac conversion units, as highlighted in Fig. 11.7.

$$z_{1a}i_{1a} + z_{2a}i_{2a} - z_{4a}i_{4a} - z_{3a}i_{3a} + v_{1a0_1} - v_{2a0_1} + v_{4a0_2} - v_{3a0_2} = 0 \quad (11.61)$$

$$z_{1a}i_{1a} + z_{2b}i_{2b} - z_{4b}i_{4b} - z_{3a}i_{3a} + v_{1a0_1} - v_{2b0_1} + v_{4b0_2} - v_{3a0_2} = 0 \quad (11.62)$$

$$z_{1a}i_{1a} + z_{2c}i_{2c} - z_{4c}i_{4c} - z_{3a}i_{3a} + v_{1a0_1} - v_{2c0_1} + v_{4c0_2} - v_{3a0_2} = 0 \quad (11.63)$$

$$z_{1b}i_{1b} + z_{2a}i_{2a} - z_{4a}i_{4a} - z_{3b}i_{3b} + v_{1b0_1} - v_{2a0_1} + v_{4a0_2} - v_{3b0_2} = 0 \quad (11.64)$$

$$z_{1b}i_{1b} + z_{2b}i_{2b} - z_{4b}i_{4b} - z_{3b}i_{3b} + v_{1b0_1} - v_{2b0_1} + v_{4b0_2} - v_{3b0_2} = 0 \quad (11.65)$$

$$z_{1b}i_{1b} + z_{2c}i_{2c} - z_{4c}i_{4c} - z_{3b}i_{3b} + v_{1b0_1} - v_{2c0_1} + v_{4c0_2} - v_{3b0_2} = 0 \quad (11.66)$$

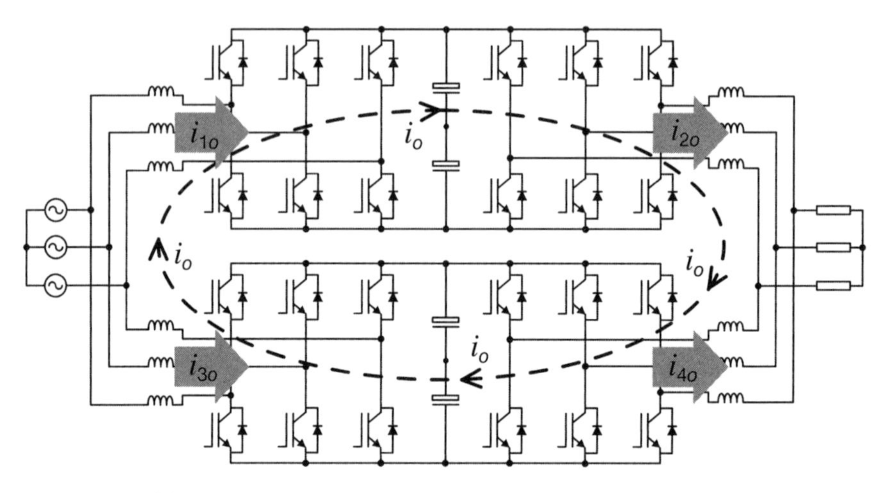

Figure 11.7 Circulating current between both back-to-back converters in parallel.

$$z_{1c}i_{1c} + z_{2a}i_{2a} - z_{4a}i_{4a} - z_{3c}i_{3c} + v_{1c0_1} - v_{2a0_1} + v_{4a0_2} - v_{3c0_2} = 0 \quad (11.67)$$

$$z_{1c}i_{1c} + z_{2b}i_{2b} - z_{4b}i_{4b} - z_{3c}i_{3c} + v_{1c0_1} - v_{2b0_1} + v_{4b0_2} - v_{3c0_2} = 0 \quad (11.68)$$

$$z_{1c}i_{1c} + z_{2c}i_{2c} - z_{4c}i_{4c} - z_{3c}i_{3c} + v_{1c0_1} - v_{2c0_1} + v_{4c0_2} - v_{3c0_2} = 0 \quad (11.69)$$

Note that equations (11.61)–(11.69) exist due to the parallel connection of the back-to-back converters. If a transformer is employed for isolation purposes, the current i_o would be zero and equations (11.61)–(11.69) would no longer exist.

An expression for the circulating voltage can be obtained by adding equations (11.61)–(11.69), to give the following expression:

$$v_o = \sum_{n=a}^{c} z_{1n}i_{1n} + \sum_{n=a}^{c} z_{2n}i_{2n} - \sum_{n=a}^{c} z_{4n}i_{4n} - \sum_{n=a}^{c} z_{3n}i_{3n} \quad (11.70)$$

with

$$v_o = -\sum_{n=a}^{c} v_{1n0_1} + \sum_{n=a}^{c} v_{2n0_1} + \sum_{n=a}^{c} v_{3n0_2} - \sum_{n=a}^{c} v_{4n0_2} \quad (11.71)$$

From the currents of the converters it is possible to define the circulating current as follows:

$$i_{1o} = \frac{1}{\sqrt{3}}(i_{1a} + i_{1b} + i_{1c}) \quad (11.72)$$

$$i_{2o} = \frac{1}{\sqrt{3}}(i_{2a} + i_{2b} + i_{2c}) \quad (11.73)$$

$$i_{3o} = \frac{1}{\sqrt{3}}(i_{3a} + i_{3b} + i_{3c}) \quad (11.74)$$

$$i_{4o} = \frac{1}{\sqrt{3}}(i_{4a} + i_{4b} + i_{4c}) \quad (11.75)$$

As presented in Fig. 11.7, the circulating current can be represented as a single current that circulates among the converters. Such a current can be written as

$$i_o = i_{1o} = -i_{3o} = i_{2o} = -i_{4o} \tag{11.76}$$

Since the parallel back-to-back converter deals with three-phase systems at both sides of the converter (i.e., grid and load sides), it is convenient to use the transformation of variables from 123 to odq as follows:

$$\overline{x}_{g123} = \overline{P}\overline{x}_{godq} \tag{11.77}$$

$$\overline{x}_{l123} = \overline{P}\overline{x}_{lodq} \tag{11.78}$$

$$\overline{x}_{kabc} = \overline{P}\overline{x}_{kodq} \tag{11.79}$$

with $\overline{x}_{g123} = [x_{g1}\ x_{g2}\ x_{g3}]^T$, $\overline{x}_{l123} = [x_{l1}\ x_{l2}\ x_{l3}]^T$ and $\overline{x}_{kabc} = [x_{ka}\ x_{kb}\ x_{kc}]^T$, where x can be either voltage or current and $k = 1, 2, 3, 4$.

Defining the voltages $v_{1n} = v_{1n0_1} - v_{g0_1}$, $v_{2n} = v_{2n0_1} - v_{l0_1}$, $v_{3n} = v_{3n0_2} - v_{g0_2}$, and $v_{4n} = v_{4n0_2} - v_{l0_2}$, with $n = a, b, c$, it is possible to determine the model of the system as follows:

$$\overline{e}_{gdq} = \overline{z}_{1odq}\overline{i}_{1odq} + \overline{v}_{1dq} \tag{11.80}$$

$$\overline{e}_{gdq} = \overline{z}_{3odq}\overline{i}_{3odq} + \overline{v}_{3dq} \tag{11.81}$$

$$\overline{e}_{ldq} = -\overline{z}_{2odq}\overline{i}_{2odq} + \overline{v}_{2dq} \tag{11.82}$$

$$\overline{e}_{ldq} = -\overline{z}_{4odq}\overline{i}_{4odq} + \overline{v}_{4dq} \tag{11.83}$$

where $\overline{e}_{gdq} = [e_{gd}\ e_{gq}]^T$, $\overline{e}_{ldq} = [e_{ld}\ e_{lq}]^T$, $\overline{i}_{kodq} = [i_{ko}\ i_{kd}\ i_{kq}]^T$, $\overline{v}_{kdq} = [v_{kd}\ v_{kq}]^T$, and
$\overline{z}_{kodq} = \frac{1}{6}\begin{bmatrix} \sqrt{2}\,(2z_{ka} - z_{kb} - z_{kc}) & (4z_{ka} + z_{kb} + z_{kc}) & \sqrt{3}(z_{kc} - z_{kb}) \\ \sqrt{6}(z_{kb} - z_{kc}) & \sqrt{3}(z_{kc} - z_{kb}) & 3(z_{kb} + z_{kc}) \end{bmatrix}$

Then it is possible to write the equation for the circulating current as follows:

$$v_o = \sqrt{3}\left[\left(\sum_{k=1}^{4} z_{ko}\right) i_o + \sum_{j=d,q} z_{1oj} i_{1j} + \sum_{j=d,q} z_{2oj} i_{2j} - \sum_{j=d,q} z_{3oj} i_{3j} - \sum_{j=d,q} z_{4oj} i_{4j}\right] \tag{11.84}$$

with $z_{ko} = \frac{2}{6}(z_{ka} + z_{kb} + z_{kc})$, $z_{kod} = \frac{\sqrt{2}}{6}(2z_{ka} - z_{kb} - z_{kc})$, and $z_{koq} = \frac{\sqrt{6}}{6}(z_{kb} - z_{kc})$

It is also convenient to write the model in dq components since: (i) the voltages $\overline{v}_{1dq}$ and $\overline{v}_{3dq}$ will be employed to control the currents $\overline{i}_{1odq}$ and $\overline{i}_{3odq}$ and consequently the grid power factor, (ii) the load variables will be controlled by the voltages $\overline{v}_{2dq}$ and $\overline{v}_{4dq}$, and (iii) the circulating current will be regulated by v_o.

Assuming a balanced case: $z_{1n} = z_{3n} = z_1$ and $z_{2n} = z_{4n} = z_2$ with $n = a, b, c$, it is possible to rewrite the model as follows:

$$\overline{e}_{gdq} = \overline{z}_1\overline{i}_{1dq} + \overline{v}_{1dq} \tag{11.85}$$

$$\overline{e}_{gdq} = \overline{z}_1\overline{i}_{3dq} + \overline{v}_{3dq} \tag{11.86}$$

$$\overline{e}_{ldq} = \overline{z}_2\overline{i}_{2dq} + \overline{v}_{2dq} \tag{11.87}$$

$$\overline{e}_{ldq} = \overline{z}_2\overline{i}_{4dq} + \overline{v}_{4dq} \tag{11.88}$$

with $\bar{z}_1 = \begin{bmatrix} z_1 & 0 \\ 0 & z_1 \end{bmatrix}$ and $\bar{z}_2 = \begin{bmatrix} z_2 & 0 \\ 0 & z_2 \end{bmatrix}$

From equations (11.80), (11.81) it is possible to obtain dq grid current:

$$v_{gd} = \frac{v_{1d} + v_{3d}}{2} = e_{gd} - \frac{z_1}{2} i_{gd} \tag{11.89}$$

$$v_{gq} = \frac{v_{1q} + v_{3q}}{2} = e_{gq} - \frac{z_1}{2} i_{gq} \tag{11.90}$$

By following the same strategy, v_{ld} and v_{lq} are given by:

$$v_{ld} = \frac{v_{2d} + v_{4d}}{2} = e_{ld} - \frac{z_2}{2} i_{ld} \tag{11.91}$$

$$v_{lq} = \frac{v_{2q} + v_{4q}}{2} = e_{lq} - \frac{z_2}{2} i_{lq} \tag{11.92}$$

Defining $z_g = z_1/2$ and $z_l = z_2/2$, yields

$$v_{gd} = e_{gd} - z_g i_{gd} \tag{11.93}$$

$$v_{gq} = e_{gq} - z_g i_{gq} \tag{11.94}$$

$$v_{ld} = e_{ld} - z_l i_{ld} \tag{11.95}$$

$$v_{lq} = e_{lq} - z_l i_{lq} \tag{11.96}$$

11.4.2 PWM

As described in the model of the three-phase back-to-back converter, the reference voltages v_{kd}^*, v_{kq}^*, and v_o^* $(k = 1, 2, 3, 4)$ are furnished by the controllers to regulate the input, output, and circulating variables, respectively. By employing a transformation of variables, it is possible to write the 123 voltages as a function of odq as follows:

$$v_{ka}^* = \sqrt{\frac{2}{3}} v_{kd}^* \tag{11.97}$$

$$v_{kb}^* = \sqrt{\frac{2}{3}} \left(\frac{1}{2} v_{kd}^* + \frac{\sqrt{3}}{2} v_{kq}^* \right) \tag{11.98}$$

$$v_{kc}^* = \sqrt{\frac{2}{3}} \left(-\frac{1}{2} v_{kd}^* - \frac{\sqrt{3}}{2} v_{kq}^* \right) \tag{11.99}$$

The reference pole voltages employed for the PWM techniques can be defined as a function of auxiliary voltages as follows:

$$v_{1a0_1}^* = v_{1a}^* + v_x^* \tag{11.100}$$

$$v_{1b0_1}^* = v_{1b}^* + v_x^* \tag{11.101}$$

$$v_{1c0_1}^* = v_{1c}^* + v_x^* \tag{11.102}$$

$$v_{2a0_1}^* = v_{2a}^* + v_y^* \tag{11.103}$$

$$v_{2b0_1}^* = v_{2b}^* + v_y^* \tag{11.104}$$

$$v_{2c0_1}^* = v_{2c}^* + v_y^* \tag{11.105}$$

$$v_{3a0_2}^* = v_{3a}^* + \frac{v_o^*}{3} - v_y^* + v_z^* + v_x^* \tag{11.106}$$

$$v_{3b0_2}^* = v_{3b}^* + \frac{v_o^*}{3} - v_y^* + v_z^* + v_x^* \tag{11.107}$$

$$v_{3c0_2}^* = v_{3c}^* + \frac{v_o^*}{3} - v_y^* + v_z^* + v_x^* \tag{11.108}$$

$$v_{4a0_1}^* = v_{4a}^* + v_z^* \tag{11.109}$$

$$v_{4b0_1}^* = v_{4b}^* + v_z^* \tag{11.110}$$

$$v_{4c0_1}^* = v_{4c}^* + v_z^* \tag{11.111}$$

The voltages v_x^*, v_y^* and v_z^* can be defined by following the same approach as in equations (11.23) and (11.24).

11.4.3 Control Strategies

Figure 11.8 shows the control block diagram of the parallel three-phase back-to-back converter. The dc-link voltages v_{C1} and v_{C2} are regulated by the controllers R_{c1} and R_{c2}, respectively. These controllers furnish the amplitudes of the reference currents I_1^* and I_3^* required at the input side of the converters 1 and 3. The power factor control, in turn, is obtained by synchronization of the reference grid current i_{1abc}^* $(i_{1a}^*, i_{1b}^*, i_{1c}^*)$

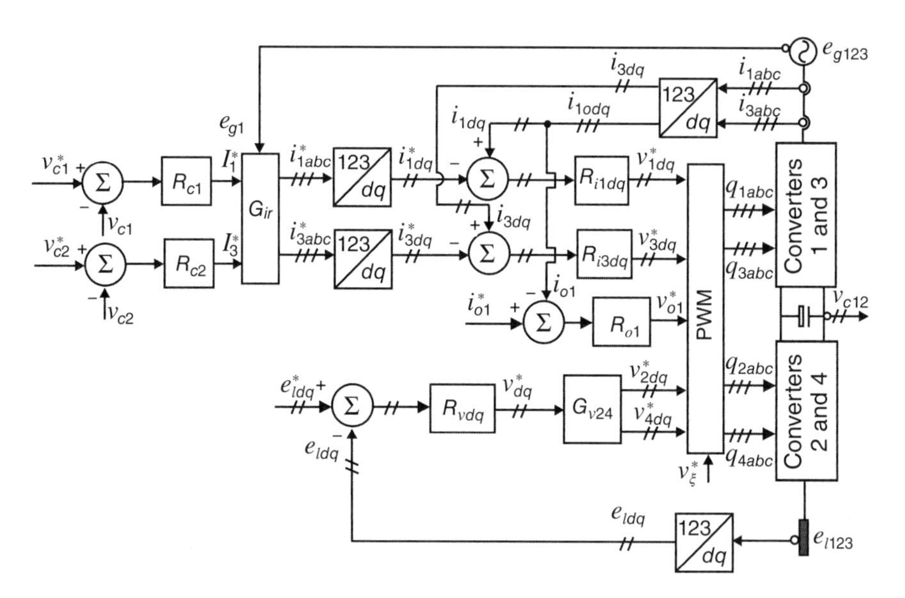

Figure 11.8 Control block diagram of the parallel three-phase converter.

and i^*_{3abc} (i^*_{3a}, i^*_{3b}, i^*_{3c}) with the three-phase voltages. Such a synchronization is done in block G_{ir}, which employs a PLL-based technique.

The output of block G_{ir} are instantaneous reference currents that are converted to dq variables through the transformation 123/dq, based on the block dq. As a consequence, the output of these blocks are the reference currents i^*_{1dq} (i^*_{1d}, i^*_{1q}) and i^*_{3dq} (i^*_{3d}, i^*_{3q}). A double-sequence type of controller is employed on R_{i1dq} and R_{i3dq} to regulate the sinusoidal currents i^*_{1dq} and i^*_{3dq}. The output of the controllers R_{i1dq} and R_{i3dq} define the reference voltages v^*_{1dq} and v^*_{3dq}.

The circulating current is regulated by R_{o1} with $i^*_{o1} = 0$. The output of block R_{o1} is circulating voltage v^*_{o1}. The load voltage controller is obtained via block R_{vdq}. The output of this controller defines v^*_{dq} (v^*_d and v^*_q).

The voltage applied for each converter can be considered as half of the voltage v_{dq}, the block G_{v24} gives $v^*_{2dq} = v^*_{4dq} = v^*_{dq}/2$. Finally the reference voltages generated by the controllers v^*_{1dq}, v^*_{3dq}, v^*_{o1}, v^*_{2dq}, and v^*_{4dq}, and the auxiliary voltages (v^*_ξ with $\xi = x, y$, and z) are applied to the PWM block. The output of this block generates the state of the switches.

11.5 TOPOLOGIES WITH INCREASED NUMBER OF SWITCHES (CONVERTERS IN SERIES)

The series connection (i.e., cascade connection) of the three-phase back-to-back converters can be obtained as shown in Fig. 11.9(a) for the power blocks representation, while Fig. 11.9(b) shows the same conversion system with power switches. Note that the line-to-line voltages viewed by both grid and load sides will be two connected H-bridge converter series. The same advantages as described for the single-phase back-to-back converter are observed in this topology. Also, it is necessary to regulate the circulating current among the conversion units to increase the efficiency of the converter.

11.6 OTHER BACK-TO-BACK CONVERTERS

In the power distribution systems, the single-phase grid has been considered as an alternative for either rural or remote areas, due to its lower cost feature, especially when compared with the three-phase solution. Hence, the single-phase voltage available as the power supply is quite common where there is a large area to be covered. On the other hand, loads connected in a three-phase arrangement present some advantages when compared to single-phase loads. This is especially true when a motor is the load connected to the converter, due to its reduced size, constant torque and constant power at balanced conditions. In this scenario, there is a need for single-phase to three-phase power conversion systems. The back-to-back converter, able to interface a three-phase load and a single-phase grid, is presented in Fig. 11.10.

Such a single-phase to three-phase power conversion presents an inherent asymmetry, that is, constant power at the output converter side (three-phase load)

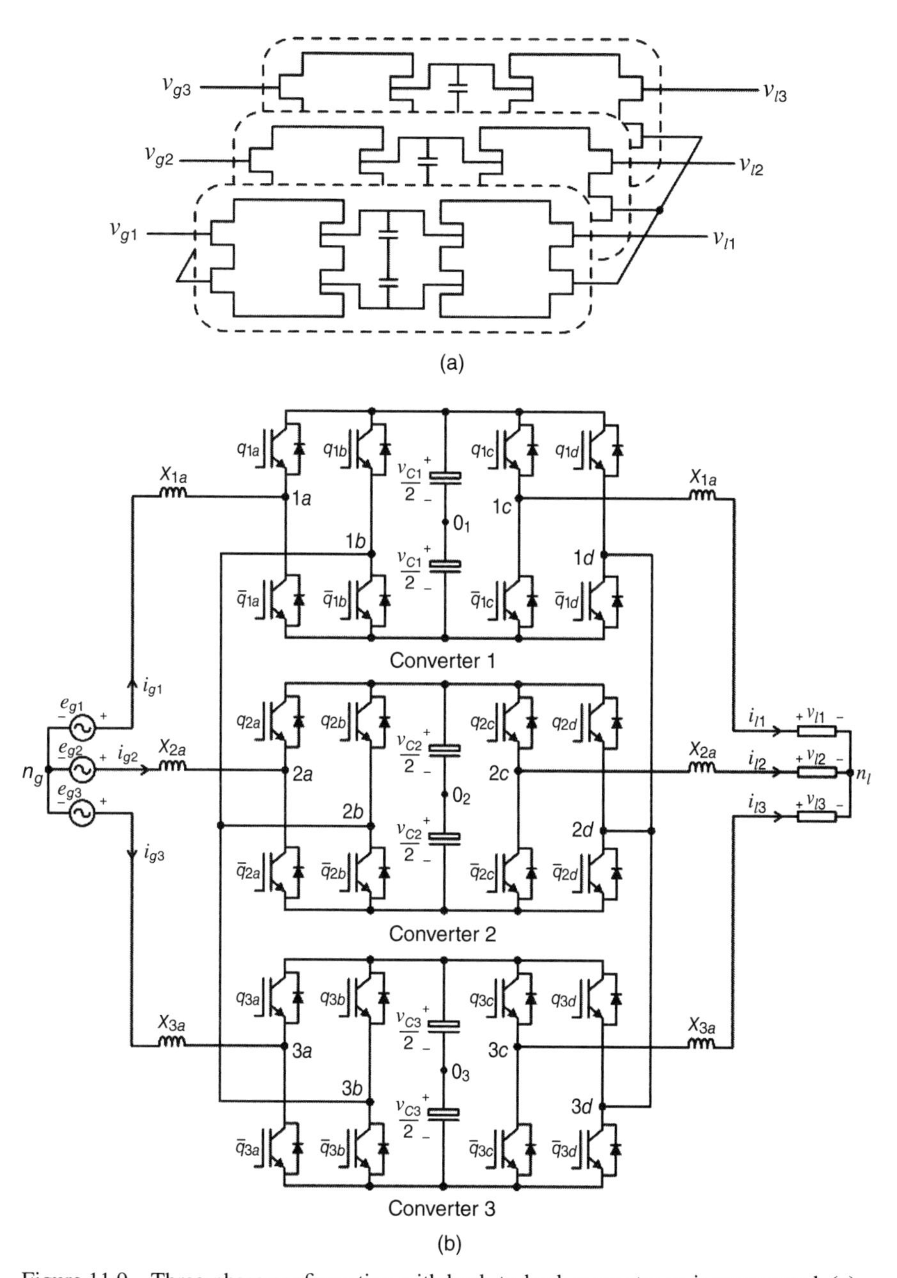

Figure 11.9 Three-phase configuration with back-to-back converter series connected: (a) block and (b) switch representation.

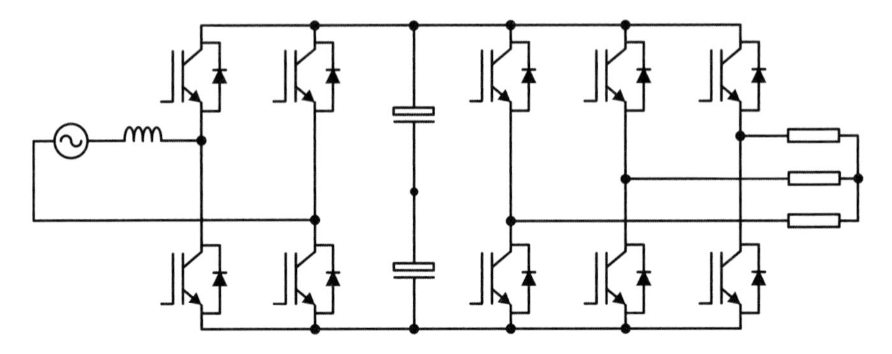

Figure 11.10 Single-phase to three-phase back-to-back converter.

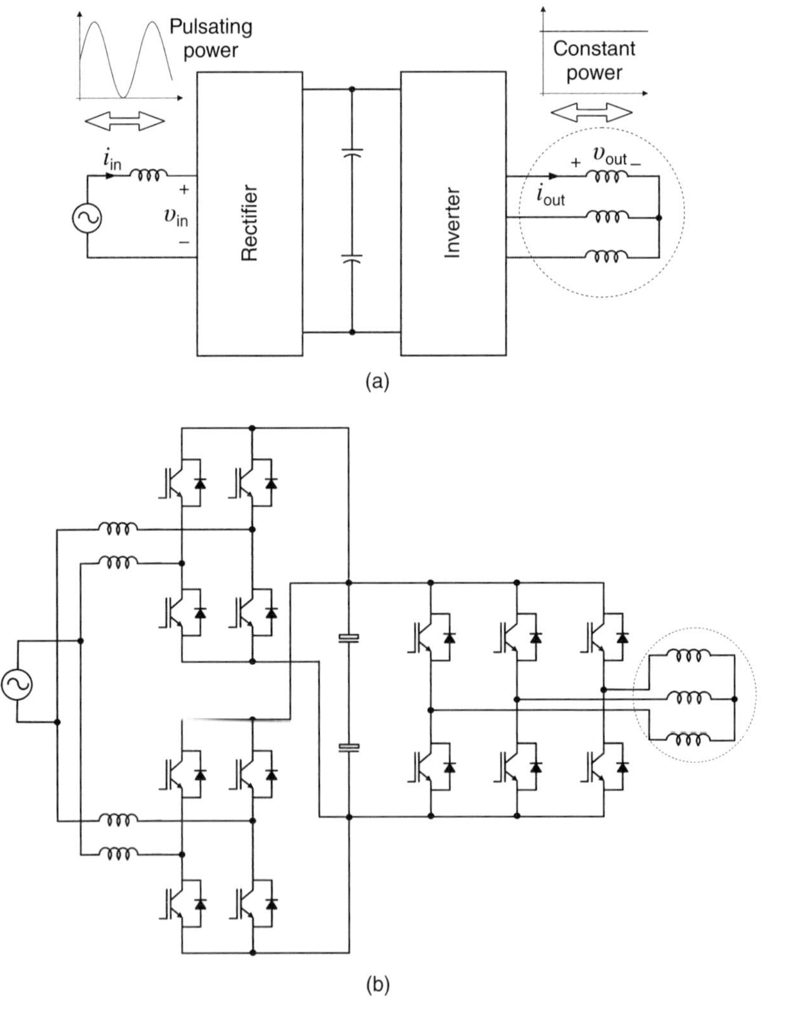

Figure 11.11 (a) Power asymmetry of the single-phase to three-phase converter (b) a potential solution to deal with the asymmetry.

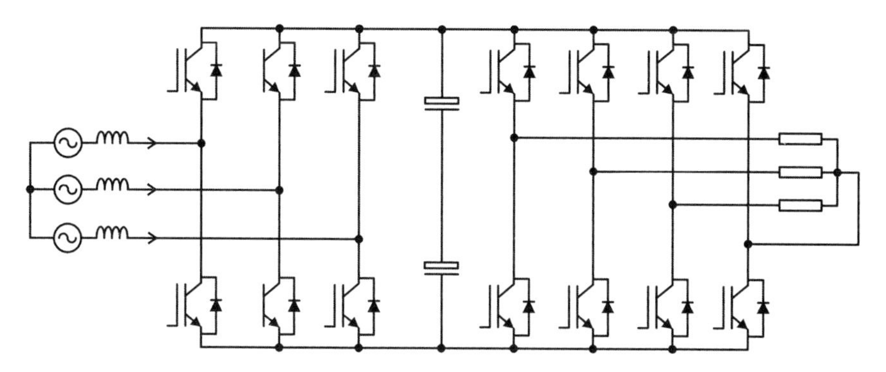

Figure 11.12 Three-phase to three-phase four-wire back-to-back converter.

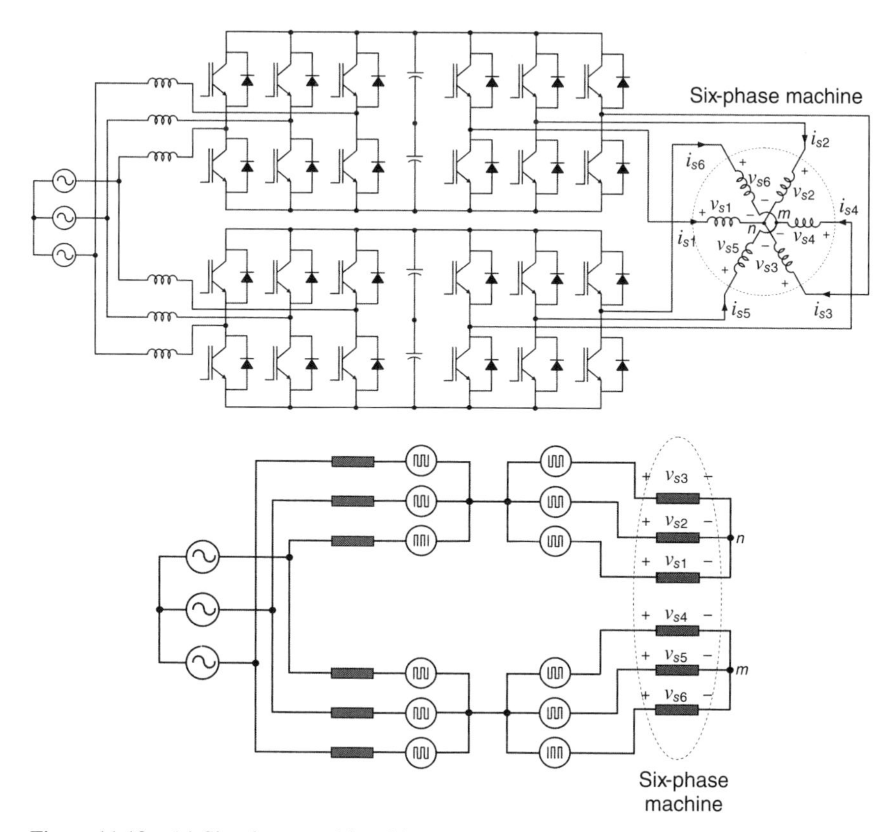

Figure 11.13 (a) Six-phase machine drive system based on three-phase back-to-back converters (b) model of the motor drive system.

and a pulsating power at the input converter side (single-phase grid), as highlighted in Fig. 11.11(a). The direct consequence of this asymmetry is the low frequency voltage oscillation observed in the dc-link capacitors; also, the power switches of the rectifier and inverter operate with different current ratings. Normally, the three-phase motor rated voltage is higher than that furnished by the single-phase grid, which means that the rectifier circuit must boost the grid voltage to guarantee the motor rated voltage. Consequently, the current relationship between input and output converter sides also implies converter asymmetry with the input current higher than the output current. There are different ways to deal with this challenge; one way is to employ the topology presented in Fig. 11.11(b) with a parallel converter used at the rectifier stage.

Another example of three-phase back-to-back converters is presented in Fig. 11.12, which deals with a load demanding a three-phase system with four wires. Note that this configuration is able to generate a regulated three-phase voltage at the load side where both three-phase and single-phase loads can be connected.

Lastly, Fig. 11.13(a) shows a six-phase machine drive system based on ac–dc–ac converter topologies connected in paralleled without isolation transformer. Note that unlike the previous parallel configurations presented in this chapter, the motor drive system in Fig. 11.13(a) does not have circulating current since the neutrals (n and m) of the six-phase machine are not connected between themselves, as highlighted in the model in Fig. 11.13(b). The six-phase machine is in fact constituted by two sets of three-phase windings, which present benefits such as reduced ripple on the torque. Another advantage of the six-phase motor drive system compared to the three-phase one is the fact that the currents processed by each switch of the inverter are lower.

11.7 SUMMARY

This chapter dealt with three-phase back-to-back converters. Section 11.2 presented the full-bridge ac–dc–ac converter, its model, PWM, and control strategies. Topologies with component count reduction were presented in Section 11.3, while Sections 11.4 and 11.5 showed the model, PWM, and control for configurations with an increased number of semiconductor devices. Shared-leg and dc-link capacitor midpoint connections were also considered for reducing the number of semiconductor devices, while connecting back-to-back converters in parallel and series. Other three-phase arrangements were considered in Section 11.6. Readers can find more details from References 1 to 8.

REFERENCES

[1] H. W. van der Broeck and J. D. van Wyk, A comparative investigation of a three-phase induction machine drive with a component minimized voltage-fed inverter under different control options, IEEE Trans Ind Appl, vol. IA-20, no. 2, pp. 309–320, Mar./Apr. 1984.

[2] P. Enjeti and A. Rahman, A new single phase to three phase converter with active input current shaping for low cost AC motor drives, IEEE Trans Ind Appl, vol. 29, no. 4, pp. 806–813, Jul./Aug. 1993.

[3] Blaabjerg F, Freysson S, Hansen HH, and Hansen S. Comparison of a space-vector modulation strategy for a three phase standard and a component minimized voltage source inverter. Proceedings EPE; 1995. p. 1806–1813.

[4] Kim GT and Lipo TA. VSI-PWM rectifier/inverter system with a reduced switch count. Conference. Rec. IEEE IAS Annual Meeting; 1995. p. 2327–2332.

[5] C. B. Jacobina, M. B. R. Correa, E. R. C. da Silva, and A. M. N. Lima, Induction motor drive system for low-power applications, IEEE Trans Ind Appl, vol. 35, no. 1, pp. 52–61, Jan./Feb. 1999.

[6] F. Blaabjerg, D. O. Neacsu, and J. K. Pedersen, Adaptive SVM to compensate dc-link voltage ripple for four-switch three-phase voltage-source inverters, IEEE Trans Power Electron, vol. 14, no. 4, pp. 743–752, Jul. 1999.

[7] E. Ledezma, B. McGrath, A. Muñoz, and T. A. Lipo, Dual AC-drive system with a reduced switch count, IEEE Trans Ind Appl, vol. 37, no. 5, pp. 1325–1333, Sep./Oct. 2001.

[8] J. R. Rodríguez, J. W. Dixon, J. R. Espinoza, J. Pontt, and P. Lezana, PWM regenerative rectifiers: State of the art, IEEE Trans Ind Electron, vol. 52, no. 1, pp. 5–22, Feb. 2005.

INDEX

Advanced Power Electronics Converters: PWM Converters Processing AC Voltages, Forty Fifth Edition. Euzeli Cipriano dos Santos Jr. and Edison Roberto Cabral da Silva.

R

S

T

U

IEEE Press Series on Power Engineering

Series Editor: M. E. El-Hawary, Dalhousie University, Halifax, Nova Scotia, Canada

The mission of IEEE Press Series on Power Engineering is to publish leading-edge books that cover the broad spectrum of current and forward-looking technologies in this fast-moving area. The series attracts highly acclaimed authors from industry/academia to provide accessible coverage of current and emerging topics in power engineering and allied fields. Our target audience includes the power engineering professional who is interested in enhancing their knowledge and perspective in their areas of interest.

1. *Principles of Electric Machines with Power Electronic Applications, Second Edition*
M. E. El-Hawary

2. *Pulse Width Modulation for Power Converters: Principles and Practice*
D. Grahame Holmes and Thomas Lipo

3. *Analysis of Electric Machinery and Drive Systems, Second Edition*
Paul C. Krause, Oleg Wasynczuk, and Scott D. Sudhoff

4. *Risk Assessment for Power Systems: Models, Methods, and Applications*
Wenyuan Li

5. *Optimization Principles: Practical Applications to the Operations of Markets of the Electric Power Industry*
Narayan S. Rau

6. *Electric Economics: Regulation and Deregulation*
Geoffrey Rothwell and Tomas Gomez

7. *Electric Power Systems: Analysis and Control*
Fabio Saccomanno

8. *Electrical Insulation for Rotating Machines: Design, Evaluation, Aging, Testing, and Repair, Second Edition*
Greg Stone, Edward A. Boulter, Ian Culbert, and Hussein Dhirani

9. *Signal Processing of Power Quality Disturbances*
Math H. J. Bollen and Irene Y. H. Gu

10. *Instantaneous Power Theory and Applications to Power Conditioning*
Hirofumi Akagi, Edson H. Watanabe, and Mauricio Aredes

11. *Maintaining Mission Critical Systems in a 24/7 Environment*
Peter M. Curtis

12. *Elements of Tidal-Electric Engineering*
Robert H. Clark

13. *Handbook of Large Turbo-Generator Operation and Maintenance, Second Edition*
Geoff Klempner and Isidor Kerszenbaum

14. *Introduction to Electrical Power Systems*
Mohamed E. El-Hawary

15. *Modeling and Control of Fuel Cells: Distributed Generation Applications*
M. Hashem Nehrir and Caisheng Wang

16. *Power Distribution System Reliability: Practical Methods and Applications*
Ali A. Chowdhury and Don O. Koval

17. *Introduction to FACTS Controllers: Theory, Modeling, and Applications*
Kalyan K. Sen and Mey Ling Sen

18. *Economic Market Design and Planning for Electric Power Systems*
James Momoh and Lamine Mili

19. *Operation and Control of Electric Energy Processing Systems*
James Momoh and Lamine Mili

20. *Restructured Electric Power Systems: Analysis of Electricity Markets with Equilibrium Models*
Xiao-Ping Zhang

21. *An Introduction to Wavelet Modulated Inverters*
S.A. Saleh and M.A. Rahman

22. *Control of Electric Machine Drive Systems*
Seung-Ki Sul,

23. *Probabilistic Transmission System Planning*
Wenyuan Li

24. *Electricity Power Generation: The Changing Dimensions*
Digambar M. Tigare

25. *Electric Distribution Systems*
Abdelhay A. Sallam and Om P. Malik

26. *Practical Lighting Design with LEDs*
Ron Lenk and Carol Lenk

27. *High Voltage and Electrical Insulation Engineering*
Ravindra Arora and Wolfgang Mosch

28. *Maintaining Mission Critical Systems in a 24/7 Environment, Second Edition*
Peter Curtis

29. *Power Conversion and Control of Wind Energy Systems*
Bin Wu, Yongqiang Lang, Navid Zargari, and Samir Kouro

30. *Integration of Distributed Generation in the Power System*
Math H. Bollen and Fainan Hassan

31. *Doubly Fed Induction Machine: Modeling and Control for Wind Energy Generation Applications*
Gonzalo Abad, Jesus Lopez, Miguel Rodrigues, Luis Marroyo, and Grzegorz Iwanski

32. *High Voltage Protection for Telecommunications*
Steven W. Blume

33. *Smart Grid: Fundamentals of Design and Analysis*
James Momoh

34. *Electromechanical Motion Devices, Second Edition*
Paul C. Krause, Oleg Wasynczuk, and Steven D. Pekarek

35. *Electrical Energy Conversion and Transport: An Interactive Computer-Based Approach, Second Edition*
George G. Karady and Keith E. Holbert

36. *ARC Flash Hazard and Analysis and Mitigation*
J. C. Das

37. *Handbook of Electrical Power System Dynamics: Modeling, Stability, and Control*
Mircea Eremia and Mohammad Shahidehpour

38. *Analysis of Electric Machinery and Drive Systems, Third Edition*
Paul Krause, Oleg Wasynczuk, S. D. Sudhoff, and Steven D. Pekarek

39. *Extruded Cables for High-Voltage Direct-Current Transmission: Advances in Research and Development*
Giovanni Mazzanti and Massimo Marzinotto

40. *Power Magnetic Devices: A Multi-Objective Design Approach*
S. D. Sudhoff

41. *Risk Assessment of Power Systems: Models, Methods, and Applications, Second Edition*
Wenyuan Li

42. *Practical Power System Operation*
Ebrahim Vaahedi

43. *The Selection Process of Biomass Materials for the Production of Bio-Fuels and Co-Firing*
Najib Altawell

44. *Electrical Insulation for Rotating Machines: Design, Evaluation, Aging, Testing, and Repair, Second Edition*
Greg C. Stone, Ian Culbert, Edward A. Boulter, Hussein Dhirani

45. *Principles of Electrical Safety*
Peter E. Sutherland

46. *Advanced Power Electronics Converters: PWM Converters Processing AC Voltages*
Euzeli Cipriano dos Santos Jr. and Edison Roberto Cabral da Silva